Cadmium Toxicity and Tolerance in Plants

Cadmium Toxicity and Tolerance in Plants

Editors
Nafees A. Khan
Samiullah

Alpha Science International Ltd.
Oxford, U.K.

Editors
Nafees A. Khan
Samiullah
Department of Botany
Aligarh Muslim University
Aligarh, India

Copyright © 2006

Alpha Science International Ltd.
7200 The Quorum, Oxford Business Park North
Garsington Road, Oxford OX4 2JZ, U.K.

www.alphasci.com

All rights reserved. No part of this publication may be reproduced, stored in a retrieval system, or transmitted in any form or by any means, electronic, mechanical, photocopying, recording or otherwise, without prior written permission of the publisher.

Printed from the camera-ready copy provided by the Editors.

ISBN 1-84265-317-2

Printed in India

Preface

Increasing environmental pollution due to indiscriminate human activities is a worldwide problem. The contamination of the environment by heavy metal pollution, particularly by cadmium is a major problem. Cadmium is a toxic element even at low concentration, and cadmium-contaminated soil poses serious threat to the sustainable agriculture and human health. It enters the environment mainly through industrial processes, large scale use of phosphatic fertilizers and sewage sludge application as well as atmospheric deposition. Because of its high plant-soil mobility, cadmium can accumulate in plant tissues and cause inhibition of growth and physiological functions and even plant death. Plants adaptation to cytotoxic amount of cadmium is associated with their ability to form metal-binding complexes and immobilization, exclusion, chelation and compartmentalisation of the metal ion and the expression of stress proteins. Its presence triggers defence mechanisms which can remove, neutralize or scavenge oxy-radicals and their intermediates by eliciting enzymatic as well as non-enzymatic processes. The understanding of physiological and molecular bases for adaptive and defence mechanisms may be useful in developing plants as agents for phytoremediation of contaminated sites.

    The intent of the book is to summarize and provide an update on our understanding of the role of antioxidants, techniques of phytoremediation and phytoextraction, metal chelation and physiological approach to alleviate cadmium toxicity. I am grateful to all those who extended support in bringing the book in its form. We also extend our apologies for any mistakes or omissions.

List of Contributors

Ana Isabel Gusmao Lima
    Centre for Cell Biology, Biology Department, University of Aveiro, Universidad de Aveiro, Portugal

Andon Vassilev
    Agricultural University of Plovdiv, Department of Plant Physiology and Biochemistry, 12 Mendeleev St. 4000, Plovdiv, Bulgaria

B. K. Dube
    Botany Department, Lucknow University, Lucknow 226007, India

C. Chatterjee
    Botany Department, Lucknow University, Lucknow 226007, India

Celine Masclaux-Daubresse
    Institut National de Recherche Agronomique: Unité de Nutrition, Azotée des Plantes, INRA Versailles, Route deSaint-Cyr, 78026 Versailles CEDEX-France

Chiraz Chaffei
    Unité de Recherche: Nutrition et Métabolisme, Azotés et Protéines de Stress (UR/09), Département de Biologie, Faculté des Sciences de Tunis

Fabricio S. Delite
    Departamento de Genética, Escola Superior de Agricultura Luiz de Queiroz, Universidade de São Paulo, Piracicaba CEP 13418-900, Brasil

Fei Chen
    Agronomy Department, Huajia Chi Campus, Zhejiang University, Hangzhou, China

Fei Dai
    Agronomy Department, Huajia Chi Campus, Zhejiang University, Hangzhou, China

Feibo Wu
    Agronomy Department, Huajia Chi Campus, Zhejiang University, Hangzhou, China

Figueira Eetelvina Maria De Almeida Paula
    Centre for Cell Biology, Biology Department, University of Aveiro, Universidad de Aveiro, Portugal

Guoping Zhang
    Agronomy Department, Huajia Chi Campus, Zhejiang University, Hangzhou, China

Honda Gouia
: Unité de Recherche: Nutrition et Métabolisme, Azotés et Protéines de Stress (UR/09), Département de Biologie, Faculté des Sciences de Tunis

Hung-Yu Lai
: Graduate Institute of Agricultural Chemistry, National Taiwan University, Taipei 106-17, Taiwan

Ivan Yordanov
: Acad. M. Popov Institute of Plant Physiology, Acad G. Bonchev St., Bl.21, 1113 Sofia, Bulgaria

Jaco Vangronsveld
: Environmental Biology, Center for Environmental Sciences, Limburgs Universitair Centrum, Universitaire Campus, B-3590 Diepenbeek, Belgium

Jung Yang
: Agronomy Department, Huajia Chi Campus, Zhejiang University, Hangzhou, China

Kang Wei
: Agronomy Department, Huajia Chi Campus, Zhejiang University, Hangzhou, China

Mohamed Habib Ghorbel
: Unité de Recherche: Nutrition et Métabolisme, Azotés et Protéines de Stress (UR/09), Département de Biologie, Faculté des Sciences de Tunis

Pallavi Sharma
: Department of Biochemistry, Faculty of Science, Banaras Hindu University, Varanasi 221005, India

Peter J. Lea
: Department of Biological Sciences, University of Lancaster, Lancaster LA1 4YQ, United Kingdom

Priscila L. Grato
: Departamento de Genética, Escola Superior de Agricultura Luiz de Queiroz, Universidade de São Paulo, Piracicaba CEP 13418-900, Brasil

Qin Dong
: Agronomy Department, Huajia Chi Campus, Zhejiang University, Hangzhou, China

R.S. Dubey
: Department of Biochemistry, Faculty of Science, Banaras Hindu University, Varanasi 221005, India

Ricardo A. Azevaedo
: Departamento de Genética, Escola Superior de Agricultura Luiz de Queiroz, Universidade de São Paulo, Piracicaba CEP 13418-900, Brasil

List of Contributors

Rui A. Gomes-Junior
> Departamento de Genética, Escola Superior de Agricultura Luiz de Queiroz, Universidade de São Paulo, Piracicaba CEP 13418-900, Brasil

Zueng-Sang Chen
> Graduate Institute of Agricultural Chemistry, National Taiwan University, Taipei 106-17, Taiwan

Contents

Preface	*v*
List of Contributors	*vii*
Antioxidant Stress Responses of Plants to Cadmium Priscila L. Gratão, Rui A. Gomes-Junior, Fabrício S. Delite, Peter J. Lea and Ricardo A. Azevedo	1
Phytoremediation Techniques of Cd-Contaminated Soils: Toxicity, Enhanced Uptake Techniques, and Mechanism Hung-Yu Lai and Zueng-Sang Chen	35
Cadmium Uptake and its Toxicity in Higher Plants Pallavi Sharma and R. S. Dubey	63
Cadmium Stress in Higher Plants Guoping Zhang, Feibo Wu, Kang Wei, Qin Dong, Fei Dai, Fei Chen and Jung Yang	87
Cadmium Phytoextraction from Contaminated Soils Andon Vassilev, Ivan Yordanov and Jaco Vangronsveld	103
Cadmium Detoxification in Roots of *Pisum Sativum* Seedlings: The Role of Phytochelatins in Metal Stress Coping Ana Isabel Gusmão Lima and Figueira Eetelvina Maria De Almeida Paula	115
Purification of Glutamate Dehydrogenase Isoenzymes from Control and Cadmium Treated Tomato Leaf Chiraz Chaffei, Celine Masclaux-Daubresse, Houda Gouia and Mohamed Habib Ghorbel	137
Cadmium-A Metal-An Enigma: An Overview C. Chatterjee and B. K. Dube	159
Index	179

Cadmium Toxicity and Tolerance in Plants
Editors: Nafees A. Khan and Samiullah
Copyright © 2006, Narosa Publishing House, New Delhi, India

Antioxidant Stress Responses of Plants to Cadmium

Priscila L. Gratão[1], Rui A. Gomes-Junior[1], Fabrício S. Delite[1], Peter J. Lea[2] and Ricardo A. Azevedo[1]*

[1] Departamento de Genética, Escola Superior de Agricultura Luiz de Queiroz, Universidade de São Paulo, Piracicaba CEP 13418-900, Brasil.
[2] Department of Biological Sciences, University of Lancaster, Lancaster LA1 4YQ, United Kingdom.

1. INTRODUCTION

Pollution is a major worldwide problem. Different types of pollutants are affecting different areas of the globe according to human activities, which are indiscriminate and the main reason for the increase in pollution. The contamination of the environment by heavy metal pollution, mainly by cadmium (Cd) which is considered one of the more toxic, is generated by mining and industrial activities, and the application of sewage sludge and phosphate fertilizers in agriculture. The latter contains Cd in its composition and with continuous agricultural utilization an increase in the heavy metal level in soils, plants and water sources have been observed, which may contribute to the bioaccumulation and biomagnification in the food chain. The heavy metal Cd causes damage to plant growth and several phytotoxic effects have been identified. A group of peptides termed phytochelatins (PCs) has been characterised that are capable of binding heavy metal ions, including Cd However, oxidative damage often occurs due to the generation of oxygen reactive species (ROS) induced by heavy metals. Antioxidant enzymes have been reported to be involved in the scavenging of ROS. Plants exposed to heavy metals may exhibit alteration in enzyme activity, which may consequently alter their metabolism. The study of such responses may allow the evaluation of the tolerance of distinct plant species to different levels of pollution in the environment. These data may be useful in breeding programs or biotechnological alternatives to produce and/or select tolerant plants that may be used in phytoremediation in order to reduce the amount of heavy metals in contaminated areas. Another important aspect is the identification or selection of plants that accumulate very low amounts of Cd in the parts that are used for animal and human food.

2. HEAVY METAL POLLUTION

In the last few decades there has been a great deal of concern over environmental pollution mainly due to the direct effects on human health and decreases in agricultural yields (Jarup, 2003). Scientists have begun to recognise that certain chemical pollutants are capable of remaining in the environment for a significant period of time and accumulating at levels harmful to humans. Some of the natural sources of pollutants include volcanic eruptions, forest fires, and evolution from anaerobic wetlands, although, the major sources are related to fossil fuels and agricultural activities, generating pollutants such as

---

* Corresponding Author (raazeved@carpa.ciagri.usp.br)

heavy metals, airborne nitrogen, sulphur and carbon oxides (Rusek and Marshall, 2000). Heavy metals can be included in the main category of pollutants, which plants have to cope with (Selim and Kingery, 2003). Their effects are highly variable depending on species sensitivity, the intensity and duration of the exposure (Rusek and Marshall, 2000). Cd is a non essential element, which is the most toxic heavy metal, being ranked at number 7 among the top 20 main toxins (Al-Khedhairy et al., 2001).

The term "heavy metal" has been widely used, however, its definition may vary according to the science branch. For instance, in toxicology, the heavy metals are defined on the basis of the toxicity of the chemical element to animal and humans. For an agronomist, the definition would be related to the toxicity of the element to plants (phytotoxicity) and the way in which soil contamination can decrease plant yields. In chemistry, the definition would be related to the atomic density (higher than 5g cm^{-3}). Therefore, the use of this term has been made in reference to metals and semi-metals more generally, which are associated with pollution and toxicity, but also include some elements that are essential for the cell in low concentrations.

Heavy metals are found in the soil and in the water and also in toxic gasses formed in the atmosphere by photochemical reactions. Heavy metals are strongly retained in the soil, with little leaching and may remain in the soil for thousands of years (Selim and Kingery, 2003). Since heavy metals do not break down, they can affect the biosphere for long periods of time and can be leached through the soil layers leading to the contamination of the water table. Consequently, the use of plants contaminated with high levels of heavy metals for food, might pose a serious risk to human and animal health (Wang et al., 2003).

Metals are present in the environment and a number of them are essential for the growth of animals and plants. They are present in the solid phase and in solution, as free ions, or adsorbed on to soil colloidal particles. Heavy metals presence in topsoil can be the result of soil-forming processes, as well as human and agricultural activities (Berglund et al., 1984). Several metals and semi-metals are essential for plants and animals: sodium (Na), magnesium (Mg), potassium (K), calcium (Ca), chromium (Cr), iron (Fe), cobalt (Co), copper (Cu), zinc (Zn), selenium (Se), Nickel (Ni), Manganese (Mn) and molybdenum (Mo). The nonessential metals, mercury (Hg), lead (Pb), cadmium (Cd), and the semi-metal arsenic (As) are recognized as health hazards as a result of environmental pollution (Berglund *et* al., 1984).

Heavy metals have been considered to be major environmental pollutants and their phytotoxicity is well established (Prasad, 1997). Cd is a toxic pollutant for humans, animals and plants (Sandalio et al., 2001) even at low doses (Balestrasse et al., 2001). Bioconcentrations of Cd in agricultural, horticultural, and silvicultural plants and weeds, and in aquatic macrophytes are of special concern in the interest of human welfare, due to its impact on human health through the food chain (Prasad, 1997). The concentration of Cd in soil is generally low, but low metal concentrations can result in significant accumulation in plant tissues (Gallego et al., 1996). Several vegetable, fruit, and cereal crops have been reported to accumulate Cd (Prasad, 1997). Cd was translocated to the rice grain, where 22 to 24% of the total metal content in the rice biomass was concentrated (Wang et al., 2003). It has recently been hypothesized that Cd accumulation in developing fruits could occur via phloem-mediated transport (di Toppi and Gabbrielli, 1999). Therefore, an understanding of the behaviour of heavy metals in the soil is fundamental to avoid toxic effects, when these metals are incorporated into the agroecosystem.

High Cd concentrations have been found to be carcinogenic, mutagenic and teratogenic for a large number of animal species (di Toppi and Gabrielli, 1999). Cd can accumulate in human with more than a 10 year half life (Salt et al., 1995). In humans Cd can cause renal disfunctions, bone demineralization (Ghoshroy et al., 1998) and cancer (Rossi et al., 1998). In addition it has been suggested that the disease itai-itai was caused by rice contaminated with Cd in Toyama city, Japan (Prasad, 1995).

Cd uptake is harmful to most plants (Gallego et al., 1996); it can reduce growth (Boussama et al., 1999), and cause death in extreme cases. Cd can also interfere with photosynthesis (Sandalio et al., 2001; Prasad, 1995), respiration, water relations, and reproduction, and cause changes in organelles by disruption of membrane structure and functions (Prasad, 1997; Sandalio et al., 2001), oxidative damage and lipid peroxidation (Balestrasse *et al.*, 2001) and alteration in chlorophyll contents (Lagriffoul et al., 1998).

According to the information described above, the increased contamination and subsequent accumulation of heavy metals in the soil may have serious consequences for agriculture, since the concentration of heavy metals may reach high levels in the soil and become a limiting factor for normal plant growth and productivity of field crops and thus affect human health (Dudka and Miller, 1999).

3. THE HEAVY METAL Cd

Cd is a toxic element that normally occurs in low concentrations in soils. The regulatory limit of Cd in agricultural soils is 100-600 mg kg^{-1}, but regionally this threshold is exceeded (Salt et al., 1998). Natural amounts of Cd are normally low, however the concentration can be significantly increased by activities such as zinc mining, iron foundries, the use of sewage sludge as a fertiliser in agriculture (Lombi et al., 2000), the combustion of fossil fuels, pesticides (Gimeno-García et al., 1996; Lagriffoul et al., 1998; Yang et al., 2002), application of phosphate fertilizers (Iretskaya et al., 1998), fertilizer impurities (Schickler and Caspi, 1999), industrial processes (Malan and Farrant, 1998), and applications where high stability and resistance to heat, cold, and light are required (Prasad, 1997). All these sources can lead to an increase in the Cd available to plants, which is not due to de novo synthesis, but to inappropriate processes leading to accumulation in the environment. Furthermore, Cd is taken up rapidly by the roots (Vitória et al., 2001). The degree to which plants are able to take up Cd depends on its concentration in the soil and its bioavailability, modulated by the presence of organic matter, pH, redox potential, temperature and concentration of other elements (di Toppi and Gabbrielli, 1999). As soon as Cd enters the roots, it can reach the xylem through an apoplastic and/or a symplastic pathway (Salt et al., 1995). Normally Cd ions are mainly retained in the roots and only small amounts are translocated to the upper parts of the plants (Schützendübel et al., 2001; Vitória et al., 2001).

Cd can cause inhibition of shoot and root growth (Schützendübel et al., 2001), disorganisation of the grana structures and reduction in chlorophyll synthesis (Siedlecka and Krupa, 1999; Somasshekaraiah et al., 1992). Moreover, Cd can also inhibit the activity of several groups of enzymes such as those of the photosynthetic Calvin cycle (Sandalio et al., 2001), nitrogen metabolism (Boussama et al., 1999), sugar metabolism (Verma and Dubey, 2001), and sulphate assimilation (Lee and Leustek, 1999), whereas leaf senescence is accelerated (Siedlecka and Krupa, 1999). Cd toxicity is also correlated with disturbances in the uptake and distribution of macro and micronutrients in plants (Sandalio et al., 2001). Cd can affect hormone balance and water movement by a reduction in the size and number of xylem vessels (Poschenrieder and Barceló, 1999) and can also differentially affect the concentrations of betaine, putrecine and spermine, suggesting a protection against metal-induced oxidative stress by polyamines (Bergmann et al., 2001). The appearance of a red-brown coloration of leaf margins and veins (Schickler and Caspi, 1999) and chlorosis of leaves has been described as one of the most characteristic symptoms of Cd toxicity in plants, which are due to the fact that Cd can compete with Fe (Siedlecka and Krupa, 1999), Ni (Baccouch et al., 1998) and Cu (Mocquot et al., 1996) uptake through the plasma membrane, leading to a deficiency in these essential metal ions.

Certain heavy metals like Cu and Fe can be toxic by their participation in redox cycles producing ROS (Schützendübel et al., 2001). By contrast with these metals, Cd is a non-redox metal unable to participate in Fenton-type reactions yielding activated oxygen species (Sandalio et al., 2001).

Cd does not appear to generate free radicals, but it does elevate lipid peroxidation and decrease GSH content, which results indirectly in ROS production (Gallego et al., 1996). The depletion of GSH is apparently a critical step in Cd sensitivity, since plants with improved capacities for GSH synthesis displayed higher Cd tolerance (Schützendübel and Polle, 2002). On the other hand, Cd stress can inhibit or stimulate the activity of several antioxidant enzymes (Gallego et al., 1996; Al-Khedhairy et al., 2001). Available data suggest that Cd, when not detoxified rapidly enough, may trigger, via the disturbance of the redox control of the cell, a sequence of reactions leading to growth inhibition, stimulation of secondary metabolism, lignification, and finally cell death that is in contrast to the idea that Cd results in unspecific necrosis (Schützendübel and Polle, 2002).

The role of Cd with respect to cell growth is still under debate (Von Zglinicki et al., 1992). A number of reports demonstrate that, at concentrations above 1 µM, Cd inhibits cell growth and DNA synthesis in a wide variety of cell types. It has been shown that Cd incorporation is proportional to the Cd concentration in the medium and to the incubation time, however, its effects are very distinct depending on the concentration. For instance, very low Cd concentrations have been shown to stimulate the growth of in vitro cell cultures of plants and *Aspergillus nidulans*, but following increase concentrations and accumulation of the metal a strong inhibitory effect was observed (Fornazier et al., 2002; Guelfi et al., 2003). Moreover, an even much lower concentration of Cd has also shown to stimulate DNA synthesis (Von Zglinicki et al., 1992).

A distinct group of plants have been shown to tolerate high metal concentrations. These hyperaccumulating plants are capable of sequestering metals in their tissues at remarkably high concentrations that would be toxic to most organisms, being important for the phytoremediation of metal-polluted soils. The majority of the research carried out so far has focused on the physiological mechanisms of metal uptake, transport and sequestration, but little is known about the genetic basis of hyperaccumulation. There are known cases of major genetic polymorphisms in which some members of a species are capable of hyperaccumulation and others are not. This is in contrast to the related phenomenon of metal tolerance, in which plant species that posses any metal tolerance are polymorphic, evolving tolerance only in local populations on metalliferous soils (Pollard et al., 2002). Although some degree of hyperaccumulation occurs in all members of the species that can hyperaccumulate, there is evidence of quantitative genetic variation in the ability to hyperaccumulate, both between and within populations (Pollard et al., 2002). The genetic basis of Cd tolerance and hyperaccumulation investigated in *Arabidopsis halleri*, has demonstrated that Cd tolerance may be governed by more than one major gene (Bert et al., 2003). Cd tolerance and accumulation are independent characters and Cd and Zn tolerance co-segregate suggesting that they are under a pleiotropic genetic control (Bert et al., 2003). In *Thlaspi caerulescens* the Cd and Zn accumulation is governed by multiple genes, but Cd tolerance and accumulation are independent traits (Zha et al., 2004). Tolerance (adaptive) is supported by (constitutive) detoxification mechanisms, which in turn rely on (constitutive) homeostatic processes (di Toppi and Gabbrielli, 1999). The mechanisms of Cd tolerance and hyperaccumulation in *Thlaspi caerulescens* hairy roots have been shown to be due to the ability of this plant species to withstand the effects of plasma membrane depolarisation (Boominathan and Doran, 2003).

Various clones of *Salix* spp. have contrasting characteristics of accumulation and translocation of Cd to shoots (Lux et al., 2004). Clones exhibiting low accumulation of Cd and high tolerance had smaller meristematic zones and more extensive vacuolation of cells in the root apices, than the clones characterized by high accumulation of and high sensitivity to Cd (Lux et al., 2004). The proportion of root apoplastic barriers, exodermis and endodermis as well as epidermis was significantly increased in clones with a higher tolerance to Cd ions, indicating that these tissues are involved in the protection of the root against the toxic effects of Cd (Lux et al., 2004). *Sedum alfredii* Hance has been identified as a new Zn hyperaccumulating plant species that also exhibited an extraordinary ability to tolerate and hyperaccumulate Cd (Yang et al., 2004). The finding of Cd/Zn hyperaccumulation in *S. alfredii* Hance provides an important new plant material for the understanding of the mechanisms of Cd/Zn co-hyperaccumulation and for phytoremediation of the heavy metal contaminated soils (Yang et al., 2004).

4. OXIDATIVE STRESS

The accumulation of dioxygen (O_2) in the Earth's atmosphere allowed the evolution of aerobic organisms that use O_2 as the terminal electron acceptor, thus providing a higher yield of energy compared with fermentation and anaerobic respiration. In aerobic metabolism, the complete breakdown of one molecule of glucose yields a total of 38 molecules of ATP, whereas the anaerobic breakdown of glucose to ethanol and CO_2 yields only 8 ATP (Foyer and Noctor, 2000). In its ground state, molecular O_2 is relatively unreactive, but the production of reactive oxygen species (ROS) such as superoxide ($O_2^{\bullet-}$), hydrogen peroxide (H_2O_2), hydroxyl radicals ($^{\bullet}OH$) and singlet oxygen (O_2^1), is an unavoidable consequence of aerobic metabolism. In plants, ROS can be produced in reactions occurring in mitochondria, chloroplasts and peroxisomes (Foyer and Noctor, 2000) in all organs including nitrogen-fixing nodules, with ROS being used as a weapon against invading pathogens in the oxidative burst (Moller, 2001). However, little is known about the ROS scavenging properties of the nucleus, which might contain redox sensitive transcription factors (Delaunay et al., 2000).

In general, $O_2^{\bullet-}$ can arise when electrons are misdirected and donated to oxygen. An estimated 1% of the total O_2 consumption of a plant tissue goes to ROS production (Moller, 2001). Mitochondrial electron transport, for example, is a well-documented source of $O_2^{\bullet-}$ radicals, as is the electron transport chain of the photosynthetic apparatus within the chloroplasts. An additional problem for chloroplasts is the transfer of excitation energy from chlorophyll to oxygen, which can generate O_2^1 (Bowler et al., 1992). Chloroplasts are at risk to oxygen toxicity since molecular O_2 can be photo-reduced to $O_2^{\bullet-}$ and subsequently to H_2O_2 in thylakoids (Mehler, 1951). The photoreducing site of O_2 in chloroplasts is PSI, and its primary product is the $O_2^{\bullet-}$ radical. Therefore, in thylakoids H_2O_2 is photoproduced via disproportionation of $O_2^{\bullet-}$, but not directly through the two-electron reduction of O_2 (Asada, 1999). Actively respiring plant mitochondria produce ROS at high rates, a more reduced electron transport chain gives more ROS, and the main sites of production are the respiratory complexes I and III (Bartoli et al., 2004). In mammalian cells the mitochondria are the major source of ROS, which has also been suggested for nonphotosynthesizing plant cells (Moller, 2001).

The redox cascades of the photosynthetic and respiratory electron transport chains not only provide the driving forces for metabolism but also generate redox signals (Foyer and Noctor, 2003; Apel and Hirt, 2004).

H_2O_2 function as a signalling molecule in plants and it is also involved in the regulation of gene expression by abiotic stresses (Neill et al., 2002). H_2O_2 is also a product of the microbody-associated β-oxidation of fatty acids and peroxisomal photorespiration reactions (Scandalios, 1993). Peroxisomes are subcellular organelles with an essentially oxidative type of metabolism. Like chloroplasts and mitochondria, plant peroxisomes also produce $O_2^{\bullet-}$ and there are, at least, two sites of $O_2^{\bullet-}$ generation: one in the organelle matrix, the generating system being xanthine oxidase, and another site in the peroxisomal membranes dependent on NAD(P)H. In peroxisomal membranes, three integral polypeptides with molecular masses of 18, 29 and 32 kDa, from a small electron transport chain, are involved in the generation of $O_2^{\bullet-}$ radicals and under certain conditions of plant stress, the release of peroxisomal membrane-generated $O_2^{\bullet-}$ can be enhanced, producing oxidative stress (del Río et al., 2002). Peroxisomes connect biosynthetic and oxidative metabolic routes and compartmentalize lethal steps of metabolism such as the formation of ROS and glyoxylate, thus preventing poisoning of the cell and futile recycling (Igamberdiev and Lea, 2002). In the last decade, several studies have revealed new sources of ROS in plants, including NADPH oxidases, amine oxidases and cell-wall-bound peroxidases, which participate in the production of ROS during pathogen defence and programmed cell death (PCD) (Lam, 2004; Laloi et al., 2004). Moreover, the presence of nitric oxide synthase (NOS) in plant peroxisomes indicates the possible function of these organelles in plant cells as a source of signal molecules like nitric oxide (NO), $O_2^{\bullet-}$, H_2O_2 and S-nitrosoglutathione (GSNO) (del Rio et al., 2002). All ROS are extremely reactive and cytotoxic to all organisms (Scandalios, 1993). It is generally assumed that the $^{\bullet}OH$ and O_2^1 are so reactive that their production must be minimized. Several Calvin-cycle enzymes within chloroplasts are extremely sensitive to H_2O_2, and high levels of H_2O_2 directly inhibit CO_2 fixation. H_2O_2 has also been shown to be active with mixed function oxidases in marking several types of enzymes for proteolytic degradation (Scandalios, 1993). However, $O_2^{\bullet-}$ and H_2O_2 are relatively unreactive in comparison with other ROS, but in the presence of metal ions (such as Fe) a series of cascade reactions are initiated resulting in the production of the $^{\bullet}OH$ in the Haber-Weiss reaction and other destructive species such as lipid peroxides (Bowler et al., 1992).

Haber-Weiss reaction:

$$O_2^{\bullet-} + H_2O_2 \xrightarrow{Fe^{2+},\ Fe^{3+}} OH^- + O_2 + {}^{\bullet}OH$$

$^{\bullet}OH$ (and derivatives) are the most potent oxidants known, as they are able to rapidly and indiscriminately attack virtually all macromolecules, leading to serious damage in cellular components (Scandalios, 1993), causing lipid peroxidation, denaturation of proteins, mutation of the DNA (Bowler et al., 1992), DNA lesions, often leading to irreparable metabolic dysfunction and cell death (Scandalios, 1993). In addition, O_2^1, which is formed when excitation energy is transferred to O_2, also produces deleterious effects (Bowler et al., 1992). In higher plants, this energy is readily obtained from light quanta via such transfer molecules as chlorophyll (Scandalios, 1993). $^1O_2$ can transfer its excitation energy to other biological molecules or react with them, forming endoperoxides or hydroperoxides (Bowler et al., 1992).

As already mentioned, under normal growth conditions the production of ROS in cells occurs but is low, however, adverse environmental factors that disrupt the cellular homeostasis enhance the production of ROS. Such factors as UV light and other forms of radiation, herbicides (e.g. paraquat, diquat), pathogens (e.g. *Cercospora*), certain injuries, hyperoxia, ozone, temperature fluctuations, and various other stresses are known to induce free radical formation in most aerobic organisms (Scandalios, 1994). Cd stress is another agent that causes oxidative stress as observed in scots pine roots. Cd treatment enhanced H_2O_2 production, that simultaneously inhibited the systems involved in its removal, i.e. glutathione (GSH), glutathione reductase (GR), catalase (CAT) and ascorbate peroxidase (APX), and resulted in elevated activities of superoxide dismutase (SOD) (Schützendübel et al., 2001).

Oxidative stress is often related with other metabolic effects. For example, Cd stress lead to stomatal closure that limits CO_2 availability for photosynthetic carbon assimilation (Sandalio et al., 2001). Under such condition and high light, excess $O_2^{\bullet-}$ production in the chloroplast can result in photoinhibition and photooxidation damage (Scandalios, 1993). The rapid increase in ROS concentration is termed the oxidative burst (Apostol et al., 1989). It has become evident that ROS generated during pathogen attack and abiotic stress situations are recognized by plants as a signal for triggering defence responses, cell death, development, redox-regulated gene expression and the action of kinases and phosphatases in redox signal transduction (Vranová et al., 2002).

Oxidative stress occurs when there is a serious imbalance in any cell compartment between production of ROS and antioxidant defence, leading to damage. Different species show different responses to metal toxicity (Gallego et al., 1996). Although information focused on the relationship of heavy metals and oxidative stress in plants has been available in more recent years, it is still difficult to draw a general conclusion about critical toxic metal concentrations in soils. The occurrence of ROS and symptoms of oxidative injury also been observed in plants exposed to Cd (Romero-Puertas et al., 1999; Schützendübel et al., 2001).

The induction of lipid peroxidation is one the most damaging effects of ROS and an indication of their production due to their reaction with unsaturated fatty acids, causing peroxidation of essential membrane lipids in the plasmalemma or intracellular organelles (Scandalios, 1993). The ROS production process as well as some of its by-products may severely affect the functional and structural integrity of biological membranes, resulting in increase of the plasma membrane permeability, which lead to leakage of K^+ ions and other solutes and may finally cause cell death (Chaoui et al., 1997). Peroxidation damage of the plasmalemma also leads to leakage of cellular contents and rapid desiccation, whereas damage to intracellular membrane can affect respiratory activity in mitochondria, cause pigment breakdown, and cause loss of carbon-fixing ability in chloroplasts (Scandalios, 1993).

Oxidation of polyunsaturated lipids involves an allylic hydrogen abstraction followed by insertion of molecular oxygen; the resulting peroxyl radicals abstract hydrogens to form lipid hydroperoxides (LOOHs). Malondialdehyde (MDA) is one of several low molecular weight end products formed via the decomposition of certain primary and secondary lipid peroxidation products (Liu et al., 1997). MDA originates from fatty acids containing more than two methylene-linked double bonds, showing that tri-unsaturated fatty acids are the in vivo source of up to 75% of MDA. The abundance of the combined pool of free and reversibly bound MDA does not change dramatically during stress, although a significant increase in the free MDA pool under oxidative conditions has been observed (Weber et al., 2004). The half-life of infiltrated MDA indicated rapid metabolic turnover/sequestration. Exposure of *A. thaliana* plants to low levels of MDA, using a recently developed protocol, powerfully upregulated many genes in a microarray system, with a bias towards those implicated in abiotic/environmental stress (e.g. ROF1 and XERO2). In contrast to the activities of other reactive electrophile species (i.e. small vinyl ketones), none of the pathogenesis-related (PR) genes tested responded to MDA (Weber et al., 2004). The use of structural mimics of MDA isomers suggested that the propensity of the molecule to act as a cross-linking/modifying reagent might contribute to the activation of gene expression. Changes in the concentration/localization of unbound MDA in vivo could strongly affect stress-related transcription (Weber et al., 2004).

O_2, although essential for the existence and survival of aerobic life, presents living organisms with a variety of physiological challenges collectively termed oxidative stress. These challenges may be greater for plants in relation to other eukaryotes because of their stationary lifestyle under constantly changing environments and because plants consume O_2 during respiration and generated it during photosynthesis. Furthermore, among all organisms, the cellular concentration of O_2 is highest in plants, which subject plants to potentially higher production of active O_2 and, consequently, oxidative damage (Scandalios, 1993).

4.1. Plant Responses to Cd Stress

The concept of stress in plants can be defined as any unfavourable condition, or substance that affects or blocks metabolism, growth or development. There is a difference between short-term and long-term stress effects as well as between low stress events, which can be partially compensated by acclimation, adaptation and repair mechanisms, and high stress or chronic stress events, which can cause considerable damage and may eventually lead to cell and plant death (Lichtenthaler, 1996). Before exposure to stress, plants are assumed to be in a certain standard condition, which is an optimum within the limits set by the growth, light and mineral supply conditions of the environment. Three distinct phases among the plant stress responses were suggested by Larcher (1987), to which a fourth has been added by Lichtenthaler and Rinderle (1988):

- Response Phase (beginning of stress): alarm reaction or general alarm syndrome (GAS) (deviation from the functional norm, decline of vitality and catabolic processes exceed anabolism);
- Restitution Phase (continuing stress): stage of resistance (adaptation processes, repair processes and reactivation);
- End Phase (long-term stress): stage of exhaustion (stress intensity too high, overcharge of the adaptation capacity and chronic disease or death);
- Regeneration Phase: partial or full regeneration of the physiological function when the stressor is removed and the damage was not too high.

At the beginning of stress when plants are suddenly confronted with a critical situation, with a consequent decline of several physiological functions, they will activate metabolic pathways, repair processes and long-term metabolic and morphological adaptations of the GAS (McKersie and Leshem, 1994), that may also stand for general acclimation syndrome or general adaptation syndrome. Thus, the GAS seems to represent a generalized effort by the organism to adapt itself to new conditions (Leshem and Kuiper, 1996). In tissues affected by stress, a "local adaptation syndrome" (LAS) develops, which has a close interaction with GAS and chemical alarm signals are sent out by stressed tissues inducing adaptive hormones. Plants under drought stress by "split-root" experimentation display a LAS reaction which leads to a GAS (Davies and Zhang, 1991).

The mechanisms conferring resistance to toxic metal ions in plants include: physical avoidance of contaminated areas, exudation of complexing agents into the rhizosphere, binding in the cell wall, efflux of metal ions from the symplasm, prevention of upward transport of metal ions into above-ground parts, complexation with various ligands in the symplasm, transport of metal-ligand complexes into the vacuole, storage of metal ions in the vacuole by complexation with vacuolar ligands and formation of metal-resistant enzymes to minimize the internal injury caused by toxicity (Prasad, 1997).

Due to the different metabolic processes described earlier that can produce ROS, enhanced by Cd stress, various defence mechanisms have emerged. These defence systems can remove, neutralize or scavenge oxy-radicals and their intermediates (Foyer and Noctor, 2003). Plants have therefore developed several mechanisms to prevent or alleviate the damage that oxygen radicals can cause in different tissues.

5. ANTIOXIDANT SYSTEMS

The first line of defence against oxidative stress in plant is the avoidance of ROS production. Thus, in plant mitochondria the electron transport chain is adequately oxidized by maintaining a balance between substrate availability and ATP requirement; activation of alternative oxidase; activation of uncoupler proteins and activation of rotenone-insensitive NAD(P)H dehydrogenases (Moller, 2001). Once formed, ROS must be detoxified as efficiently as possible to minimize damage, thus, the detoxification mechanisms constitute the second line of defence against the detrimental effects of ROS (Moller, 2001). The term antioxidant describes any compound capable of quenching ROS without itself undergoing conversion to a destructive radical. Antioxidant enzymes are considered as those that either catalyse such reactions or are involved in the direct processing of ROS. Hence, antioxidants and antioxidant enzymes function to interrupt the cascades of uncontrolled oxidation (Noctor and Foyer, 1998). Finally, the third line of defence against ROS is the repair of the damage (Moller, 2001), which is outside the scope of this chapter.

5.1. Non-Enzymatic Antioxidant System

The plant cell non-enzymatic antioxidant system is essentially composed of relatively high concentrations of ascorbate, glutathione, and α-tocopherol, which are efficient oxyradical scavengers (Bowler et al., 1992). Ascorbate is a major primary antioxidant (Smirnoff et al., 2001) that reacts with ROS (Buettner and Jurkiewicz, 1996). Glutathione (GSH), is another major redox buffer in most aerobic cells, and plays an important role in physiological functions, including redox regulation, conjugation of metabolites, detoxification of xenobiotics and homeostasis and cellular signalling that trigger adaptive responses. These functions depend of the concentration and/or the redox state of leaf GSH pools (Noctor et al., 2002).

The equilibrium between different antioxidants must be tightly controlled, as it has been shown that in cells with enhanced GSH biosynthesis in chloroplasts, there is oxidative stress damage, due to changes in the overall redox state of the chloroplasts (Apel and Hirt, 2004).

5.1.1. *Ascorbate*

All plants can synthesize ascorbate, which can accumulate in both photosynthetic and nonphotosynthetic tissues. Ascorbate is a major primary antioxidant, reacting directly with $^{\bullet}OH$, $O_2^{\bullet-}$ and O_2^1. Ascorbate is also a powerful secondary antioxidant, reducing the oxidized form of α-tocopherol, an important antioxidant in nonaqueous phases. In addition to its importance in the ascorbate-glutathione cycle, ascorbate plays a role in preserving the activities of enzymes that contain prosthetic transition metal ions (Noctor and Foyer, 1998).

The structure of the ascorbic acid was described in 1933, but its biosynthesis in plants remained elusive until recently. The Smirnoff-Wheeler pathway of ascorbate biosynthesis in plants was discovery in 1998, which is based on the description in 1996 of an *A. thaliana* mutant deficient in ascorbate (Smirnoff et al., 2001). In this pathway, the ascorbate is generated from glucose and proceeds through the intermediates GDP-D-mannose, L-galactose, and L-galactono-1, 4-lactone (Smirnoff et al., 2001). The penultimate step of ascorbate synthesis is catalysed by L-galactose dehydrogenase (L-GalDH), an enzyme that oxidizes L-galactose to L-galactono-1,4-lactone (L-GalL). The enzyme has been purified from pea seedlings and cloned from *A. thaliana*. Overexpression of the *GalDH* gene in tobacco had no effect, whilst antisense suppression in *A. thaliana* caused a reduction in ascorbate content (Gatzek et al., 2002). The final step in ascorbate synthesis involves the oxidation of L-galactono-1, 4-lactone catalysed by L-galactono-1,4-lactone dehydrogenase (GalLDH). The enzyme is an integral protein of the inner mitochondrial membrane and cytochrome c is the electron acceptor for the GalLDH reaction. It has been proposed that the rate of mitochondrial respiration could control the rate of ascorbate synthesis (Millar et al., 2003). Since ascorbate is synthesized in both green and non-green tissues, its formation is not dependent on photosynthesis (Rautenkranz et al., 1994). However, the amount of ascorbate is upregulated by light (Smirnoff et al., 2001). Vacuolar ascorbate concentrations are relatively low (Rautenkranz et al., 1994), whilst high values for chloroplastic and cytosolic compartments have been calculated (Foyer and Lelandais, 1996). Ascorbate is also the major and probably the only antioxidant buffer in the apoplast, where it is considered important in the scavenging of ROS (Luwe, 1996). Low concentrations of ascorbate can inhibit oxidation reactions of phenoxyl radicals, because of their rapid reactions with ascorbate (Takahama, 1993). The apoplastic enzyme ascorbate oxidase (AO) regulates the redox state of the apoplastic ascorbate pool and the oxidative burst and AO can modify the apoplastic redox state in such a way as to modify receptor activity and signal transduction involved in defence and growth (Pignocchi and Foyer, 2003). The ascorbate redox system consists of L-ascorbate, monodehydroascorbate (MDHA) and dehydroascorbate (DHA). Both oxidized forms of ascorbate are unstable in aqueous environments (Smirnoff et al., 2001) and they can be catabolized to two and four carbon products such as oxalate and tartrate, which accumulate to high concentrations (Loewus, 1988). Cd stress has been shown to lead to a decrease in ascorbate content in soybean roots and nodules (Balestrasse et al., 2001), in leaves of *A. thaliana* (Skorzynska-Polit et al., 2003) and cucumber chloroplasts (Zhang et al., 2003), whereas ascorbate remained unaffected in poplar roots treated with Cd (Schützendübel et al., 2002). Similarly, ascorbate/dehydroascorbate ratios were diminished in soybean roots and nodules (Balestrasse et al., 2001). However, in sunflower plants, the increase in ascorbate based detoxification mechanisms did not provide complete protection against the oxidative stress imposed by combined Cd and ozone treatments because a photosynthetic decrease was observed due to lipid peroxidation and protein oxidation (Di Cagno et al., 2001).

5.1.2. Glutathione

The tripeptide glutathione (γ-glutamylcysteinylglycine, GSH) is the major source of non-protein thiols in most plant cells (Xiang and Oliver, 1998). GSH is involved in many important pathways (Noctor and Foyer, 1998), which includes a major role in sulphur transport, where it regulates sulfate uptake in the root (Creissen et al., 1994), and a central role in plant defence mechanisms. GSH is the precursor for the synthesis of phytochelatins, metal-binding peptides involved in heavy metal tolerance and sequestration. GSH is the substrate for glutathione S-transferase, enabling neutralization of potentially toxic xenobiotics (Beck et al., 2003). Moreover, GSH has an important function in the removal of H_2O_2 from plant cells, via the ascorbate-glutathione cycle. In addition to its effects on the expression of defence genes, GSH may also be involved in the redox control of cell division (Noctor and Foyer, 1998).

GSH is considered essential in plant cells, although some plants contain tripeptide homologs of GSH, in which the carboxy terminal glycine is replaced by other amino acids, such as homoglutathione (γ-glutamylcysteinylalanine), hydroxymethylglutathione (γ-glutamylcysteinylserine), that may partly or totally replace GSH (Creissen et al., 1994). Another GSH homologue, γ-glutamylcysteinylglutamate, was discovered in maize seedlings exposed to cadmium (Prasad, 1997). The oxidized forms of GSH homologues can be reduced by yeast glutathione reductase (GR), suggesting similar physiological and biochemical roles to the more widespread GSH (Noctor and Foyer, 1998).

The pathway of GSH synthesis has been elucidated and appears to be common to all organisms that contain GSH (Noctor and Foyer, 1998). GSH is synthesized from glutamate, cysteine and glycine by a two-step ATP-dependent reaction. The first reaction forms γ-glutamylcysteine (γ-EC) from glutamate and cysteine by the enzyme γ-glutamylcysteine synthetase (γ-ECS), which is encoded by the *gsh1* gene. GSH is then synthesized by the ligation of γ-EC and glycine in a reaction catalysed by the enzyme GSH synthetase (GS), which is encoded by the *gsh2* gene (Xiang and Oliver, 1998). GSH can be synthesized in the cytosol, chloroplast and mitochondria (Moran et al., 2000) and the cellular concentration can vary considerably between species and in different organs or development stages in the same plant (Noctor and Foyer, 1998). Chloroplasts contain a high GSH content due the role in the ascorbate-glutathione cycle (Creissen et al., 1994).

When GSH is oxidized as part of its antioxidant activity, it forms glutathione disulfide (GSSG), the oxidized form of GSH (Xiang and Oliver, 1998). In response to stress, plants can increase the activity of GSH biosynthetic enzymes (Vernoux et al., 2000) and GSH levels (Noctor et al., 2002). However, the mechanism behind H_2O_2-induced GSH synthesis is unclear (Xiang and Oliver, 1998). Regulation of thiol contents and alteration of thiol pools in maize exposed to Cd could be regulated by intermediates and effectors of GSH synthesis (Prasad, 1997). GSH synthesis may also be stimulated by induction of enzymes involved in sulphate uptake and reduction (Heiss et al., 1999). The modulation of GSH contents transmits information through diverse signalling mechanisms that include the release of Ca^{2+} to the cytosol and the establishment of an appropriate redox potential for thiol/disulphide exchange (Gomez et al., 2004). These results support the idea that the GSH concentration is controlled at multiple levels.

The role of GSH in the antioxidative defence system provides a rationale for its use as a stress marker, although responses of GSH concentrations and redox states have not been consistent among the available publications (Tausz et al., 2004). Several limitations have to be highlighted about the suggested stress response concept because of the importance of the GSH system relative to other components of the photoprotective and antioxidative defence system. Thus, within such response patterns, the GSH system is a valuable stress marker in ecophysiological studies (Tausz et al., 2004).

GSH synthesis is increased with the elevation of cysteine concentration, as the rate limiting step for GSH and consequently phytochelatin (PC) synthesis is considered to be the availability of reduced sulfur for cysteine synthesis (Goldsbrough, 1998). Cysteine is the first committed molecule in plant metabolism containing both sulphur and nitrogen and the regulation of its biosynthesis via the sulfur reduction pathway involving at least 6 enzymes, is critically important. Cysteine itself is required for the production of an abundance of key metabolites in diverse pathways (Hesse et al., 2004). Upon heavy metal stress, some genes involved in the pathway are transcriptionaly activated, resulting in an elevation of enzyme activity (Harada et al., 2001). Furthermore, since supplying glutamate has little effect on GSH content, cysteine may regulate the rate of GSH and phytochelatin synthesis (Noctor and Foyer, 1998).

Another fundamental GSH synthesis control point is the feedback inhibition of γ-ECS by GSH. *In vitro* studies with the enzymes from tobacco and parsley cells showed that the plant γ-ECS was inhibited by GSH (Noctor and Foyer, 1998). Under stress conditions, the oxidation of GSH to GSSG, or the synthesis of phytochelatin would decrease the GSH concentration and thus stimulate γ-ECS activity (Noctor and Foyer, 1998). Cd-resistant tomato cells containing increased levels of GSH were shown to possess two-fold higher extractable γ-ECS activities than susceptible cells. Exposure of roots or cultured cells to Cd resulted in an increased rate of GSH synthesis and concomitant increases in γ-ECS and GS activities (Noctor and Foyer, 1998).

Cd accumulation and tolerance have been correlated with the expression level of *gshII* with high concentrations of GSH and phytochelatin being found in Cd-treated GS plants when compared to wild-type plants (Zhu et al., 1999a). These data suggested that in the presence of Cd the GS enzyme is rate limiting for the biosynthesis of GSH and PCs, and that the overexpression of GS and genes involved in GSH and PC syntheses offers a promising strategy for the production of plants with superior phytoremediation capacity (Zhu et al., 1999a).

Antioxidant activity in the leaves and chloroplasts of *Phragmites australis* Trin. (Cav.) ex Steudel was associated with a large pool of GSH, protecting the activity of many photosynthetic enzymes against the thiophilic binding of Cd, exerting a direct important protective role in the presence of Cd (Pietrini et al., 2003). GSH concentrations also increased with increasing Cd in romaine lettuce (Maier et al., 2003). However, Cd stress lead to a decrease in GSH in soybean roots (Balestrasse et al., 2001), sunflower leaves (Gallego et al., 1996), maize seedlings (Rauser, 1990), *Pisum sativum* (Ruegsegger et al., 1990), scots pine roots (Schützendübel et al., 2001), cucumber chloroplasts (Zhang et al., 2003), poplar roots (Schutzebdubel et al., 2002) and rice leaves (Hsu and Kao, 2004), whilst there was no change in soybeans nodules (Balestrasse et al., 2001). The reduction in GSH content in Cd treated plants is probably due to the induction of phytochelatin synthesis, which consumes GSH.

5.1.3. Phytochelatins (PCs)

Cd may be detoxified in plants by a family of sulphur rich peptides termed phytochelatins (PCs) that are able to bind Cd and some other heavy metals (Cobbett, 2000). PC peptides have a higher affinity for Cd than GSH (Schmöger et al., 2000). The chelation of Cd by PC is important to avoid the accumulation of free Cd ions inside plant cells, and thus preventing their toxicity. The chemical structure of PCs suggests that they are not formed as a direct result of the expression of a metal tolerance gene, but rather a product of a biosynthetic pathway (Cobbett, 2000). The peptides are structurally related to GSH and contain a varying number (normally 2-5) of glutamate and cysteine, linked through the γ-carboxyl group of glutamate. Synthesis of these PCs can be induced very rapidly in tissue culture cells and roots, and is accompanied by a fall in the concentration of GSH, following the addition of Cd or other heavy metals (Grill et al., 1987).

Phytochelatin synthase (PCS) has been characterized as a specific γ-glutamyl cysteine dipeptidyl transpeptidase (EC 2.3.2.15) (Vatamaniuk et al., 2004), which carries out the conversion of GSH to PCs, and has been shown to be stimulated by Cd (Cobbett, 2000). In most organisms, PCS is expressed constitutively and the accumulation of PCs occurs within a few minutes of initial exposure to Cd or other heavy metals, indicating that PCS is primarily regulated by enzyme activation of a pre-formed protein (Cobbett and Goldsbrough, 2002). *Arabidopsis thaliana* cad2 mutants lacking γ-glutamylcysteine synthetase are sensitive to Cd and are unable to synthesize GSH, whilst the cad1 mutants lacking PCS are sensitive to Cd and accumulate GSH (Howden et al., 1995; Cobbett et al., 1998).

Exposure of *Brassica juncea* (Heiss et al., 2003), *Phragmites australis* roots (Ederli et al., 2004), *Cuscuta reflexa* (Srivastava et al., 2004) to Cd have been shown to stimulate the synthesis of PCs, which rapidly form a "low molecular weight" (LMW) complex with Cd (Vögelli-Lange and Wagner, 1996). These complexes acquire acid labile sulphur (S^{2-}) at the tonoplast, and form a "high molecular weight" (HMW) complex (Speiser et al., 1992) with a higher affinity towards Cd ions. Free Cd ions can also enter the vacuole by means of a $Cd^{2+}/2H^+$ antiport system (Gries and Wagner, 1998). In the vacuole, the HMW complex dissociates and the released Cd can be complexed by vacuolar organic acids (Krotz et al., 1989).

The majority of the experimental studies have argued that PCs play a role only at high levels of Cd exposure, which are not normally found in natural environments. For instance, a PC-deficient mutant of *Arabidopsis thaliana* was shown to be sensitive to concentrations of Cd as low as 0.6 µM in agar medium (non-polluted soils contain Cd concentrations ranging up to 0.3 µM) (Howden et al., 1995).

In animals, cyanobacteria and fungi, Cd and other heavy metals can be complexed and detoxified by metallothioneins, a group of gene-encoded cysteine-rich (about 30%) peptides (di Toppi and Gabbrielli, 1999). A gene in the ascomycete fungus *Magnaporthe grisea*, *Magnaporthe metallothionein 1* (*MMT1*), which is highly expressed throughout growth and development by *M. grisea* and encodes an unusual 22-amino acid metallothionein-like protein containing only six Cys residues has been identified. The MMT1-encoded protein shows a very high affinity for Zn and can act as a powerful antioxidant and it may play a novel role in fungal cell wall biochemistry that is required for fungal virulence (Tucker et al., 2004). The expression of a metallothionein gene (*OsMT2b*) was synergically down-regulated by OsRac1 and *Oryza sativa* blast-derived elicitors. Transgenic plants overexpressing *OsMT2b* showed increased susceptibility to bacterial blight and blast fungus (Wong et al., 2004). OsMT2b-overexpressing cells showed reduced elicitor-induced H_2O_2 production. Homozygous OsMT2b::Tbs17-inserted mutant and OsMT2b-RNAi-silenced transgenic cells showed significantly higher elicitor-induced H_2O_2 production than the wild-type cells. The recombinant

OsMT2b protein possessed $O_2^{\bullet-}$ and $^{\bullet}OH$ radical-scavenging activities. These results showed that OsMT2b is an ROS scavenger and its expression is down-regulated by OsRac1, thus potentiating ROS, which function as signals in resistance response (Wong et al., 2004). Metallothioneins are not typical of plant species, and there is no clear indication in the literature of the existence in higher plants of metallothioneins induced by Cd. However, in *A. thaliana* two Cu-induced metallothioneins were isolated using polyclonal antibodies (Murphy et al., 1997).

5.1.4. Others Non-Enzymatic Antioxidants

Cysteine, hydroquinones, mannitol, vitamin E, some alkaloids and β-carotene, are also important antioxidants. Carotenoids, which are essential components of thylakoid membranes, can effectively quench the excited triplet state of chlorophyll and/or O_2^1 (Scandalios, 1993). In contrast to ascorbate and GSH, ROS detoxification in plants by flavonoids and carotenoids has not been investigated in great detail. In *A. thaliana*, the overexpression of β-carotene hydroxylase lead to increased amounts of xanthophyll in the chloroplasts and resulted in tolerance towards oxidative stress induced by high light (Davison et al., 2002). The phytotoxic effects of Cd inhibited the synthesis of carotenoids in mustard seedlings (Fargasova, 2004) and leaves of *Amaranthus lividus* seedlings (Bhattacharjee and Mukherjee, 2003).

Anthocyanins have been considered important to mitigate photooxidative injury in leaves, by shielding chloroplasts from excess high energy quanta and scavenging ROS (Neill and Gould, 2003). According to Neill and Gould (2003), anthocyanins offer effective and versatile protection to leaves without significantly compromising photosynthesis, as observed in *Lactuca sativa*. Krupa et al. (1996) have demonstrated using the first leaves of rye seedlings, that the effect of Cd, Ni, Pb and Zn on the content of anthocyanins is not directly related to the individual metal toxicity, since less anthocyanins were accumulated following treatment with more toxic metals (Krupa et al., 1996). Moreover, the level of chlorophyll and total carotenoids is postulated as a simple and reliable indicator of heavy metal toxicity for higher plants (Krupa et al., 1996).

5.2. Enzymatic Antioxidant System

Chloroplasts and thylakoids have been identified to be the photogenerating site of O_2 from water, but no O_2 uptake was considered to occur in them (Asada, 1999). Efficient destruction of $O_2^{\bullet-}$ and H_2O_2 requires the action of several antioxidant enzymes acting in synchrony. The major ROS scavenging mechanisms of plants include the enzymes, superoxide dismutase (SOD), catalase (CAT), ascorbate peroxidase (APX), dehydroascorbate reductase (DHAR), monodehydroascorbate reductase (MDHAR), glutathione reductase (GR) and glutathione peroxidase (GPX). Cell oxidative stress levels are determined by the amounts of $O_2^{\bullet-}$, H_2O_2, and $^{\bullet}OH$ radicals. H_2O_2 can be directly metabolized by peroxidases, particularly those of the cell wall and by CAT in the peroxisome (Polidoros and Scandalios, 1999). In the chloroplast, $O_2^{\bullet-}$ is converted by SOD into H_2O_2, which is then detoxified to H_2O and O_2 by the glutathione-ascorbate cycle, which involves the action of several enzymes, including GR (Noctor et al., 2002). In addition, some of the alternative systems channelling electrons in the electron transport chains of the chloroplasts and mitochondria, can decrease ROS production through alternative oxidases (AOXs) that reduce O_2 to water (Maxwell et al., 1999).

The activity and expression of genes encoding antioxidant enzymes have been shown to change in some plants when subjected to environmental conditions such as chilling stress (Pinhero et al., 1997), light intensity and type (Willekens et al., 1994), salt stress (Fadzilla et al., 1997), pathogens (Williamson and Scandalios, 1992), herbicides (Donahue et al., 1997) and several gaseous pollutants

(Van Camp et al., 1994; Conklin and Last 1995; Torsethaugen et al., 1997; Azevedo et al., 1998). The responses of antioxidants to heavy metal induced oxidative stress have provided variable and controversial results (Schützendübel et al., 2001; Pinto et al., 2003) for three reasons :- (i) the heavy metals exert different mechanisms of stress induction; (ii) the cell antioxidant stress is compartmentalized and consequently there are differences between the responses of organelles, cells and tissues; (iii) the detoxification and complexation of heavy metals can reduce their stimulatory effects on oxidative stress.

The numerous ROS sources and a complex system of oxidant scavengers provide the flexibility necessary for anti-oxidative metabolism. Given the mechanisms utilised by plants to detoxify ROS, it is clearly important to establish whether exposing plants to Cd causes a detrimental or stimulatory effect on the enzymes involved in this detoxification process. Somewhat surprisingly, the limited number of experiments that have been carried out on the subject, have frequently produced contradictory results.

The main ROS scavenging pathways of plants are:- (1) Water-water cycle in chloroplasts that includes SOD; (2) Ascorbate-glutathione cycle in chloroplasts, cytosol, mitochondria, apoplast and peroxisomes; (3) Glutathione peroxidase (GPX) and CAT in the peroxisomes. Compensatory mechanisms will be induced if the balance of scavenging enzymes is changed. For instance, when CAT activity was reduced in plants, other ROS scavenging enzymes such as APX, GPX and mitochondrial AOX were shown to be upregulated (Yoshimura et al., 2000).

Intact chloroplasts photoreduce O_2 and no H_2O_2 accumulates, in the absence of CAT. These results indicate that chloroplasts have a system to reduce O_2 to water using the electrons derived from water. The stoichiometry of the evolution and uptake of O_2 and the absence of CAT in chloroplasts indicate that the H_2O_2 derived from the $O_2^{\bullet-}$ photogenerated in PSI is reduced to water via a peroxidase reaction using the reductant generated in PSI as the electron donor. The occurrence of an ascorbate-specific peroxidase in spinach chloroplasts, and dehydroascorbate reductase (DHAR) indicate that the electron donor for the peroxidase is ascorbate (Asada, 1999).

The water-water cycle is composed of the following reactions (Asada, 1999):
[1] $2 H_2O \rightarrow 4 [e^-] + 4 H^+ \, O_2$ [photooxidation of water in PSII]
[2] $2 O_2 + 2 [e^-] \rightarrow 2 O_2^{\bullet-}$ [photoreduction of O_2 in PSI]
[3] $2 O_2^{\bullet-} + 2 H^+ \rightarrow H_2O_2 + O_2$ [SOD-catalyzed disproportionation of $O_2^{\bullet-}$]
[4] $H_2O_2 + 2$ ascorbate $\rightarrow 2 H_2O + 2$ MDA [APX-catalyzed reduction of H_2O_2 by ascorbate]
[5] 2 MDA (or DHA) $+ 2 [e^-] + 2 H^+ \rightarrow 2$ ascorbate (or 1 ascorbate) [reduction of oxidized ascorbate]

The most important function of this cycle is the scavenging of $O_2^{\bullet-}$ and H_2O_2 at the site of its generation, shortening their lifetimes to suppress the production of $^{\bullet}OH$ and its interaction with the target molecules. Moreover, the water-water cycle can dissipate excess excitation energy under environmental stress (Asada, 1999).

The ascorbate-glutathione cycle, also called Halliwell-Foyer-Asada cycle (Fig. 1), is an efficient way for plant cells to dispose of the H_2O_2 in certain cellular compartments where this metabolite is produced and CAT is not present (del Rio et al., 2002). This cycle makes use of the non-enzymatic antioxidants ascorbate and GSH in a series of reactions catalysed by four antioxidative enzymes: APX, which has a much higher affinity for H_2O_2 than CAT, and monodehydrosacorbate reductase (MDHAR) or DHAR and GR (Foyer et al., 1997). APX uses two molecules of ascorbate to reduce H_2O_2 to water, with the concomitant generation of two molecules of monodehydroascorbate (MDHA). MDHA is a radical with a short lifetime that, if not rapidly reduced, disproportionates to ascorbate and dehydroascorbate (DHA) (Noctor and Foyer, 1998).

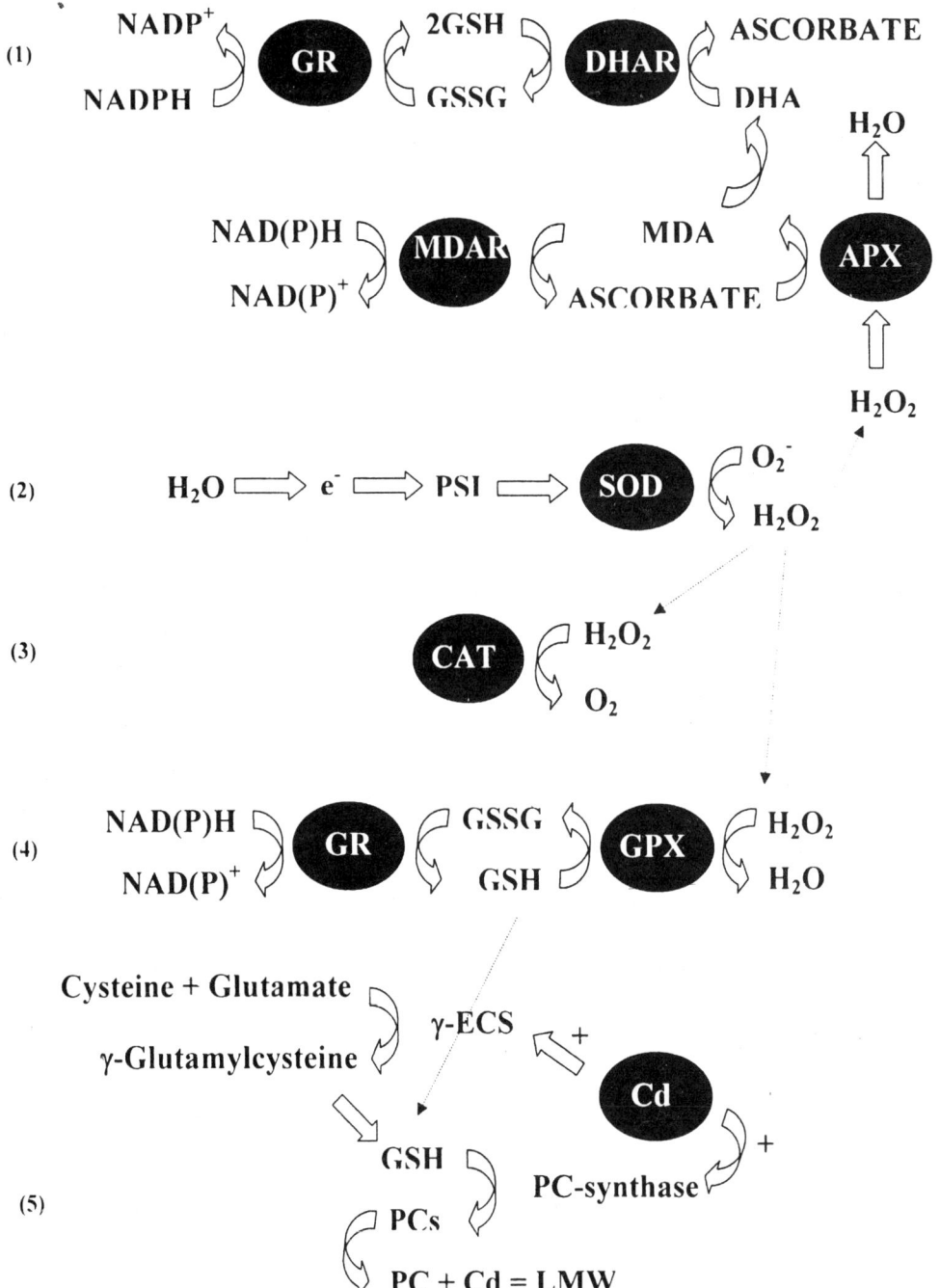

Fig. 1. Pathway for ROS scavenging: (1) ascorbate-glutathione cycle, (2) water-water cycle, (3) CAT, (4) GPX cycle and (5) mechanisms involved in Cd chelation (PCs). LMW – low molecular weight.

The absence of CAT in chloroplasts precludes a role in the protection of the thiol-regulated enzymes of the Calvin cycle. The plant cell can synthesize ascorbate via several routes, but the main route includes a mitochondrial enzyme in the last step. Mitochondria apparently lack the enzymes for GSH synthesis and, therefore, must import GSH from the cytosol (Moller, 2001). During both biotic and abiotic stresses, the mitochondria can be damaged by oxidative stress, because it is very susceptible to oxidative inhibition of function (Moller, 2001, Millar et al., 2003). The ascorbate-glutathione cycle has also been demonstrated in the cytosol and peroxisomes (del Rio et al., 2002).

5.2.1. *Superoxide Dismutase (SOD)*

Superoxide dismutase (SOD, EC 1.15.1.1) was first isolated by Mann and Keilin in 1938 from bovine blood as a green copper protein whose biological function was believed to be Cu storage. Over the years, the enzyme has been variably referred to as erythrocuprein, indophenol oxidase, and tetrazolium oxidase (Scandalios, 1993). The catalytic function of SOD was discovered by McCord and Fridovich in 1969, using a Cu/Zn protein from bovine erythrocytes. Similar Cu/Zn proteins with SOD activity were subsequently isolated from various eukaryotic sources. Mn containing proteins with SOD activity were found in prokaryotes and in the mitochondria of eukaryotes and later, Fe proteins from *Escherichia coli* and algae were shown to possess SOD activity. SOD enzymes are ubiquitous, being widely distributed among O_2-consuming organisms, aerotolerant anaerobes, and some obligate anaerobes (Olmos et al., 2003).

SODs are considered to be the first line of defence against $O_2^{\bullet-}$ and their reaction products are H_2O_2 and O_2 (Alscher et al., 2002). Among the enzymatic mechanisms, the dismutation of $O_2^{\bullet-}$ by SOD has been studied extensively (Milone et al., 2003). No enzymes for the reduction of $O_2^{\bullet-}$ to H_2O_2 or the oxidation of $O_2^{\bullet-}$ to O_2 have been found to date in either plants or other organisms (Apel and Hirt, 2004). The enzyme SOD is unique in that its activity determines the concentration of $O_2^{\bullet-}$ and H_2O_2, the two Haber-Weiss reaction substrates, and it is therefore likely to be central in the defence mechanism, thereby preventing the formation of $^{\bullet}OH$ radical (León et al., 2002).

The catalytic cycle of SOD for the disproportionation of $O_2^{\bullet-}$ is shown below:

$$SOD\text{-}Cu(II) + O_2^{\bullet-} \rightarrow SOD\text{-}Cu(I) + O_2$$
$$SOD\text{-}Cu(I) + O_2^{\bullet-} \rightarrow SOD\text{-}Cu(II) + H_2O_2$$

Where SOD-Cu(II) and SOD-Cu(I) represent oxidized and reduced enzymes, respectively. Three classes of SODs have been reported in plants, which have been classified according to their metal co-factor, manganese (Mn), copper/zinc (Cu/Zn) or iron (Fe) (Pereira et al., 2002). Mn-SOD is located in the mitochondria, although it has also been reported in the chloroplasts of some plants (Azevedo et al., 1998), whereas Fe-SOD, although observed in a more limited number of plant species, is associated with chloroplasts (Vitória et al., 2001), while the abundant Cu/Zn-SODs are generally found in the cytosol of eukaryotic cells and chloroplasts (Azevedo et al., 1998). Chloroplastic SOD is generally the most abundant SOD in green leaves, while in germinating seedlings and in etiolated material, the cytoplasmic mitochondrial SODs are prevalent. This distribution presumably reflects changes occurring in the subcellular sites of oxyradical formation, i.e. during the greening process photosynthetic reactions become more dominant in cell metabolism, necessitating an increase in chloroplastic SOD (Bowler et al., 1994). The presence of Mn-SOD and Cu/Zn-SOD isoforms in isolated plant peroxisomes was reported in pea, watermelon, cucumber, tomato and other plant species (del Rio et al., 2002).

SODs are dimeric enzymes with two identical subunits (Chen and Liu, 1996). Cu/Zn-SOD is a homodimer and contains one atom each of Cu and Zn in each subunit. Cu/Zn-SOD is a stable enzyme that is resistant to denaturating stress including increasing the temperature up to 80°C because of its βbarrel structure, with a low content of α-helix formations. Fe-SOD is not as thermostable as Cu/Zn-SOD, but is stable up to 50°C (Asada, 1999). Sequence data suggest that the three types of SOD fall into phylogenetic families. The Fe-SODs and Mn-SODs are closely related in protein structure and other chemical characteristics, but are unrelated to Cu/Zn-SODs. Hence, Fe-SOD and Mn-SOD may have evolved from a unique ancestor, whereas Cu/Zn-SOD evolved independently and more recently (Chen and Liu, 1996).

Both Cu/Zn-SOD and Fe-SOD are inactivated by H_2O_2, but Mn-SOD is not. This inactivation is due the generation of a hydroxyl radical at the reaction center of the enzyme, after the reverse reaction of the catalytic cycle. The hydroxyl radical generated oxidizes the metal-ligated His residue to 2-oxohistidine, resulting in irreversible inhibition (Asada, 1999). Furthermore, Cu/Zn-SOD is inhibited reversibly by KCN, whereas the other SOD isoenzymes are not (Mallick and Mohn, 2000).

Unlike most other organisms, plants have multiple enzymatic forms (isoenzymes) of SOD. The existence of SOD isoenzymes, their location within cells, tissues, or organelles, and any changes they may undergo during development, or in response to various signals, imply separate metabolic roles for each of SOD isoenzymes (Scandalios, 1993). The number of isozymes of each type of SOD varies greatly from plant to plant, as does the relative abundance of each one. All of the isoenzymes appear to be nuclear encoded and, where necessary, are transported to their organellar locations by means of an amino-terminal targeting sequences (Bowler et al., 1992). Nine nuclear genes (*Sod1*, *Sod2*, *Sod3.1*, *Sod3.2*, *Sod3.3*, *Sod3.4*, *Sod4*, *Sod4A* and *Sod5*) encode the corresponding differentially compartmentalized SOD isoenzymes: SOD-1 (chloroplastic), all SOD-3s (mitochondrial), and SOD-2, SOD-4, SOD-4A and SOD-5 (cytosolic) (Acevedo and Scandalios, 1996; Kliebenstein et al., 1998). Regulation of SOD genes also appears to be very sensitive to environmental stress, presumably as a consequence of increased oxygen radical formation (Bowler et al., 1992). It is conceivable that high level of oxidative stress may result in high SOD protein turnover, resulting in the requirement of new SOD enzyme synthesis to maintain SOD levels sufficiently high for effective protection (Scandalios, 1993). The ubiquity of $O_2^{\bullet-}$ and H_2O_2 suggests that they do not themselves direct the diverse profiles of SOD gene expression (Alscher et al., 2002).

Living cells must maintain a delicate balance between the rates of $O_2^{\bullet-}$ generation and removal. To maintain such a balance, organisms involved elaborate regulatory mechanisms to control the synthesis of SODs in response to different oxidative stimuli (Scandalios, 1993). Mitochondrial Mn-SOD responds to increased oxyradical formation in mitochondria while chloroplastic Fe-SOD responds to such an event occurring in the chloroplasts and cytosolic Cu/Zn-SOD responds to cytosol-localized reactions. The effect of a particular stress on SOD gene expression is thus likely to be governed by the subcellular sites at which oxidative stress is generated (Bowler et al., 1992). For instance, overexpression of mitochondrial Mn-SOD from *Nicotiana plumbagnifolia* in *Nicotiana tabacum* mitochondria (Bowler et al., 1991) and the overexpression of Mn-SOD in maize (Kingston-Smith and Foyer, 2000), protected the transgenic plants produced from oxidative damage.

The SOD activity response to Cd metal stress varies considerably depending on plant species, tissue, stage of development and exposure time. SOD activity in radish leaves exhibited increases in activity in response to Cd treatment, whereas in roots there was no significant variation in SOD activity (Vitoria et al., 2001). Activity staining for SOD in soybean revealed seven isoenzymes in leaves and eight in roots, corresponding to Mn-SOD and Cu/Zn-SOD isoenzymes, and although a clear effect of Cd on plant growth was observed, the activities of the SOD isozymes were unaltered (Ferreira et al., 2002). In sugarcane seedlings several isoenzymes have been observed, but growth in the presence of Cd did not result in any significant SOD activity alteration (Fornazier et al., 2002). In pea plants, a strong reduction of chloroplastic and cytosolic Cu/Zn-SODs by Cd was verified and to a lesser extent for Fe-SOD, while Mn-SOD was only affected by the highest Cd concentration tested, showing that Mn-SOD was the isoenzyme more resistant to Cd (Sandalio et al., 2001). In pea leaf peroxisomes, the Mn-SOD activity did not change in response to Cd treatment (Romero-Puertas et al., 1999). Contrary to these later reports, increases in total SOD activity were detected following the application of Cd in pea plants (Dalurzo et al., 1997), *S. tuberosum* plants (Stroinski and Kozlowska, 1997), *Hordeum vulgare* (Guo et al., 2004), *A. thaliana* (Skorzynska-Polit et al., 2003), rice leaves (Hsu and Kao, 2004) and hyperaccumulator plants of the genus *Alyssum* (Schickler and Caspi, 1999). SOD activity remained constant in *H. annuus* (Gallego et al., 1996, 1999) and declined in *Amaranthus lividus* (Bhattacharjee, 1998), *Phragmitis australis* (Iannelli et al., 2002), sunflower leaves (Gallego et al., 1996) and pepper (Leon et al., 2002) in response to Cd exposure. In clonal, hydroponically grown poplar plants (*Populus* x *Canescens*, a hybrid of *Populus tremula* x *Populus alba*) (Schützendübel et al., 2002) and *A. lividus* (Bhattacharjee, 1998), exposure to Cd resulted in inhibition of SOD activity.

Since the action of SOD results in the formation of H_2O_2, it is also intimately linked with the activity of other peroxidases, thus maintaining an interaction with these other enzymes and antioxidants to guarantee a highly optimized balance in order to reduce the risk of damage.

5.2.2. *Catalase (CAT)*

Catalase (CAT, EC 1.11.1.6) is a tetrameric Fe porphyrin protein that catalyses the conversion of H_2O_2 to water and O_2 (Frugoli et al., 1996), the enzyme is abundant in peroxisomes. The function of CAT in the peroxisome is to metabolize the H_2O_2 liberated following the conversion of glycolate to glyoxylate during photorespiration (Igamberdiev and Lea, 2002) and to decompose the H_2O_2 formed during the β-oxidation of fatty acids in the glyoxysomes of lipid-storing tissues (Holtman et al., 1994). CAT has an extremely high maximum catalytic rate, but a low substrate affinity, since the reaction requires the simultaneous access of two H_2O_2 molecules to the active site. There is no doubt that the contribution of CAT to protect organisms from the damage caused by ROS is essential (Zámocký and Koller, 1999). CAT is also thought to protect SOD against inactivation by higher levels of H_2O_2 (Fridovich, 1995). Initially, it was thought that the main function of peroxisomes was the removal by CAT of toxic H_2O_2 generated by different oxidases. In recent years, it has been demonstrated that peroxisomes are involved in a range of important cellular functions in almost all eukaryotic cells (del Rio et al., 2002). In plants, there are several types of peroxisomes which are specialized in certain metabolic functions. Glyoxisomes are specialized peroxisomes of lipid storing tissues, where CAT decomposes H_2O_2 formed during the β-oxidation of fatty acids (Holtman et al., 1994). Glyoxysomes have the glyoxylate cycle enzymes to convert the seed reserve lipids into sugars, which are used for germination and plant growth (del Rio et al., 2002). Unspecialised peroxisomes have also been found in organs that contain neither glyoxysomes nor leaf peroxisomes. The rapid proliferation of peroxisomes in plants under natural and abiotic stress conditions has been confirmed in pea leaves of plants subjected to Cd (Romero-Puertas et al., 1999; Lopez-Huertas et al., 2000). Moreover, the induction of peroxisome biogenesis genes (PEX) by H_2O_2 has also been demonstrated in both plant and animal cells, indicating that H_2O_2 is the signal molecule leading to peroxisome proliferation (del Rio et al., 2002).

Three genetically distinct CAT isoenzymes have been characterized in maize plants (Scandalios, 1994), whilst only two have been identified in barley (Skadsen et al., 1995). CAT is encoded by a multigene family in *Arabidopsis thaliana* that includes three genes encoding individual subunits *CAT1*, *CAT2* and *CAT3*, which associate to form at least six distinct isozymes (Frugoli et al., 1996). As might be suspected from the different functions of catalases, many plants contain multiple catalase isozymes, from two isoenzymes in barley (Azevedo et al., 1998) to as many as 12 in mustard (Frugoli et al., 1996). Isoenzymes of CAT have been studied extensively in higher plants (Polidoros and Scandalios, 1999). One specific class of CAT is located in vascular tissues and may be involved in the protection against environmental stress (Willekens et al., 1994). CAT isoenzymes have been shown to be regulated temporally and spatially and may respond differentially to light (Willekens et al., 1994; Skadsen et al., 1995). CAT is not present in chloroplasts (Asada, 1999), but has been reported to be present in the matrix of rat heart mitochondria, but whether it is present in plant mitochondria is still an open question. (Moller, 2001).

CAT activity has been shown to be variable in presence of Cd. CAT activity declined in sunflower leaves (Gallego et al., 1996), *Phaseolus vulgaris* (Chaoui et al., 1997), *Phaseolus aureus* (Shaw, 1995), *Pisum sativum* (Dalurzo et al., 1997), *Lemna minor* (Mohan and Hossetti, 1997), *Amaranthus lividus* (Bhattacharjee, 1998), soybean roots (Balestrasse et al., 2001), *Phragmitis australis* (Iannelli et al., 2002) and pepper (Leon et al., 2002) when submitted to Cd treatment. On the other hand, in the presence of Cd, CAT activity increased considerably in *Agropyron repens* (Brej, 1998), *Helianthus annuus* (Gallego et al., 1999), soybean nodules (Balestrasse et al., 2001), rice leaves (Hsu and Kao, 2004) in tolerant varieties of *Solanum tuberosum* (Stroinski and Kozlowska, 1997), in roots of *Raphanus sativus* seedlings (Vitória et al., 2001) and sugarcane *in vitro* callus culture (Fornazier et al., 2002). Moreover, in roots of the transgenic CAT deficient tobacco line (CAT1AS), the DNA damage induced by Cd was higher than in wild type tobacco (SR1) roots (Gichner et al., 2004). However, CAT activity was shown to remain unaltered under Cd stress in soybean leaves (Ferreira et al., 2002), and in *Phaseolus vulgaris*, although decreases in CAT activity were detected in roots and leaves, there was no effect on the stem enzyme (Chaoui et al., 1997).

5.2.3. *Ascorbate Peroxidase (APX)*

Ascorbate is a key plant compound involved in the defence against oxidative and photoxidative damage, through its action as a substrate for the H_2O_2 degrading enzyme ascorbate peroxidase (APX, EC 1.11.1.11). Moreover, analyses indicated the importance of APX for the removal of H_2O_2 in photosynthetic tissues, when non-green tissues were compared to green tissues. The elucidation of cellular location and function of cytosolic APX in non photosynthetic tissue of potatoes tubers, demonstrated the role of APX in protection against oxidative damage in non-photosynthetic cells (Asada, 1992).

APX is a heme peroxidase with an amino acid sequence indicating that APX belongs to the same superfamily of heme-peroxidase (Class I) as cytochrome c peroxidase (CPX). APX occurs also in mammals and is inhibited by cyanide, azide, thiol-modifying reagents, hydroxyurea, p-aminophenol, and thiols via a suicide mechanism, whereas the elicitor salicylic acid does not inhibit the enzyme activity (Asada, 1999). The catalytic cycle of APX is a peroxidase ping-pong mechanism; a two-electron oxidized intermediate of APX is formed, which then oxidizes ascorbate, successively producing two molecules of MDHA, being reduced back to the resting ferric state (Asada, 1999):

$$APX\text{-}Fe(III)\text{-}R + H_2O_2 \rightarrow APX\text{-}Fe(IV)\text{=}O\text{-}R^+ + H_2O$$
$$APX\text{-}Fe(IV)\text{=}O\text{-}R^+ + AsA \rightarrow APX\text{-}Fe(IV)\text{=}O\text{-}R + MDHA$$
$$APX\text{-}Fe(IV)\text{=}O\text{-}R + AsA \rightarrow APX\text{-}Fe(III)\text{-}R + MDHA + H_2O$$

APX and CAT belong to two different classes of H_2O_2 scavenging enzymes because of their different affinities, with APX in the µM range and CAT in the mM range. Thus, while APX might be responsible for the fine modulation of ROS for signalling, CAT might be responsible for the removal of excess ROS during stress (Mittler, 2002).

The peroxidase (APX) family consists of at least five different isoforms (Asada, 1992) including thylakoid, soluble stromal, cytosolic, peroxisomal and apoplastic isoenzymes (Miyake and Asada, 1992; Jiménez et al., 1997). Thylakoid membrane-bound APX (tAPX) is a limiting factor of antioxidant systems under photo-oxidative stress in chloroplasts and the enhanced activity of tAPX functions to maintain the redox status of ascorbate under stress conditions (Yabuta et al., 2002). The various APX isoforms respond differentially to metabolic and environmental signals (Kubo et al., 1995). Increases in transcript abundance may not necessarily be accompanied by corresponding increases in enzyme activities (Mittler and Zilinskas, 1992). In subsequent studies, a glyoxysomal membrane APX was shown to be a constitutive enzyme in the different types of cucumber seedling peroxisomes, probably protecting the boundary membrane from toxic H_2O_2 produced by the metabolic pathways in the differentiating peroxisomes (Corpas and Trelease, 1998). The reduction of cytosolic APX gene expression is an important event in heat shock-induced programmed cell death in tobacco bright-yellow 2 cells (BY-2 cells) (Vacca et al., 2004).

Chloroplasts contain APX in two isoforms, thylakoid-bound (tAPX) and soluble stromal enzymes (sAPX). At least one half of the chloroplastic APX is tAPX, but the ratio of tAPX/sAPX varies according to the plant species and, possibly, leaf age, but the biosynthetic ratio of the two APXs is controlled by alternative splicing (Asada, 1999). tAPX binds to the stroma thylakoids where the PSI complex is located, while sAPX is though to be localized in the stroma (Asada, 1999). Plants also contain the cytosolic isoforms of APX (cAPX), which has a different amino acid sequence when compared to the chloroplastic APXs, but participating in the scavenging of H_2O_2 in compartments other than chloroplasts. cAPX is a homodimer and its electron donor is not so specific for ascorbate, unlike tAPX and sAPX (Asada, 1999).

The activity of APX has been shown to increase following treatment with Cd in *Ceratophyllum demersum* (Aravind and Prasad, 2003), peas leaves (Romero-Puertas et al., 1999), *Phaseolus aureus* (Shaw, 1995), in the upper parts of *Phaseolus vulgaris* plants (Chaoui et al., 1997), and in green and greening barley seedlings, however, in roots the APX activity was reduced in high concentration of Cd (Hegedus et al., 2001). APX activity was increased in leaves of salt-sensitive wheat under combined NaCl and Cd stress, indicating that both abiotic factors in combination enhanced the production of oxygen radicals and H_2O_2, especially in leaves of the salt sensitive genotype (Muhling and Lauchli, 2003). Moreover, the APX in *Ceratophyllum demersum* showed a very high increase in activity in Cd + Zn treated plants as compared to Cd or Zn alone treated plants, indicating efficient antioxidant and ROS scavenging activities by Zn against Cd induced free radicals and oxidative stress (Aravind and Prasad, 2003). Low-concentration Cd treatments led to an increase in APX in soybean roots and nodules, but the activity was lower in higher concentrations of Cd (Balestrasse et al., 2001). This lower activity was also observed in cucumber chloroplasts with increasing Cd concentration in the nutrient solution (Zhang et al., 2003). Opposite effects of Cd exposure in which APX activity was inhibited have been also demonstrated in clonal, hydroponically grown poplar plants (Schützendübel et al., 2002) and sunflower leaves (Gallego et al., 1996). APX appears to have an important role in detoxification of H_2O_2, under abiotic stressful conditions, but its activity seems to be dependent on the Cd concentrations applied.

5.2.4. *Monodehydroascorbate Reductase (MDHAR)*

Monodehydroascorbate reductase (MDHAR, EC 1.6.5.4) is a FAD enzyme that is present as chloroplastic and cytosolic isoenzymes, sharing similar enzymatic properties. MDHAR shows a high specificity for MDHA as the electron acceptor. The enzyme prefers NADH rather than NADPH as the electron donor. The catalytic cycle of MDHA reductase is represented as follows (Asada, 1999):

$$E\text{-}FAD + NAD(P)H + H^+ \rightarrow E\text{-}FADH_2\text{-}NAD(P)^+$$
$$E\text{-}FADH_2\text{-}NAD(P)^+ + MDA \rightarrow E\text{-}FADH\text{-}NAD(P)^+ + ascorbate + H^+$$
$$E\text{-}FADH\text{-}NAD(P)^+ + MDA + H^+ \rightarrow E\text{-}FAD + ascorbate + NAD(P)^+$$

The first step is the reduction of the enzyme-FAD to form a charge transfer complex. The reduced enzyme donates electrons successively to MDHA, producing two molecules of ascorbate via a semiquinone form (E-FAD-NAD(P)$^+$) (Asada, 1999). The MDHA radical is directly reduced prior to its spontaneous disproportionation by photoreduced ferredoxin (redFd) in the thylakoids. The redFd competes between MDA and NADP$^+$ in thylakoids, but the reduction rate of MDHA is 34-fold higher than that of NADP$^+$ (Asada, 1999). Since redFd can more effectively reduce MDHA than NADP$^+$, MDHAR cannot participate in the reduction of MDHA in the thylakoidal scavenging system. Therefore, MDA reductase would function at a site where NAD(P)H is available, but redFd is not. In chloroplasts, Fd associates with the basic domains of psaD and psaE of the PSI complex via ionic interaction. The diffusion of Fd from the PSI complex is likely very limited, given the occurrence of MDHAR in the vicinity of thylakoid membranes (Asada, 1999).

MDHAR is also located accompanying APX in peroxisomes and mitochondria, and a scavenging system similar to that in chloroplasts seems to operate for further removal of H_2O_2, which has escaped from break down by peroxisomal CAT (del Rio et al., 2002).

In contrast to other antioxidant enzymes, MDHAR has not been studied in detail in response to oxidative stress induced by heavy metals. MDHAR activity in Cd grown *Pinus sylvestris* plants exhibited an enhanced activity as a response to the stress induced by Cd (Schützendübel et al., 2001). On the other hand, MDHAR activity was shown to be reduced in the roots of poplar hybrids (*Populus* x *canescens*) exposed to Cd (Schützendübel et al., 2002).

5.2.5. Dehydroascorbate Reductase (DHAR)

The MDHA produced in the lumen cannot be reduced by either ferredoxin or NAD(P)H, so MDHA is spontaneously disproportionated to dehydroascorbate (DHA) and ascorbate. DHA is a non-dissociated compound and easily penetrates cell membranes such as the plasma membrane, but can penetrate through the thylakoid membranes, whilst the anionic MDHA cannot (Asada, 1999). The rate of spontaneous disproportionation of MDHA is higher at lower pH, so that in the lumen where the pH is 5-6, the MDHA is rapidly converted to DHA and ascorbate, but in the stroma, where the pH is 7-8, the life span of MDHA is longer and it may be reduced by either redFd or NAD(P)H (Asada, 1999).

Spontaneous disproportionation of MDHA:

DHA + MDHA → ascorbate + DHA

For the photoreduction of DHA the addition of GSH is required, thus, GSH-dependent dehydroascorbate reductase (DHAR, EC 1.8.5.1) is the most likely enzyme for the reduction of DHA in chloroplasts. DHAR using GSH as the electron donor has been purified from leaf and other tissues. Its molecular size is 23 kDa and a thiol group participates in the reaction (Asada, 1999).

Reduction of DHA:

DHA + 2 GSH → ascorbate + GSSG

Whilst the effect of Cd on DHAR has been investigated, the available information is still very limited such as for the MDHAR enzyme. Low Cd concentrations led to an increase in DHAR extracted from soybean roots and nodules, but the activity was shown to be lower when higher Cd concentrations were applied (Balestrasse et al., 2001). Cd stress induced a decrease in DHAR activity in sunflower leaves (Gallego et al., 1996), whereas no significant changes were observed in DHAR activity in *Pinus sylvestris* (Shcützendübel et al., 2001).

5.2.6. Glutathione Reductase (GR)

Glutathione reductase (GR, EC 1.6.4.2) is a flavoprotein that catalyses the NADPH-dependent reduction of oxidized GSSG to the reduced form GSH. The enzyme protein although synthesised in the cytoplasm, can be targeted to both the chloroplast and mitochondria (Mullineaux and Creissen, 1997). In higher plants, GR is involved in defence against oxidative stress, whereas GSH plays an important role within the cell system, which includes participation in the ascorbate-glutathione cycle, maintenance of the sulfhydryl groups of cysteine in a reduced form, store of reduced sulphur and a substrate for glutathione-S-transferases (Noctor et al., 2002).

Two genes encoding GR have been identified in *A. thaliana*, *gr1* and *gr2*, which encode cytosolic and plastidic GR isoenzymes, respectively (Xiang and Oliver, 1998), however in plants most of the GR activity is located in chloroplasts (Creissen et al., 1994). Although pea leaf GR can be resolved into one single band of 55 kDa on SDS-PAGE, two-dimensional PAGE revealed the presence of up to 8 isoenzyme whilst in tobacco leaves at least 6 GR isoenzymes were detected on native PAGE stained for GR activity (Creissen et al., 1994).

It has been suggested that the GSSG inhibition of NADPH-consuming enzymes other than those involved in ROS-detoxification, is a result of the need to conserve NADPH (Moller, 2001). The majority of the studies determining the response of GR to Cd exposure have shown that GR activity increases as part of the defence against Cd stress, an alteration which has been shown to be often dose dependent and variable over time. Such a result suggests that GR is responding to Cd stress by maintaining glutathione in the reduced form prior to incorporation into PCs, and/or the activation of the ascorbate-glutathione cycle for the removal of H_2O_2. For instance, GR activity increased in the presence of Cd in *Raphanus sativus* (Vitória et al., 2001), *Crotalaria juncea* (Pereira *et al.*, 2002), pepper plants (Leon et al., 2002), soybean (Ferreira et al., 2002), sugarcane leaves (Fornazier et al., 2002), *Phaeodactylum tricomutum* (Morelli and Scarano, 2004). Enhancement of GR activity was shown to be higher at 40 µM Cd than at 4 µM Cd in roots and leaves of pea (Dixit et al., 2001). In *Alyssum* species, GR activity increased at 0.02 mM Cd, but decreased when the Cd concentration was elevated to 0.05 mM (Schickler and Caspi, 1999). In leaves of *A. thaliana*, GR activity was higher than in control plants (no Cd) but decreased as the Cd concentration was increased and was accompanied by gradual increase in the content of SH-groups (Skorzynska-Polit et al., 2003). A similar decrease in GR activity with increasing Cd concentration, was also observed in cucumber chloroplasts (Zhang et al., 2003) A decrease in GR activity following the application of Cd has also been reported previously in pea (Dalurzo et al., 1997), in a similar manner to *S. tuberosum* (Stroinski and Kozlowska, 1997), *H. annuus* (Gallego et al., 1999) and two *Alyssum* species (Schickler and Caspi, 1999). In *Vicia faba* exposed to Cd in solution, the GR activity in shoots increased without a concentration response relationship, but in roots, after an initial increase, GR activity showed a negative concentration response relationship (Rosa et al., 2003).

5.2.7. *Glutathione Peroxidase*

Glutathione peroxidases (GPX, EC 1.11.1.9) comprise another family of enzymes using GSH to reduce H_2O_2, lipid hydroperoxides and other hydroperoxides. GPX has been found in mammalian mitochondria, where it is the main enzyme for the removal of H_2O_2. Although several homologues have been identified in plants and one enzyme purified and characterized, no information is available for plant mitochondria. One member of the GPX superfamily is the phospholipidhydroperoxide glutathione peroxidase (EC 1.11.1.12), which can act directly on lipid hydroperoxide without the need for release of the hydroperoxy fatty acid, indicating that GPX can also contribute to repair ROS damage, the third line of defence against ROS (Moller, 2001).

Plant GPX are not constitutive but are induced in response to stress. Plant GPX can metabolise H_2O_2 at rates which are low compared with the high rates of H_2O_2 generation in plants (Noctor et al., 2002). GPX may be important in other areas of oxidant metabolism such as the removal of lipid and alkyl peroxides (Eshdat et al., 1997). A family of seven related proteins named AtGPX1 -

AtGPX7 in *A. thaliana* was identified in the cytosol, chloroplast, mitochondria and endoplasmic reticulum (Rodriguez Milla et al., 2003). Analysis of the upstream region of the *AtGPX* genes revealed the presence of multiple conserved motifs and some of them resembled antioxidant-responsive elements found in plant and human promoters (Rodriguez Milla et al., 2003). Using a conditional life or death screen in yeast, it was possible to isolate a tomato gene encoding a phospholipid hydroperoxide GPX (LePHGPX) that can protect plant tissue from a variety of stresses (Chen et al., 2004). A cDNA encoding a rice GPX (*OsGPX1*) has been considered a stress-inducible gene of the rice GPX family that protects cells against both metabolic and environmental oxidative stresses (Kang et al., 2004). Pepper plants that have been exposed to Cd exhibited increases in GPX activity (Leon et al., 2002), while the opposite result was observed for the enzyme activity of pea roots, whilst the activity remained almost unaltered in pea leaves (Dixit et al., 2001).

5.2.8. Guaiacol Peroxidase (GOPX)

Guaiacol peroxidases (GOPX, EC 1.11.1.7) isolated from plants, are distinguishable from APX in both amino acid sequence and in physiological function (Chen et al., 1992). GOPX participates in metabolic reactions such as the biosynthesis of lignin, decomposition of indoleacetic acid (IAA), and defence against pathogens, but not in the scavenging of H_2O_2. The peroxidase ping-pong mechanism of GOPX, is the same as APX, but GPX prefers aromatic electron donors such as guaiacol and pyrogallol, and oxidizes ascorbate usually at a rate of around 1% that of guaiacol (Asada, 1999). However, a high ascorbate oxidizing peroxidase with an amino acid sequence similar to GPX has been found in *Camellia sinensis* (Kvaratskhelia et al., 1997). GOPX has four disulfide bridges linking the conserved eight cysteine residues and amino acid sequences for the binding of carbohydrate and Ca^{+2}. APX does not contain carbohydrate and has no corresponding sequences for glycosylation and Ca^{2+}-binding (Asada, 1999).

Cd treatment has been shown to increased GOPX activity in *Phaseolus aureus* (Shaw, 1995), *Phaseolus vulgaris* (Chaoui et al., 1997) and wheat (Milone et al., 2003). In pepper plants, Cd concentrations higher than 0.5 mM produced an increase in the activity of GPOX (Leon et al., 2002). However, decreases in GOPX activity as a response of Cd stress have also been observed in soybean root and nodules (Balestrasse et al., 2001), pea (Sandalio et al., 2001) and sunflower (Gallego et al., 2002). In *A. thaliana*, GOPX activity was particularly high at the lowest and highest Cd concentrations (Skorzynnska-Polit et al., 2003). These data indicate that GOPX activity is variable, and depends on the applied Cd concentration and exposure time.

5.2.9. Glutathione-S-Transferase (GST)

Glutathione-S-transferase (GST, EC 2.5.1.18) is an enzyme that catalyzes the conjugation of GSH to a variety of hydrophobic, electrophylic and cytotoxic substrates (Marrs, 1996). GST is a dimeric soluble protein, mainly cytosolic with a molecular mass of 50 kDa with two subunits of 23 and 29 kDa (Dixon et al., 2002). The subunits have ligant sites for GSH (G site) and for adjacent eletrophilic substrate (H site). The G site specificity is high for GSH, whereas the H site has a wide specificity for electrophilic substrates (Marrs, 1996).

Plant GST was originally identified due its association with resistance mechanisms for the herbicide atrazine in maize (Marrs, 1996). Other plant GSTs were identified in response to several biotic and abiotic stresses, including heavy metals (Dixon et al., 2002). GSTs can also use GSH to reduce peroxides and form a large heterogeneous family of proteins that catalyse the nucleophylic attack of the sulphur atom of GSH. They remove compounds that are potentially genotoxic or cytotoxic because of their reaction with DNA, RNA and proteins (Noctor et al., 2002). Glutathione-S-conjugates are generally metabolized by removal of the carboxylterminal glycine residue of the GSH to give rise to the S-glutamylcysteinyl derivative (Beck et al., 2003). GSTs can also function as flavonoid-binding proteins for efficient anthocyanin export from the cytosol in petunia (Mueller and Goodman, 2000).

Investigations using the human *Omega* GST identified two outlying groups of the GST superfamily in *A. thaliana* which differed from all other plant GSTs by containing a cysteine in place of a serine at the active site (Dixon et al., 2002). Both groups have roles as antioxidant enzymes, inferred from the induction of the respective genes following exposure to chemicals and oxidative stress (Dixon et al., 2002). A plant GST (BI-GST) has been identified as a potent inhibitor of Bax lethality in yeast associated with oxidative stress and the disruption of mitochondrial functions (Kilili et al., 2004). Screening of a tomato two-hybrid library for BI-GST interacting proteins identified five homologous Tau class GSTs, which participate in a network of catalytic and regulatory functions involved in the oxidative stress response (Kilili et al., 2004). In the *A. thaliana* genome 48 genes with GST homology have been identified (Dixon et al., 2002), whereas in maize and soybean, 29 and 20 genes encoding GST, respectively, have been identified (McGonigle et al., 2000). The amino acid sequence identity of GSTs is about 25 to 30%, but this is higher in the amino terminal group. Conserved amino acids are probably in the catalytic active site of GSH ligation (Dixon et al., 2002).

The induction of GST genes by Cd may occur through the generation of oxidative stress (Marrs and Walbot, 1997). Heavy metals have been shown to induce the *NT107*, *NT103* and *Bz-2* genes (Droog et al., 1993; Marrs and Walbot, 1997; Takahashi et al., 1995) and the wheat GST25 and GST26 proteins (Medhy, 1994). GST activity has been shown to be considerably increased by Cd-induced stress in both roots and leaves of pea plants (Dixit et al., 2001) and roots of rice seedling (Moons, 2003). Spruce cells cultures were incubated with $CdSO_4$, Na_2HAsO_4 or $PbCl_2$ and demonstrated a strong induction of GST subunits, although no novel subunit was expressed, the appearance of a new GST isoform occurred (Schoroder et al., 2003). Moreover, in the roots of *Phragmites australis* plants treated with Cd, an increased GST activity was observed (Iannelli et al., 2002). The GST activity can be associated with the induction of the detoxification processes in response to increased Cd concentrations.

6. MOLECULAR APPROACH TO Cd TOXICITY AND TOLERANCE IN PLANTS

Transgenic plants overexpressing genes encoding proteins and enzymes involved in stress tolerance have been produced and tested against distinct types of stress. For instance, the overexpression of the mitochondrial Mn-SOD from *Nicotiana plumbagnifolia* in *Nicotiana tabacum* mitochondria protected the latter from oxidative damage (Bowler et al., 1991). More recently several transgenic plants with increased heavy metal tolerance have been developed for the purpose of phytoremediation (Lee et al., 2003). The *PvSR2* stress related gene of *Phaseolus vulgaris* has been cloned from French bean and shown to be expressed specifically upon heavy metal treatment (Chai et al., 2003). Transgenic tobacco seedlings expressing the *PvSR2* coding sequence exhibited a higher tolerance to Cd compared with the wild-type under Cd exposure (Chai et al., 2003). Other tobacco transgenic lines ALR1/5 and ALR1/9 were less susceptible to Cd induced stress, suggesting that the introduction and overexpression of the alfafa aldose/aldehyde reductase genes may generally induce higher stress tolerance (Hegedus et al., 2004). Lee et al. (2003) have demonstrated that transgenic *A. thaliana* plants overexpressing a PC synthase encoding gene, exhibited an increased tolerance to Cd stress.

To identify the limiting factors for heavy metal accumulation and tolerance, and to develop transgenic plants with an increased capacity to accumulate and/or tolerate heavy metals, the *Escherichia coli gshII* gene encoding GS was overexpressed in the cytosol of Indian mustard (*Brassica juncea*) (Zhu et al., 1999a). The transgenic GS plants accumulated significantly more Cd than the wild type plants when grown in the presence of Cd and also exhibited higher concentrations of GSH, PCs, thiol, S, and Ca than wild type plants (Zhu et al., 1999a). In the presence of Cd, the GS enzyme is rate limiting for the biosynthesis of GSH and PCs, and the overexpression of GS offers a promising strategy for the production of plants with superior heavy metal phytoremediation capacity (Zhu et al., 1999b). In another study, a bacterial GR was expressed *in Brassica juncea*, targeted to the cytosol (cytGR) or the plastids (cpGR) (Pilon-Smits et al., 2000). cpGR plants exhibited enhanced Cd tolerance at the chloroplast level, however the Cd tolerance of the whole plant was not affected. The lower Cd stress experienced by the chloroplastic cpGR may be the result of reduced Cd uptake and/or translocation, since Cd concentrations in shoots of cpGR plants were only half those determined in wild type shoots. These differences in Cd tolerance and accumulation may result from increased root GSH concentrations, which were up to 2-fold higher in cpGR plants than in the wild type (Pilon-Smits et al., 2000). Experiments with transgenic poplar plants overexpressing γ-ECS, in the presence of Cd in a hydroponic system revealed a significant accumulation of Cd in the root tissue when compared to the wild type. The use of transgenic poplar lines with enhanced glutathione production capacity seems to be of particular advantage in highly polluted soils (Koprivova et al., 2002).

A recent study has shown that overexpressing the yeast vacuolar transporter YCF1 increased Pb and Cd tolerance and consequently increased the accumulation of these metals in the shoots of transgenic *A. thaliana* plants even though expression levels of YCF1 were relatively low (Tong et al., 2004). Following *Agrobacterium* mediated transformation in *Nicotiana glauca*, the induction and overexpression of a wheat gene encoding PCS increased its tolerance to metals such as Pb and Cd, in developing seedling roots by 160% when compared to wild type plants (Gisbert et al., 2003). Moreover, the seedlings of transformed plants grown in mining soils containing high levels of Pb accumulated twice the concentration of this heavy metal when compared to the wild type (Gisbert et al., 2003). The bacterial arsenate reductase gene (*ArsC*), that conferred Cd resistance to *E. coli*, was overexpressed in tobacco and *A. thaliana*, with both transgenic plant species showing greater Cd toterance than wild type controls (Dhankler et al., 2003).

7. CONCLUDING COMMENTS

Unlike animals, plants are sedentary and are vulnerable to varying concentration of metals. Therefore, plants must adapt themselves to the prevailing conditions for their survival, resulting in the acquisition of a wide range of metal-tolerance mechanisms, both among species and among genotypes within a species (Prasad, 1997). In spite of this, a significant increase in research involving all aspects related to Cd contamination of the environment has been investigated.

Thus, it is expected that the information available in the literature will improve our understanding of the basic events of the phytotoxicity caused by Cd. The antioxidant systems have the capacity to interrupt the cascades of uncontrolled oxidation, some of which have been discussed in this chapter. Therefore, Cd stress cannot be understood without establishing the integrated processes involving other cycles and metabolic pathways that have not been mentioned in this chapter.

8. FUTURE PROSPECTS

The past decade has witnessed a rapid evolution of our knowledge of the biological roles of antioxidant defences in plants. The antioxidant systems must keep ROS under tight control to prevent damage. Understanding the responses of plants to heavy metal stress and the regulation of the enzymes and other compounds involved will lead to indications as to their usefulness in detoxification or phytoremediation programs of polluted environments by heavy metals and in particular, Cd.

Although many questions about ROS metabolism have not been answered, the rapid progress of proteomics and metabolomics and their interaction may be important tools to help define these roles more clearly in the near future. Such information complemented by using mutants, genetic engineering and the development of cellular markers, will allow the manipulation of the cellular redox status and identify traits for plant breeding programs.

Thus, the "omics" technologies could reveal the non-targeted identification of all gene products (mRNA transcripts, proteins and metabolites) in a specific biological sample, which could be followed by a refined analysis of quantitative dynamics in biological systems. In addition, metabolomics can provide biochemical and physiological knowledge about network organization in plants subject to Cd stress. Understanding these processes may considerably increase our knowledge of Cd phytotoxicity and induced oxidative stress.

References

Acevedo A, Scandalios JG (1996) Antioxidant gene (Cat/Sod) expression during the process of accelerated senescence in silks of the maize ear shoot. Plant Physiol Biochem 34: 539-545.

Al-Khedhairy AA, Al-Rokayan AS, Al-Misned FA (2001) Cadmium toxicity and cell stress response. Pakistan J Biol Sci 4: 1046-1049.

Alscher RG, Erturk N, Heath LS (2002) Role of superoxide dismutases (SODs) in controlling oxidative stress in plants. J Exp Bot 53: 1331-1341.

Apel K, Hirt H (2004) Reactive oxygen species: metabolism, oxidative stress, and signal transduction. Ann Rev Plant Biol 55: 373-399.

Apostol I, Heinsten PF, Low PS (1989) Rapid stimulation of an oxidative burst during elicitation of cultured plant cells: role in defense and signal transduction. Plant Physiol 90: 109-116.

Aravind P, Prasad MNV (2003) Zinc alleviets cadmium-induced oxidative stress in *Ceratophyllum demersum* L.: a free floating freshwater macrophyte. Plant Physiol Biochem 41: 391-397.

Asada K (1992) Ascorbate peroxidase: A hydrogen peroxide scavenging enzyme in plants. Physiol Plant 85: 235-241.

Asada K (1999) The water cycle in chloroplast: Scavenging of active oxygens and dissipation of excess photons. Ann Rev Plant Physiol Plant Mol Biol 50: 601-639.

Azevedo RA, Alas RM, Smith RJ, Lea PJ (1998) Response of antioxidant enzymes to transfer from elevated carbon dioxide to air and ozone fumigation, in leaves and roots of wild-type and a catalase-deficient mutant of barley. Physiol Plant 104: 280-292.

Baccouch S, Chaoui A, El Ferjani E (1998) Nickel-induced oxidative damage and responses in *Zea mays* shoots. Plant Physiol Biochem 36: 689-694.

Balestrasse KB, Gardey L, Gallego SM, Tomaro ML (2001) Response of antioxidant defence system in soybean nodules and roots subjected to cadmium stress. Australian J Plant Physiol 28: 497-504.

Bartoli CG, Gomez F, Martinez DE, Guiamet JJ (2004) Mitochondria are the main target for oxidative damage in leaves of wheat (*Triticum aestivum* L.). J Exp Bot 55: 1663-1669.

Beck A, Lendzian K, Oven M, Christmann A, Grill E (2003) Phytochelatin synthase catalyses key steps in turnover of glutathione conjugates. Phytochemistry 62: 423-431.

Berglund S, Davis RD, L'Hermite P (1984) Utilization of Sewage Sludge on Land: Rates of Application and long-Term Effects of Metals. Reidel Publishing, 216.

Bergmann H, Machelett B, Lippmann B, Friedrich Y (2001) Influence of heavy metals on the accumulation of trimethylglycine, putrescine and spermine in food plants. Amino Acids 20: 325-329.

Bert V, Meerts P, Saumitou-laprade P, Salis P, Gruber W, Verbruggen N (2003) Genetic basis of Cd tolerance and hyperaccumulation in *Arabidopsis halleri*. Plant Soil 249: 9-18.

Bhattacharjee S (1998) Membrane lipid peroxidation, free radical scavengers and ethylene evolution in *Amaranthus* as affected by lead and cadmium. Biol Plant 40: 131-135.

Boominathan R, Doran PM (2003) Organic acid complexation, heavy metal distribution and the effect of ATPase inhibition in hairy roots of hyperaccumulator plant species. J Biotechnol 101: 131-146.

Boussama N, Ouariti O, Ghorbal MH (1999) Cd-stress on nitrogen assimilation. J Plant Physiol 155: 310-317.
Bowler C, Slooten L, Vandenbranden S, Derycke R, Botterman J, Sybesma C, Vanmontagu M, Inzé D (1991) Manganese superoxide-dismutase can reduce cellular-damage mediated by oxygen radicals in transgenic plants. EMBO J 10: 1723-1732.
Bowler C, Van Camp W, Van Montagu M, Inzé D (1994) Superoxide dismutase in plants. Critical Rev Plant Sci 13: 199-218.
Bowler C, Vanmontagu M, Inzé D (1992) Superoxide dismutase and stress tolerance. Ann Rev Plant Physiol Plant Mol Biol 43: 83-116.
Brej T (1998) Heavy metal tolerance in *Agropyron repens* (L.) P. Bauv. Populations from the Legnica copper smelter area, Lower Silesia. Acta Soc Bot Pol 67: 325-333.
Buettner GR, Jurkiewicz BA (1996) Handbook of Antioxidants. N.Y.
Chai TY, Chen Q, Zhang YX (2003) Cadmium resistance in transgenic tobacco plants enhanced by expressing bean heavy metal-responsive gene *PvSR2*. Science in China Series C-Life Sciences 46: 623-630.
Chaoui A, Mazhoudi S, Ghorbal MH, El Ferjani E (1997) Cadmium and zinc induction of lipid peroxidation and effects on antioxidant enzymes activities in bean (*Phaseolus vulgaris*). Plant Sci 127: 139-147.
Chen GX, Sano S, Asada K (1992) The amino acid sequence of ascorbate peroxidase from tea has high degree of homology to that of cytochrome c peroxidase from yeast. Plant Cell Physiol 33: 109-116.
Chen HY, Liu WY (1996) The molecular evolution of superoxide dismutase based on its distribution and structure. Progress Biochem Biophys 23: 408-413.
Chen SR, Vaghchhipawala Z, Li W, Asard H, Dickman MB (2004) Tomato phospholipid hydroperoxide glutathione peroxidase inhibits cell death induced by Bax and oxidative stresses in yeast and plants. Plant Physiol 135: 1630-1641.
Cobbett C, Goldsbrough P (2002) Phytochelatins and metallothioneins: roles in heavy metal detoxification and homeostasis. Ann Rev Plant Biol 53: 159-182.
Cobbett CS, May MJ, Howden R, Rolls B (1998) The glutathione-deficient, cadmium-sensitive mutant, cad2-1, of *Arabidopsis thaliana* is deficient in gamma-glutamylcysteine synthetase. The Plant J 16: 73-78
Cobbett CS (2000) Phytochelatins and their roles in heavy metal detoxification. Plant Physiol 123: 825-832.
Conklin PL, Last RL (1995) Differential accumulation of antioxidant mRNAs in *Arabidopsis thaliana* exposed to ozone. Plant Physiol 109: 203-212.
Corpas FJ, Trelease RN (1998) Differential expression of ascorbate peroxidase and a putative molecular chaperone in the boundary membrane of differentiating cucumber seedling peroxisomes. J Plant Physiol 153: 332-338.
Creissen GP, Edwards EA, Mullineaux PM (1994) Glutathione reductase and ascorbate peroxidase. In: Foyer CH, Mullineaux, PM (eds.) Causes of Photooxidative Stress and amelioration of defense systems in plants. CRC Press, Boca Raton, FL, pp. 343-364.
Dalurzo HC, Sandalio LM, Gomez M, Del Rio LA (1997) Cadmium infiltration of detached pea leaves: effect on its activated oxygen metabolism. Phyton (Austria) 37: 59-64.
Davies WJ, Zhang J (1991) Roots signals and the regulation of growth and development of plants in drying soil. Ann Rev Plant Physiol Plant Mol Biol 76: 42-55.
Davison PA, Hunter CN, Horton P (2002) Overexpression of β-carotene hydroxylase enhances stress tolerance in *Arabidopsis*. Nature 418: 203-206.
del Rio LA, Corpas FJ, Sandalio LM, Palma, JM, Gomez M, Barroso JB (2002) Reactive oxygen species, antioxidant systems and nitric oxide in peroxisomes. J Exp Bot 53: 1255-1272.
Delaunay A, Isnard, AD, Toledano MB (2000) H_2O_2 sensing through oxidation of the Yap1 transcription factor. EMBO J 19: 5157-5166.
Dhankler OP, Shasti NA, Rosen BP, Fuhrmann M, Meagher RB (2003) Increased cadmium tolerance and accumulation by plants expressing bacterial arsenate reductase. New Phytol 159: 431-441.
Di Cagno R, Guidi L, De Gara L, Soldatini GF (2001) Combined cadmium and ozone treatments photosynthesis and ascorbate-dependent defences in sunflower. New Phytol 151: 627-636.
di Toppi LS, Gabbrielli R (1999) Response to cadmium in higher plants. Environ Exp Bot 41: 105-130
Dixit V, Pandey V, Shyam R (2001) Differential antioxidative responses to cadmium in roots and leaves of pea (*Pisum sativum* L. cv. Azad). J Exp Bot 52: 1101-1109.

Dixon DP, Davis BG, Edwards R (2002) Functional divergence in the glutathione transferase superfamily in plants: identification of two classes with putative functions in redox homeostasis in *Arabidopsis thaliana*. J Biol Chem 277: 30859-30869.

Droog FNJ, Hooykaas PJJ, Libbenga KR, van der Zaal EJ (1993) Proteins encoded by an auxin-regulated gene family of tobacco share limited but significant homology with glutathione S-transferases and one member indeed shows *in vitro* GST activity. Plant Mol Biol 21: 965-972.

Dudka S, Miller WP (1999) Accumulation of potentially toxic elements in plants and their transfer to the human food chain. J Environ Sci 34: 681-708.

Ederli L, Reale L, Ferranti F, Pasqualini S (2004) Responses induced by high concentration of cadmium in *Phragmites australis* roots. Physiol Plant 121: 66-74.

Eshdat Y, Holland D, Faltin Z, Ben-Hayyim G (1997) Plant glutathione peroxidases. Physiol Plant 100: 234-240.

Fadzilla NM, Finch RP, Burdon RH (1997) Salinity, oxidative stress and antioxidant responses in shoot cultures of rice. J Exp Bot 48: 325-331.

Fargasova A (2004) Toxicity comparison of some possible toxic metals (Cd, Cu, Pb, Se, Zn) on young seedlings of *Sinapis alba* L. Plant Soil Environ 50: 33-38.

Ferreira RR, Fornazier RF, Vitória AP, Lea PJ, Azevedo RA (2002) Changes in antioxidant enzyme activities in soybean under cadmium stress. J Plant Nutr 25: 327-342.

Fornazier RF, Ferreira RR, Pereira GJG, Molina SMG, Smith RJ, Lea PJ, Avezedo RA (2002) Cadmium stress in sugar cane callus cultures: Effect on antioxidant enzymes. Plant Cell Tissue Organ Cult 71: 125-131.

Foyer CH, Lelandais M (1996) A comparison of the relative rates of transport of ascorbate and glucose across the thylakoid, chloroplast and plasmalemma membranes of pea leaf mesophyll cells. J Plant Physiol 148: 391-398.

Foyer CH, Lopez-Delgado H, Dat JF, Scott IM (1997) Hydrogen peroxide- and glutathione-associated mechanisms of acclimatory stress tolerance and signalling. Physiol Plant 100: 241-254.

Foyer CH, Noctor G (2000) Oxygen processing in photosynthesis: Regulation and signalling. New Phytol 146: 359-388.

Foyer CH, Noctor G (2003) Redox sensing and signalling associated with reactive oxygen in chloroplasts, peroxisomes and mitochondria. Physiol Plant 119: 355–364.

Fridovich I. (1995) Superoxide radical and superoxide dismutases. Ann Rev Biochem 27: 97-112.

Frugoli JA, Zhong HH, Nuccio ML, McCourt P, McPeek MA, Thomas TL, McClung CR (1996) Catalase is encoded by a multigene family in *Arabdopsis thaliana* (L.) Plant Physiol 112: 327-336.

Gallego SM, Benavídes MP, Tomaro ML (1996) Effects of heavy metal ion excess on sunflower leaves: evidences for involvement of oxidative stress. Plant Sci 121: 151-159.

Gallego SM, Benavides MP, Tomaro ML (1999) Effect of cadmium ions on antioxidant defence system in sunflower cotyledons. Biol Plant 42: 49–55.

Gatzek S, Wheeler GL, Smirnoff N (2002) Antisense suppression of L-galactose dehydrogenase in *Arabidopsis thaliana* provides evidence for its role in ascorbate synthesis and reveals light modulated L-galactose synthesis. The Plant J 30: 541-553.

Ghoshroy S, Freedman K, Lartey R, Citovsky V (1998) Inhibition of plant viral systemic infection by non-toxic concentration of cadmium. The Plant J 13: 591-602.

Gichner T, Patkova Z, Szakova J, Demnerova K (2004) Cadmium induces DNA damages in tobacco roots, but no DNA damage, somatic mutations or homologous recombination in tobacco leaves. Mutation Reasearch-Genetic Toxic Environ Mut 559: 49–57.

Gimeno-García E, Andreu V, Boluda R (1996) Heavy metals incidence in the application of inorganic fertilizers and pesticides to rice farming soils. Environ Poll 92: 19-25.

Gisbert C, Ros R, De Haro A, Walker DJ, Bernal MP, Serrano R, Navarro-Avino J (2003) A plant genetically modified that accumulates Pb is especially promising for phytoremediation. Biochem Biophys Res Comm 303: 440-445.

Goldsbrough PB. 1998. Metal tolerance in plants: The role of phytochelatins and metallothioneins. In: Terry N, Banuelos GS (eds.) Phytoremediation of Contaminated Soil and Water. CRC Press, Boca Raton, FL, pp. 221-233.

Gomez LD, Noctor G, Knight MR, Foyer CH (2004) Regulation of calcium signaling and gene expression by glutathione. J Exp Bot 55: 1851-1859.

Gries GE, Wagner GJ (1998) Association of nickel versus transport of Cd and calcium in tonoplast vesicles of oat roots. Planta 204: 390-396.

Grill E, Winnacker E-L, Zenk MH (1987) Phytochelatins, a class of heavy-metal binding peptides from plants, are functionally analogous to metallothioneins. Proc Natl Acad Sci USA 84: 439-443.

Guelfi A, Azevedo RA, Lea PJ, Molina SMG (2003) Growth inhibition of the filamentous fungus *Aspergillus nidulans* by cadmium: an antioxidant enzyme approach. J Gen App Microbiol 49: 63-73.

Guo TR, Zhang GP, Zhou MX, Wu FB, Chen JX (2004) Effects of aluminum and cadmium toxicity on growth and antioxidant enzyme activities of two barley genotypes with different Al resistance. Plant Soil 258: 241–248.

Harada E, Choi YE, Tsuchisaka A, Obata H, Sano H (2001) Transgenic tobacco plants expressing a rice cysteine synthase gene are tolerant to toxic levels of cadmium. J Plant Physiol 158: 655–661.

Hegedus A, Erdei S, Horvath G (2001) Comparative studies of H_2O_2 detoxifying in green and greening barley seedlings under cadmium stress. Plant Sci 160: 1085–1093.

Hegedus A, Erdei S, Janda T, Toth E, Horvath G, Dudits D (2004) Transgenic tobacco plants overproducing alfalfa aldose/aldehyde reductase show higher tolerance to low temperature and cadmium stress. Plant Sci 166: 1329-1333.

Heiss S, Schäfer HJ, Haag-Kerwer A, Rausch T (1999) Cloning sulfur assimilation genes of *Brassica juncea* L.: Cadmium differentially affects the expression of a putative low affinity sulfate transporter and isoforms of ATP sulfurylase and APS reductase. Plant Mol Biol 39: 847–857.

Heiss S, Wachter A, Bogs J, Cobbett C, Rausch T (2003) Phytochelatin synthase (PCS) protein is induced in *Brassica juncea* leaves after prolonged Cd exposure. J Exp Bot 54: 1833–1839.

Hesse H, Nikiforova V, Gakiere B, Hoefgen R (2004) Molecular analysis and control of cysteine biosynthesis: Integration of nitrogen and sulphur metabolism. J Exp Bot 55: 1283-1292.

Holtman WL, Heistek JC, Mattern KA, Bakhuizen R, Douma AC (1994) Beta-oxidation of fatty acids is linked to the glyoxylate cycle in the aleurone but not in the embryo of germinating barley. Plant Sci 99: 43–53.

Howden R, Goldsbrough PB, Andersen CR, Cobbett CS (1995) Cadmium-sensitive, cad1 mutants of *Arabidopsis thaliana* are phytochelatin deficient. Plant Physiol 107: 1059-1066.

Hsu YT, Kao CH (2004) Cadmium toxicity is reduced by nitric oxide in rice leaves. Plant Growth Regul 42: 227–238.

Iannelli MA, Pietrini F, Fore L, Petrilli L, Massacci A (2002) Antioxidant response to cadmium in *Phragmites australis* plants. Plant Physiol Biochem 40: 977-982.

Igamberdiev AU, Lea PJ (2002) The Role of peroxisomes in the integration of metabolism and evolutionary diversity of fotosynthetic organism. Phytochemistry 60: 651-674.

Iretskaya SN, Chien SH, Menon RG (1998) Effects of acidulation of high cadmium containing phosphate rocks on cadmium uptake by upland rice. Plant Soil 201: 183-188.

Jarup L (2003) Hazards of heavy metal contamination. British Med Bull 68: 167-182.

Jiménez A, Hernández JA, del Río LA, Sevilla F (1997) Evidence for the presence of the ascorbate-glutathione cycle in mitochondria and peroxisomes of pea leaves. Plant Physiol 114: 275-284.

Kang SG, Jeong HK, Suh HS (2004) Characterization of a new member of the glutathione peroxidase gene family in *Oryza sativa*. Molecules Cells 17: 23-28.

Kilili KG, Atanassova N, Vardanyan A, Clatot N, Al-Sabarna K, Kanellopolos PN, Makris AM, Kampranis SC (2004) Differential roles of Tau class glutathione S-transferases in oxidative stress. J Biol Chem 279: 24540-24551.

Kingston-Smith AH, Foyer CH (2000) Overexpression of Mn-superoxide dismutase in maize leaves leads to increased monodehydroascorbate reductase and glutathione reductase activities. J Exp Bot 51: 1867–1877.

Kliebenstein DJ, Monde RA, Last RL (1998) Superoxide dismutase in *Arabidopsis*: An eclectic enzyme family with disparate regulation and protein localization. Plant Physiol 118: 637–650.

Koprivova A, Kopriva S, Jager D, Will B, Jouanin L, Rennenberg H (2002) Evaluation of transgenic poplars overexpressing enzymes of glutathione synthesis for phytoremediation of cadmium. Plant Biol 4: 664–670.

Krotz RM, Evangelou BP, Wagner GJ (1989) Relationships between cadmium, zinc, Cd-peptide, and organic acid in tobacco suspension cells. Plant Physiol 91: 780-787.

Krupa Z, Baranowska M, Orzot D (1996) Can anthocyanins be considered as heavy metal stress indicator in higher plants? Acta Physiol Plant 18: 147-151.

Kubo A, Saji H, Tanaka K, Kondo N (1995) Expression of *Arabidopsis* cytosolic ascorbate peroxidase gene in response to ozone or sulfur dioxide. Plant Mol Biol 29: 479–489.

Kvaratskhelia M, Winkel C, Thorneley RNF (1997) Purification and characterization of a novel class in peroxidase isoenzyme from tea leaves. Plant Physiol 114: 1237-1245.

Lagriffoul A, Mocquot B, Vangronsveld J, Mench M (1998) Cadmiun toxicity effects on growth, mineral and chlorophyll contents, and activites of stress related enzymes in young maize plants (*Zea mays* L.). Plant Soil 200: 241-250.

Laloi C, Apel K, Danon A (2004) Reactive oxygen signaling: The latest news. Curr Opn Plant Biol 7: 323–328.

Lam E (2004) Controled cell death, plant survival and devolopment. Nature Rev Mol Cell Biol 5: 305-315.

Larcher WS (1987) Stress in plants. Naturwissenschaften 74: 158-167.

Lee J, Bae H, Jeong J, Lee JY, Yang YY, Hwang I, Martinoia E, Lee Y (2003) Functional expression of a bacterial heavy metal transporter in *Arabidopsis* enhances resistance to and decreases uptake of heavy metals. Plant Physiol 133: 589-596.

Lee SM, Leustek T (1999) The effect of cadmium on sulfate assimilation enzyme in *Brassica juncea*. Plant Sci 141: 201-207.

León AM, Palma JM, Corpas FJ, Gomez M, Romero-Puertas MC, Chatterjee D, Mateos RM, del Rio LA, Sandalio LM (2002) Antioxidative enzymes in cultivars of peppers plants with different sensitivity to cadmium. Plant Physiol Biochem 40: 813-820.

Leshem YY, Kuiper PJC (1996) Is there a GAS (general adaptation syndrome) response to various types of environmental stress? Biol Plant 38: 1–18.

Lichtenthaler HK, Rinderle U (1988) The role of chlorophyll fluorescence in the detection of stress conditions in plants. CRC Critical Rev Anal Chem 19: 29-85.

Lichtenthaler HK (1996) Vegetation stress: Na introduction to the stress concept in plants. Journal of Plant Physiol 148: 4-14.

Liu JK, Yeo HC, Doniger SJ, Ames BN (1997) Assay of aldehydes from lipid peroxidation Gas chromatography mass spectrometry compared to thiobarbituric acid. Anal Biochem 245: 161-166.

Loewus FA (1988) Ascorbic acid and its metabolic products. In: Preiss J (ed.) The Biochemistry of Plants. Academic Press, NY, pp. 85-107.

Lombi E, Zhao FJ, Dunham SJ, McGrath SP (2000) Cadmium accumulation in populations of *Thlaspi caerulescens* and *Thlaspi goesingense*. New Phytol 145: 11-20.

Lopez-Huertas E, Charlton WL, Johnson B, Graham IA, Baker A (2000) Stress induces peroxisome biogenesis genes. The EMBO J 19: 6770-6777.

Luwe M (1996) Antioxidants in the apoplast and symplast of beech (*Fagus sylvatica* L) leaves: Seasonal variations and responses to changing ozone concentrations in air. Plant Cell Environ 19: 321-328.

Lux A, Sottnikova A, Opatrna J, Greger M (2004) Differences in structure of adventitious roots in *Salix* clones with contrasting characteristics of cadmium accumulation and sensitivity. Physiol Plant 120: 537-545.

Maier T, Yu C, Kullertz G, Clemens S (2003) Localization and functional characterization of metal-binding sites in phytochelatin synthases. Planta 218: 300-308.

Malan HL, Farrant JM (1998) Effects of the metal pollutants cadmium and nickel on soybean seed development. Seed Sci Res 8: 445-453.

Mallick N, Mohn FH (2000) Reactive oxygen species: response of alga cells. J Plant Physiol 157: 183–193.

Marrs KA (1996) The functions and regulation of glutathione S-transferases in plants. Ann Rev Plant Physiol Plant Mol Biol 47: 127-158.

Marrs KA, Walbot V (1997) Expression and RNA splicing of the maize glutathione S-transferase Bronze2 is regulated by cadmium and other stresses. Plant Physiol 113: 93-102.

Maxwell DP, Wang Y, McIntosh L (1999) The alternative oxidase lowers mitochondrial reactive oxygen production in plant cells. Proc Natl Acad Sci USA 96: 8271-8276.

McCord JM, Fridovich I (1969) Superoxide dismutase: An enzymatic function for erythrocuprein (hemocuprein). J Biol Chem 244: 6049-6050.

McGonigle B, Keeler SJ, Lan SMC, Koeppe MK, O'Keefe DP (2000) A genomics approach to the comprehensive analysis of the glutathione S-transferase gene family in soybean and maize. Plant Physiol 124: 1105-1120.

McKersie BD, Leshem YY (1994) Stress and Stress Coping in Cultivated Plants. Kluwer Academic Publishers, Dordrecht, pp. 1-256.

Medhy MC (1994) Active oxygen species in plant defense against pathogens. Plant Physiol 105: 467-472.

Mehler AH (1951) Studies on reactions of illuminated chloroplasts. 1. Mechanism of the reduction of oxygen and other hill reagents. Arch Biochem Biophys 33: 65–77.

Millar AH, Mittova V, Kiddle G, Heazlewood JL, Bartoli CG, Theodoulou FL, Foyer CH (2003) Control of ascorbate synthesis by respiration and its implication for stress responses. Plant Physiol 133: 443–447.

Milone MMT, Sgherri C, Clijster H, Navari-Izzo F (2003) Antioxidante responses of wheat treated with realistic concentration of cadmium. Environ Exp Bot 50: 265-276.

Mittler R, Zilinskas BA (1992) Molecular cloning and characterization of a gene encoding pea cytosolic ascorbate peroxidase. J Biol Chem 267: 21802-21807.

Mittler R (2002) Oxidative stress, antioxidants and stress tolerance. Trends Plant Sci 7: 405-410.

Miyake C, Asada K (1992) Thylakoid-bound ascorbate peroxidase in spinach chloroplasts and photoreduction of its primary oxidation product the monodehydroascorbate radicals in thylakoids. Plant Cell Physiol 33: 541-553.

Mocquot B, Vangronsveld J, Clijsters H, Mench M (1996) Copper toxicity in young maize (*Zea mays* L.) plants: Effects on growth, mineral and chlorophyll contents, and enzyme activities. Plant Soil 182: 287-300.

Mohan BS, Hossetti BB (1997) Potential phitotoxicity of lead and cadmium to *Lemna minor* grown in sewage stabilization ponds. Environ Poll 98: 233-238.

Moller IM (2001) Plant mitochondria and oxidative stress: Electron transport, NADPH turnover, and metabolism of reactive oxygen species. Ann Rev Plant Physiol Plant Mol Biol 52: 561–591.

Moons A (2003) Osgstu3 and osgstu4, encoding tau class glutathione S-transferase, are heavy metal – and hypoxic stress-induced and differentially salt stress-responsive in rice roots. FEBS Lett 553: 427-432.

Moran JF, Iturbe-Ormaetxe I, Matamoros MA, Rubio MC, Clemente MR, Brewin NJ, Becana M (2000) Glutathione and homoglutathione synthetases of legume nodules: cloning, expression, and subcellular localization. Plant Physiol 124: 1381–1392.

Morelli E, Scarano G (2004) Copper-induced changes of non-protein thiols and antioxidant enzymes in the marine microalga *Phaeodactylum tricomutum*. Plant Sci 167: 289-296.

Mueller LA, Goodman CD, Silady RA, Walbot V (2000) AN9, a petunia glutathione S-transferase required for anthocyanin sequestration, is a flavonoid-binding protein. Plant Physiol 123: 1561-1570.

Muhling KH, Lauchli A (2003) Interaction of NaCl and Cd stress on compartmentation pattern of cations, antioxidant enzymes and proteins in leaves of two wheats genotypes differing in salt tolerance. Plant Soil 253: 219-231.

Mullineaux PM, Creissen GP (1997) Glutathione reductase: regulation and role in oxidative stress. In: Scandalios JC (ed.) Oxidative Stress and the Molecular Biology of Antioxidant Defenses. Cold Spring Harbor Laboratory Press, NY, pp. 667-713.

Murphy A, Zhou J, Goldsbrough PB, Taiz L (1997) Purification and immunological identification of metallothioneins 1 and 2 from *Arabidopsis thaliana*. Plant Physiol 113: 1293-1301.

Neill SJ, Desikan R, Clarke A, Hurst RD, Hancock JT (2002) Hydrogen peroxide and nitric oxide as signalling molecules in plants. J Exp Bot 53: 1237–1247.

Neill SO, Gould KS (2003) Anthocyanins in leaves: light attenuators or antioxidants? Func Plant Biol 30: 865-873.

Noctor G, Foyer CH (1998) Ascorbate and glutatione: Keeping active oxygen under control. Ann Rev Plant Physiol Plant Mol Biol 49: 249-279.

Noctor G, Gomez L, Vanacker H, Foyer CH (2002) Interactions between biosynthesis, compartmentation and transport in the control of gluthatione homeostasis and signalling. J Exp Bot 53: 1283-1304.

Olmos E, Martinez-Solano JR, Piqueras A, Hellin E (2003) Early steps in the oxidative burst induced by cadmium in cultured tobacco cells (BY-2 line). J Exp Bot 54: 291-301.

Pereira GJG, Molina SMG, Lea PJ, Azevedo RA (2002) Activity of antioxidant enzymes in response to cadmium in *Crotalaria juncea*. Plant Soil 239: 123-132.

Pignocchi C, Foyer CH (2003) Apoplastic ascorbate metabolism its role in the regulation of cell signalling. Curr Opn Plant Biol 6: 379-389.
Pilon-Smits EAH, Zhu YL, Sears T, Terry N (2000) Overexpression of glutathione reductase in *Brassica juncea*: Effects on cadmium accumulation and tolerance. Physiol Plant 110: 455–460.
Pinhero RG, Rao MV, Paliyath G, Murr DP, Fletcher RA (1997) Changes in activities of antioxidant enzymes and their relationship to genetic and paclobutrazol-induced chilling tolerance of maize seedlings. Plant Physiol 114: 695–704.
Pinto E, Sigaud-Kutner TCS, Leitao MAS, Okamoto OK, Morse D, Colepicolo P (2003) Heavy metal-induced oxidative stress in algae. J Phycol 39: 1008–1018.
Polidoros NA, Scandalios JG (1999) Role of hydrogen peroxide and different classes of antioxidants in the regulation of catalase and glutathione S-transferase gene expression in maize (*Zea mays* L.). Physiol Plant 106: 112-120.
Pollard AJ, Powell KD, Harper FA, Smith JAC (2002) The genetics basis of metal hyperaccumulation in plants. Critical Rev Plant Sci 21: 539-566.
Prasad MNV (1995) Cadmiun toxicity and tolerance in vascular plants. Environ Exp Bot 35: 525-545.
Prasad MNV (1997) Plant Ecophysiology. John Wiley and Sons, NY, pp. 542.
Rauser WE (1990) Phytochelatins. Ann Rev Biochem 59: 61-86.
Rautenkranz AAF, Li LJ, Machler F, Martinoia E, Oertli JJ (1994) Transport of ascorbic and dehydroascorbic acids across protoplast and vacuole membranes isolated from barley (*Hordeum-vulgare* 1 cv gerbel) leaves. Plant Physiol 106: 187-193.
Rodriguez Milla MA, Maurer A, Huete Rodriguez A, Gustafson JP (2003) Glutathione peroxidase genes in *Arabidopsis* are ubiquitous and regulated by abiotic stresses through diverse signaling pathways. The Plant J 36: 602–615.
Romero-Puertas MC, MccCarthy I, Sandalio LM, Palma JM, Corpas FJ, Gomez M, del Rio LA (1999) Cadmium toxicity and oxidative metabolism of pea liaf peroxisomes. Free Radical Res 31: 25-31.
Rosa EVC, Valgas C, Souza-Sierra MM, Correa AXR, Radetski CM (2003) Biomass growth, micronucleus induction, and antioxidant stress enzyme responses in *Vicia faba* exposed to cadmium in solution. Environ Toxic Chem 22: 645–649.
Rossi C, Padilha PM, Padilha CCF (1998) Absorção de cádmio e crescimento de feijoeiro. Sci Agric 55: 332-337.
Ruegsegger A, Schmutz D, Brunold C (1990) Regulation of glutathione synthesis by cadmium in *Pisum sativum*. Plant Physiol 93: 1579-1584.
Rusek J, Marshall VG (2000) Impacts of airborne pollutants on soil fauna. Ann Rev Ecol Syst 31: 395-423.
Salt DE, Blayieck M, Kuma NPBA, DushenkovV, Ensley BD, Chet H, Raskin H (1995) Phytoremediation: A novel strategy for the removal of toxic metals from the environment using plants. Biotechnology 13: 468-474.
Salt DE, Smith RD, Raskin I (1998) Phytoremediation. Ann Rev Plant Physiol Plant Mol Biol 49: 643-668.
Sandalio LM, Dalurzo HC, Gómez MC, Romero-Puertas, MC, Del Rio LA (2001) Cadmium-induced changes in the growth and oxidative metabolism of pea plants. J Exp Bot 52: 2115-2126.
Scandalios JG (1993) Oxygen stress and superoxide dismutases. Plant Physiol 101: 7–12.
Scandalios JG (1994) Regulation and properties of plant catalases. 1994. In: Foyer CH, Mullineaux PM (eds.) Causes of Photooxidative Stress and Amelioration of Defence Systems in Plants. CRC Press, Boca Raton, pp. 275-315.
Schickler H, Caspi H (1999) Response of antioxidant enzymes to nickel and cadmium stress in hyperaccumulator plants of the genus *Alyssum*. Physiol Plant 105: 39-44.
Schmöger MEV, Oven M, Grill E (2000) Detoxification of arsenic by phytochelatins in plants. Plant Physiol 122: 793-802.
Schoroder P, Fischer C, Debus R, Wenzel A (2003) Reaction of detoxification mechanisms in suspension cultured spruce cells (*Picea abies* L. Karst.) to heavy metal in pure mixture and soil eluates. Environ Sci Poll Res 10: 225–234.
Schützendübel A, Nikolova P, Rudolf C, Polle A (2002) Cadmium and H_2O_2-induced oxidative stress in Populusxcanescens roots. Plant Physiol Biochem 40: 577–584.
Schützendübel A, Polle A (2002) A plant response to abiotic stresses: Heavy metal induced oxidative stress and protection by micorrhization. J Exp Bot 53: 1351-1365.

Schützendübel A, Schwanz P, Teichmann T, Groos K, Langenfeld-heyser R, Godbold DL, Polle A. (2001) Cadmium-induced changes in antioxidant systems, hydrogen peroxide content and differentation in scots pine roots. Plant Physiol 127: 887-898.

Selim HM, Kingery WL (2003) Geochemical and Hydrological Reactivity of Heavy Metals in Soils. Lewis Publishers, Boca Raton, FL, pp. 360.

Shaw BP (1995) Effects of mercury and cadmium on the activities of antioxidative enzymes in the seedling of *Phaseolus aureus*. Biol Plant 37: 587-596.

Siedlecka A, Krupa Z (1999) Cd/Fe interaction in higher plants - its consequences for the photosynthetic apparatus. Photosynthetica 36: 321-331.

Skadsen RW, Schulze-Lefert P, Herbst JM (1995) Molecular cloning, characterization and expression analysis of two catalase isozyme genes in barley. Plant Mol Biol 29: 1005-1014.

Skorzynska-Polit E, Drazkiewicz M, Krupa Z (2003) The activity of the antioxidante system in cadmium-treated *Arabdopsis thaliana*. Biol Plant 47: 71–78.

Smirnoff N, Conklin PL, Loewus FA (2001) Biosynthesis of ascorbic acis in plants: A renaissance. Ann Rev Plant Physiol Plant Mol Biol 52: 437-467.

Somasshekaraiah BV, Padmaja K, Prasad ARK (1992) Phytotoxicity of cadmium ions on germinating seedlings of mung bean (*Phaseolus-Vulgaris*): Involvement of lipid peroxides in chlorophyll degradation. Physiol Plant 85: 85–89.

Speiser DM, Abrahamson SL, Banuelos G, Ow DW (1992) *Brassica juncea* produces a phytochelatin-cadmium-sulfide complex. Plant Physiol 99:817-821.

Srivastava S, Tripathi RD, Dwivedi UN (2004) Synthesis of phytochelatins and modulation of antioxidants in responses to cadmium stress in *Cuscuta reflexa* – an angiospermic parasite. J Plant Physiol 161: 665-674.

Stroinski A, Kozlowska M (1997) Cadmium-induced oxidative stress in potato tuber. Acta Soc Bot Pol 66: 189–195.

Takahama U (1993) Regulation of peroxidase-dependent oxidation of phenolics by ascorbic-acid - different effects of ascorbic-acid on the oxidation of coniferyl alcohol by the apoplastic soluble and cell wall-bound peroxidases from epicotyls of *Vigna angularis*. Plant Cell Physiol 34: 809-817.

Takahashi Y, Hasezawa S, Kusaba M, Nagata T (1995) Expression of the auxin-regulated para gene in transgenic tobacco and nuclear-localization of its gene-products. Planta 196: 111–117.

Tausz M, Pilch B, Rennenberg H, Grill D, Herschbach C (2004) Root uptake, transport, and metabolism of externally applied glutathione in *Phaseolus vulgaris* seedlings. J Plant Physiol 161: 347-349.

Tong L, Nakashima S, Shibasaka M, Katsuhara M, Kasamo K (2004) A novel histidine-rich CPx-ATPase from the filamentous cyanobacterium *Oscillatoria brevis* related to multiple-heavy-metal cotolerance. J Bacteriol 184: 5027-5035.

Torsethaugen G, Pitcher LH, Zilinskas BA, Pell EJ (1997) Overproduction of ascorbate peroxidase in the tobacco chloroplast does not provide protection against ozone. Plant Physiol 114: 529-537.

Tucker SL, Thornton CR, Tasker K, Jacob C, Giles G, Egan M, Talbot NJ (2004) A fungal metallothionein is required for pathogenicity of *Magnaporthe grisea*. Plant Cell 16: 1575–1588.

Vacca RA, de Pinto MC, Valenti D, Passarella S, Marra E, De Gara L (2004) Production of reactive oxygen species, alteration of cytosolic ascorbate peroxidase, and impairment of mitochondrial metabolism are early events in heat shock-induced programmed cell death in tobacco bright-yellow 2 cells. Plant Physiol 134: 1100-1112.

Van Camp W, Van Montagu M, Inzé D (1994) Superoxide dismutases. In: Foyer CH, Mullineaux PM (eds.) Causes of Photooxidative Stress and Amelioration of Defense Systems in Plants. CRC Press, Inc., Boca Raton, FL, USA. pp. 317-341.

Vatamaniuk OK, Mari S, Lang A, Chalasani S, Demkiv LO, Rea PA (2004) Phytochelatin synthase, a dipeptidyltranferase that undergoes multisite acylation with gamma-glutamylcysteine during catalysis – Stoichiometric and site-directed mutagenic analysis of arabidopsis thaliana PCS1-catalyzed phytochelatin synthesis. J Biol Chem 279: 22449–22460.

Verma S, Dubey RS (2001) Effect of cadmium on soluble sugars and enzymes of their metabolism in rice. Biol Plant 44: 117-123.

Vernoux T, Wilson RC, Seeley KA, Reichheld JP, Muroy S, Brown S, Maughan SC, Cobbett CS, Montagu MV, Inzé D, May MJ, Sung ZR (2000) The root meristemless1/cadmium sensitive2 gene defines a glutathione-dependent pathway involved in initiation and maintenance of cell division during postembryonic root development. Plant Cell 12: 97–109.

Vitória AP, Lea PJ, Azevedo RA (2001) Antioxidant enzymes responses to cadmium in radish tissues. Phytochemistry 57: 701-710.

Vögelli-Lange R, Wagner GJ (1996) Relationship between cadmium, glutathione and cadmium-binding peptides (phytochelatins) in leaves of intact tobacco seedlings. Plant Sci 114: 11-18.

Von Zglinicki T, Edwall C, Östlund E, Lind B, Nordberg M, Ringertz NR, Wroblewski J (1992) Very low cadmium concentrations stimulate DNA synthesis and cell growth. J Cell Sci 103: 1073-1081.

Vranová E, Inzé D, Van Breusengem F (2002) Signal transduction during oxidative stress. J Exp Bot 53: 1227–1236.

Wang QR, Cui YS, Liu XM, Dong YT, Christie P (2003) Soil contamination and uptake of heavy metals at polluted sites in China. J Environ Sci Health Part A-Toxic/Hazardous Substances & Environ Engg 38: 823–838.

Weber M, Harada E, Vess CV, Roepenack-Lahaye E, Clemens, S (2004) Comparative microarray analysis of *Arabidopsis thaliana* and *Arabidopsis halleri* roots identifies nicotianamine synthase, a ZIP transporter and other genes as potencial metal hyperaccumulation factors. The Plant J 37: 269–281.

Willekens H, Langebartels C, Tiré C, Van Montagu M, Inzé D, Van Camp W (1994) Differential expression of catalase genes in *Nicotiana plumbaginifolia* (L.). Proc Natl Acad Sci USA 91: 10450-10454

Williamson JD, Scandalios JG (1992) Differential response of maize catalases and superoxide dismutases to the photoactivated fungal toxin cercosporin. The Plant J 2: 351-358.

Wong HL, Sakamoto T, Kawasaki T, Umemura K, Shimamoto K (2004) Down-regulation of metallothionein, a reactive oxygen scavenger, by the small GTPase OsRac1 in rice. Plant Physiol 135: 1447–1456.

Xiang C, Oliver DJ (1998) Glutathione metabolic genes coodinately respond to heavy metals and jasmonic acid in *Arabidopsis*. Plant Cell 10: 1539–1550.

Yabuta Y, Motoki T, Yoshimura K, Takeda T, Ishikawa T, Shigeoka S (2002) Thylakoid membrane-bound ascorbate peroxidase is a limiting factor of antioxidative systems under photo-oxidative stress. The Plant J 32: 915-925.

Yang HD, Rose NL, Battarbee RW (2002) Distribution of some trace metals in Lochnagar, a Scottish mountain lake ecosystem and its catchment. Science Total Environ 285: 197-208.

Yang XE, Long XX, Ye HB, He ZL, Calvert DV, Stoffella PJ (2004) Cadmium tolerance and hiperaccumulation in a new Zn-hiperaccumulating plant species (*Sedum alfredii* Hance). Plant Soil 259: 181-189.

Yoshimura K, Yabuta Y, Ishikawa T, Shigeoka S (2000) Expression of spinach ascorbate peroxidase isoenzymes in response to oxidative stresses. Plant Physiol 123: 223–234.

Zámocky M, Koller F (1999) Understanding the structure and function of catalases: Clues from molecular evolution and in vitro mutagenesis. Prog Biophys Mol Biol 72: 19–66.

Zha HG, Jiang RF, Zhao FJ, Vooijs R, Schat H, Barker JHA, McGrath SP (2004) Co-segregation analysis of cadmium and zinc accumulation in *Thlaspi caerulenses* interecotypic crosses. New Phytol 163: 299-312.

Zhang FQ, Shi WY, Jin ZX, Shen ZG (2003) Response of antioxidative enzymes in cucumber chloroplasts to cadmium toxicity. J Plant Nutr 26: 1779-1788.

Zhu YL, Pilon-Smits EAH, Jouanin L, Terry N (1999a) Overexpression of glutathione synthetase in indian mustard enhhances cadmium accumulation and tolerance. Plant Physiol 119: 73-79.

Zhu YL, Pilon-Smits EAH, Tarun A, Weber SU, Jouanin L, Terry N (1999b) Cadmium tolerance and accumulation in Indian mustard is enhanced by overexpressing -glutamylcysteine synthetase. Plant Physiol 121: 1169-1177.

Cadmium Toxicity and Tolerance in Plants
Editors: Nafees A. Khan and Samiullah
Copyright © 2006, Narosa Publishing House, New Delhi, India

Phytoremediation Techniques of Cd-Contaminated Soils: Toxicity, Enhanced Uptake Techniques, and Mechanism

*Hung-Yu Lai and Zueng-Sang Chen**

Graduate Institute of Agricultural Chemistry, National Taiwan University, Taipei 106-17, Taiwan.

1. THE SOURCES AND BEHAVIOR OF Cd IN THE CONTAMINATED SOILS

There are many sources for cadmium (Cd) to add into the soil, including anthropogenic emission, phosphate fertilizers, and sewage sludge, etc. (Alloway, 1995; Alloway and Steinnes, 1999). Among these sources, anthropogenic emission is the most dominate which was 3,700-21,400 tons yr^{-1} and was about 56% of the total input per year (Singh and McLaughlin, 1999). Alloway (1995) indicated that agricultural materials and atmospheric deposition were the two dominate adding sources of Cd into the soil and was 54-58% and 39-41% of the total anthropogenic sources, respectively.

The uptake of Cd was affected by plant species, growth stage, soil pH value, redox potential, organic matter, clay content, and oxides, etc. (Singh and McLaughlin, 1999). The result of Page *et al.*. (1981) showed that the Cd concentration in the leaves of Swiss chard increased (2 to 3.9-fold) when the soil pH value decreased from 7.4 to 4.5. After applying with lime, the soil pH value increased to 7.0 and thus decreased 43% and 41% of the Cd content in the cabbage and lettuce, respectively (Jackson and Alloway, 1992). The increase of soil pH value decreased the Cd concentration in both the leaves of soybean and seedling of corn (Kitagishi and Yamane, 1981). The mobility of Cd was low when the soil pH value was more than 7.5, however, with the highest mobility when the soil pH value was 4.5-5.5. With soil pH value of 4-7.7 in the soil, with the texture of sandy or loam, the adsorption of Cd increased 3-fold when the soil pH increased per unit. Some studies also indicated that the adsorption of Cd increased with the increased soil pH value until pH 8 (Farrah and Pickering, 1977).

The redox potential of soil also has great effect on the uptake of Cd by plants. The rice grown in the flooded soil of reduced condition accumulated lower concentration of Cd compared with that in the condition of oxidation (Bingham et al., 1976). This is because of the formation of CdS which thus decrease the availability of Cd. However, the availability of Cd increased with the decreased soil pH value result from the oxidation of CdS. The result of Ito and Iimura (1975) and Iimura et al. (1977) showed that the concentration of H_2S increased and thus decreased the concentration of soluble Cd when the redox potential decreased to -130 mV.

Cadmium is a nonessential element for the growth of plants, but this element can be uptake by the root or leaves of plants. The mean concentration of Cd ranged from 0.013 to 0.22 mg kg^{-1} for cereal grains, ranged from 0.07 to 0.27 mg kg^{-1} for grasses, and ranged from 0.08 to 0.28 mg kg^{-1} for legumes (Kabata-Pendias and Pendias, 2001). The capacity of plants to accumulate Cd in the plants varied with

---

*Corresponding Author (soilchen@ccms.ntu.edu.tw)

the different plants species. Davis and Calton-Smith (1980) reported that lettuce, spinach, celery, and cabbage accumulated high concentration of Cd in the tissues, but potato, corn, french soybean, and pea did not.

Other elements exist in the soil at the same time can affect the uptake of Cd by plants. Zinc has antagonistic effect on the uptake of Cd by plants when soil has lower Cd concentration, but has synergistic effect when the soil has high concentration of Cd (Page et al., 1981). The adsorption of Pb with soil particle was higher compared with Cd when Pb and Cd were existed simultaneously, and therefore increase the Cd concentration in soil solution and the uptake by plants (Adriano, 1986). Because of the competition of Zn and Cd for the adsorption sites in the soil surface, the soluble Cd concentration and the uptake of Cd increased when Zn and Cd were existed in the soil simultaneously (Kitagishi and Yamane, 1981). Wallace et al. (1980) indicated that the uptake of Cd by plants increased in the soil of low pH value and high concentration of Zn. There were some interactions between the metal in the contaminated soil and that uptake by plants when soil was contaminated with combined metals (Alloway, 1995; Kabata-Pendias and Pendias, 2001) (Fig. 1).

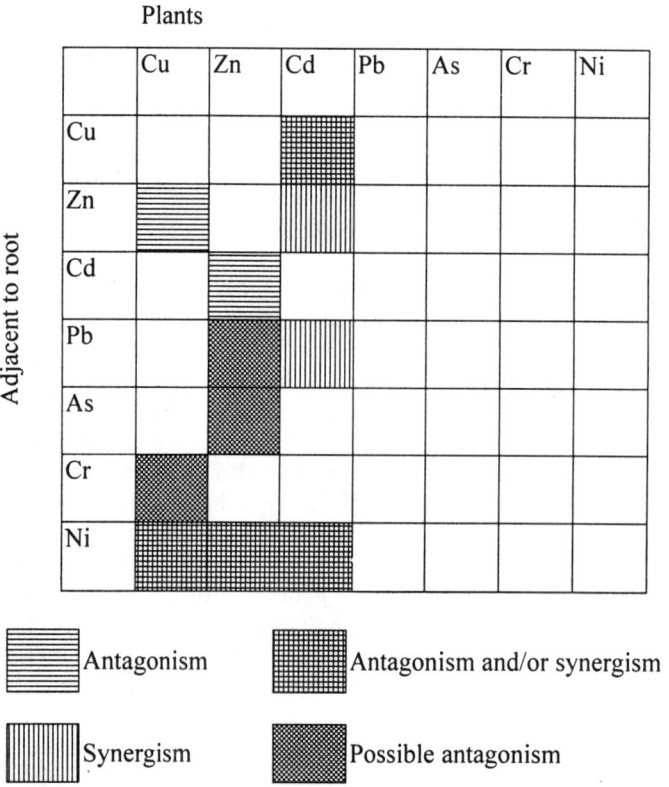

Fig. 1. The interactions between the metal in the contaminated soil and uptake by plants (Alloway, 1995; Kabata-Pendias and Pendias, 2001)

2. PHYTOREMEDIATION OF Cd-CONTAMINATED SOIL BY DIFFERENT PLANTS

Traditional techniques used to remediate the metals-contaminated soils including soil washing, soil flushing, pyrometallurgical, and electrokinetic, etc (Mulligan et al., 2000). However, most of soil characteristics changed after treating with these techniques and thus decreased the soil quality. Phytoremediation is the use of plants to remediate organic, inorganic, or nuclear pollutants in contaminated sites. The advantages of this technique including the related lower costs compared with traditional techniques, avoiding the excavation of soil, have no advised effect on the soil quality, and can be used in the contaminated soil of large area. The mechanisms of phytoremediation techniques including phytoextraction, phytostabilization, phytovolatilization, phytodegradation, and phytofiltration (Salt et al., 1995; Chaney et al., 1997; Lombi et al., 1998; Wenzel et al., 1999; Garbisu and Alkorta, 2001). Of these techniques, phytoextraction and phytostabilization were used to remediate inorganic-contaminated soil, especially for soil contaminated with heavy metals. Phytoextraction accumulated toxic metals from contaminated soil into the aboveground tissue of higher plants, which were then harvested and incinerated and/or buried (Kumar et al., 1995; Chaney et al., 1997; Wenzel et al., 1999; Garbisu and Alkorta, 2001). Phytostabilization used the dense root system of the plant to physically stabilize the contaminated soil, protecting against erosion by wind and water, thereby reducing the risk to the environment by the leaching of pollutants into groundwater (Salt et al., 1995; Chaney et al., 1997; Garbisu and Alkorta, 2001).

The plants used in the phytoremediation must accumulate more than 0.5% of contaminants and the bioconcentration factor (BCF) should bigger than 1,000 when planting in the solution (Zayed et al., 1998). There were many plants that accumulated high concentration of Cd in the different parts of plants. In the water containing 0.64 mg Cd L^{-1}, the BCF of *Lemna trisulca* was 3,594 (Huebert and Shay, 1993). *Azolla filiculoides* accumulated more than 10,000 mg Cd kg^{-1} when planting in the solution of 15 mg Cd L^{-1} (Sela et al., 1989). When duckweed (*Lemna minor* L.) was grown in the solution of 10 mg Cd L^{-1}, the Cd concentration reached 13,300 mg Cd kg^{-1} and the BCF was 1,300 (Zayed et al., 1998). *Eichhornia crassipes* accumulated 6,300 mg Cd kg^{-1} and the BCF was 2,150 (Zhu et al., 1999a). The results of Muramoto and Oki (1983) showed that the Cd in the *Eichhornia crassipes* was 10,600 and 36,000 mg kg^{-1}, respectively. *Thlaspi caerulescens* accumulated 1,140 mg Cd kg^{-1} in the shoot with the symptom of toxicity and accumulated 1,000 mg Cd kg^{-1} without showing the symptom of toxicity when planting in the solution of 200 mg Cd L^{-1} (Brown et al., 1995). Lombi et al. (2000) reported that *T. caerulescens* accumulated more than 14,100 mg Cd kg^{-1} in the shoot and more than 28,000 mg Cd kg^{-1} in the root when planting in the solution containing 500 µM of Cd. When planting in the Cd solutions of 0.5~50 µM, the Cd concentrations accumulated in the shoots and roots were 10~300 and 10~890 mg kg^{-1} in Chicory (*Cichorium intybus* L.), 20~410 and 20~1,360 mg Cd kg^{-1} in *Taraxacum officinale* Web. (Simon et al., 1996). *Taraxacum officinale* Web. accumulated As, Cu, Cr, Hg, Se, and Zn in the leaves and the Zn and Cd concentration in the leaves when planting in the different places were 1,049±22 and 15.2±3.5 mg kg^{-1}, respectively (Kuleff and Djingova (1984). Brown and Thomas (1983) reported that Cd concentration in the shoot of wheat (*Triticum aestivum*) when planting in the solution of 150 mg Cd L^{-1} was 479 mg kg^{-1}.

Thlaspi caerulescens accumulated high concentration of Cd when planting in the Cd-contaminated soil. The Cd concentration in the *T. caerulescens* was more than 1,000 mg kg^{-1} when planting in the France (Robinson et al., 1998). Lombi et al. (2000) indicated that when planting in the soil contamination with 500 mg Cd kg^{-1}, the Cd concentration in the *T. caerulescens* was 2,800 mg kg^{-1}, Baker et al. (1994) reported that the *T. caerulescens* growing in the wasted pit of northern England accumulated 164 mg kg^{-1} in the shoot of plant. Brown et al. (1994) also indicated that *T. caerulescens* accumulated 1,740 mg Cd kg^{-1} in the shoot when growing in the contaminated soil with soil pH value of 5.84-7.04, however some symptoms of toxicities occurred. The highest Cd concentration accumulated in the shoot of *T. caerulescens* was 1,020 mg kg^{-1} without showing the symptom of toxicity. From above result we can know that *T. caerulescens* accumulated 164 ~ 2,800 mg Cd kg^{-1} when grown in the Cd-contaminated soils. Some researches used sunflower for removal of heavy metals in the contaminated soils. The result of Simon (1998) showed that *Helianthus annuus* L. accumulated 114 mg Cd kg^{-1} in the kernel when planted in the 10 mg Cd kg^{-1} treated soil. Most of the Cd was accumulated in the root and the Cd concentration in the root was 13.7 mg Cd kg^{-1}. Dahmani-Muller et al. (2000) reported that the leaves of *Cardaminopsis halleri* accumulated 281Cd kg^{-1}. Table 1 shows the plants that accumulated high concentration of Cd in previous studies.

Table 1. The plants which accumulated high concentration of Cd in their tissues

Species	Concentration (mg kg^{-1})	References
Lemna trisulca	2,300	Huebert and Shay, 1993
Lemna minor L.	13,300	Zayed et al., 1998
Azolla filiculoides	10,000	Sela et al., 1989
Eichhornia crassipes	6,300	Zhu et al., 1999a
Eichhornia crassipes	10,600	Muramoto et al., 1983
Eichhornia crassipes	36,000	Muramoto et al., 1989
T. caerulescens	>1,000	Brown et al., 1995
T. caerulescens	14,100-28,000	Lombi et al., 2000
Cardaminopsis halleri	281	Dahmani-Muller et al., 2000
Taraxacum officinale Web.	20-1,360	Simon et al., 1996
Taraxacum officinale Web.	1,049	Kuleff et al., 1984
Triticum aestivum	479	Brown et al., 1983
T. caerulescens	>1,000	Robinson et al., 1998
T. caerulescens	2,800	Lombi et al., 2000
T. caerulescens	164	Baker et al., 1994
T. caerulescens	>1,020	Brown et al., 1994
Helianthus annuus L.	114	Simon, 1998
Cichorium intybus L.	10-890	Simon et al., 1996

Using garden plants to remove metals from contaminated soils is a popular and satisfactory strategy, because it recovers the contaminated sites to their natural condition and also generates economic value if appropriate flower species are cultivated. Chen and Lee (1997) reported that when rainbow pink (*Dianthus chinensis*) was grown in a contaminated site in northern Taiwan for five weeks, the Cd concentration in the shoot of the plant increased from 1.56 (under the control treatment) to 115 mg kg^{-1}, and the total Cd uptake was about 100 g ha^{-1} yr^{-1}. These findings showed that this plant can accumulate a high concentration of Cd. Moreover, the accumulated Cd concentration can reach the threshold (100 mg Cd kg^{-1}) of a Cd hyperaccumulator (Baker et al., 2000).

Over 420 plant species of hyperaccumulators from all over the world that can accumulate high concentrations of metals at contaminated sites were discovered (Baker et al., 2000), but many of them have a low growth rate and very low biomass; they therefore need much time to remove contaminants from soils. Two strategies, one involving chelating agents and the other involving genetic engineering, are being developed to increase the phytoextraction of metals with higher biomass or by transgenic plants. Synthesized chelating agents, such as EDTA (ethylenedinitrilo- tetraacetic acid), DTPA (diethylenetrinitrilopentaacetic acid), HEDTA (hydroxyl- ethylenediaminetriacetic acid), CDTA (trans-1,2-cyclohexylenedinitrilotetraacetic acid), and EGTA (ethylenebis (oxyethylenetrinitrilo) tetraacetic acid), were applied to metal-contaminated soil to increase the mobility and bioavailability of the metal in the soils and also to increase the amount of heavy metals accumulated in the upper parts of plants (Huang and Cunningham, 1996; Huang et al., 1997; Blaylock et al., 1997; Ebbs and Kochian, 1998; Wu et al., 1999). The results of these studies revealed that adding synthetic chelating agents can increase both the solubility of metal in soil solution and the concentration of metal in the shoots of plants. However, in soils contaminated with combined metals, the application of synthetic chelating agents can reduce both the biomass of the plant and the total amount of metal removed because the high concentrations of other metals in the soil solution are toxic to the plant (Chen and Cutright, 2001; Lombi et al., 2001a). Synthetic chelating agents at high concentrations can also be toxic to plants (Cooper et al., 1999). Applying synthetic chelating agents to combined metals-contaminated soils represents another challenge for chemical enhanced phytoextraction by plants.

3. THE EFFECT OF EDTA ON THE PHYTOEXTRACTION OF A COMBINED METALS-CONTAMINATED SOIL BY RAINBOW PINK

Rainbow pink (*Dianthus chinensis*) accumulated 115 mg Cd kg^{-1} when grown in the Cd-contaminated soil in northern Taiwan for 5 weeks (Chen and Lee, 1997). After planting for 50 days in an artificially combined metals-contaminated soil (total concentration of Cd, Zn, and Pb was 17.4±0.86, 458±26.3, and 938±45.5 mg kg^{-1}, respectively), the Cd, Zn, and Pb concentration in the shoot of rainbow pink was about 80, 3700, and 220 mg kg^{-1}, respectively. The result showed that this kind of garden flower also accumulated high concentration of Cd when planted in the combined metals-contaminated soil. The result of previous studies showed that applying EDTA increased the concentration of metals in the soil solutions and in the shoots of plants (Blaylock et al., 1997; Huang et al., 1997; Wu et al., 1999; Luo et al., 2000). In this study, three different concentration of 2Na-EDTA solutions were added into an artificially combined metals-contaminated soil to (i) identify the effect of EDTA on the concentration of heavy metals in soil solution when rainbow pink was grown in an artificially Cd, Zn, and Pb-contaminated soil in northern Taiwan, and (ii) to evaluate the effect of EDTA on the phytoextraction by rainbow pink.

The soil used in this study was collected from a Cd-contaminated site in northern Taiwan. The surface soil (0-20 cm) of the contaminated site was sampled, air-dried, and stored in large plastic bottles for laboratory analysis and pot experiments. To increase the concentration of metals in the innate contaminated soil, $Cd(NO_3)_2 \cdot 4H_2O$, ZnO, and $Pb(NO_3)_2$ were added to the air-dried soil at the same time to control the total concentration of the three metals as 20 mg Cd kg^{-1}, 500 mg Zn kg^{-1}, and 1,000 mg Pb kg^{-1}. The treated soils were wetted for two weeks by adding deionized water to maintain water content of 60% of the water-holding capacity (WHC) to enable the added heavy metal to reach a steady state. Then, the soils were dried at room temperature for approximately four weeks. The artificially contaminated soil was subjected to three cycles of wet and dry processes before pot experiments were conducted (Blaylock et al., 1997).

Each pot (16 cm diameter and 19 cm height) contained 3.5 kg (dry weight) of artificially contaminated soils. Two rhizon soil moisture samplers (RSMS) were put in each pot at a depth of around 10 cm below the surface of the soil. Two seedlings of rainbow pink (*Dianthus chinensis*) were planted in each pot. The pot experiments were conducted in a phytotron using a RCBD (randomized complete block design) with a number of treatments and three replicates. The soil moisture content was maintained at 60% WHC by weighing and adding deionized water every two days. Three concentrations of 2Na-EDTA solution (0 (control), 5, and 10 mmol kg^{-1}) were added to the contaminated soils at 7th day after planting. The growth conditions and observed symptoms of toxicity, displayed by the plants, were also recorded. Tested plants were harvested at 7th day after the EDTA solution was first added to the pot. The harvested plants were oven-dried at 60°C for 72 hr, weighed, and ground to 0.5 mm for analysis.

Then, 0.2 g of the plant was digested with the H_2SO_4/H_2O_2 digestion method (Harmon and Lajtha, 1999). The digested solutions were filtered through a Whatman No. 42 filter paper, and then diluted with deionized water to a volume of 50 mL in a flask. The Cd, Zn, and Pb concentrations were determined with atomic absorption spectrometer (Hitachi 180-30 type). Soil solutions were also collected directly by a RSMS before harvesting. The bioavailable concentrations of Cd, Zn, and Pb in contaminated soils were evaluated from the concentrations of heavy metals in the soil solution, which were determined with inductively coupled plasma optical emission spectrometer (ICP-OES) (Perkin Elmer 2000 DV).

3.1. The Physical and Chemical Characteristics of the Studied Soil

The studied soil texture was silty clay with a moderate cation exchange capacity of 13.9 $cmol_{(+)}$ kg^{-1}. The original total concentrations of the metals in the contaminated soil, digested with aqua regia, was 2.58±0.08 mg Cd kg^{-1}, 80.0±5.32 mg Zn kg^{-1} and 31.3±2.09 mg Pb kg^{-1}, respectively (Table 2). After the soil was spiked with solutions of heavy metals, the final total concentrations of the elements were 19.0±1.74 mg Cd kg^{-1}, 503±44.1 mg Zn kg^{-1}, and 931±54.7 mg Pb kg^{-1}, respectively.

Table 2. The physical and chemical characteristics of the study soil

pH (H_2O)	O.C.[#] (g/kg)	Texture	CEC[§] ($cmol_{(+)}$/kg)	Total concentration (mg/kg)[†]		
				Cd	Zn	Pb
4.9	16.9	silty clay[¶]	13.9	2.58±0.08	80.0±5.32	31.3±2.09

[#]O.C.: organic carbon.
[§]CEC: cation exchangeable capacity.
[†]: The total content of Cd, Zn, and Pb in the original soil was digested with aqua regia.
[¶]: sand 142 g kg^{-1}, silt 462 g kg^{-1}, and clay 396 g kg^{-1}.

3.2. Effect of EDTA on the Metal Concentration in Soil Solution

The results showed that the Cd, Zn, and especially the Pb concentrations in the soil solution of rainbow pink were significantly increased ($p< 0.05$) after adding 5 or 10 mmol EDTA kg^{-1}, except for Cd in the treatment of 10 mmol EDTA kg^{-1} (Fig. 2). In pervious studies, applying EDTA solution was efficiently to increase the soluble or exchangeable fraction of metals which was available to plants (Blaylock et al., 1997; Barona et al., 2001; Chen and Cutright, 2001). The result of this study was similar to those of pervious studies. Perhaps because of the high variation of Cd concentrations in the soil solutions after applying 10 mmol EDTA kg^{-1}, the concentration of Cd in the soil solutions was not significantly increase when 10 mmol EDTA kg^{-1} was applied. The Pb concentration in the soil solution significantly increased from 5.93±1.27 (control) to 90.4±12.5 mg L^{-1} (15-fold) and also from 5.93±1.27 to 252±182 mg L^{-1} (42-fold) when 5 and 10 mmol of EDTA kg^{-1} was added, respectively ($p< 0.001$) (Fig. 2c). These findings are consistent with those obtained following the application of EDTA treatments in previous studies (Blaylock et al., 1997; Huang et al., 1997; Wu et al., 1999; Luo et al., 2000; Chen and Cutright, 2001; Lombi et al., 2001a). These differences of the chelating effects between EDTA and metals are reflected by the formation constant (log $K_f=19.0$) of the Pb-ligand complexes, which exceeds those of the Cd-ligand complexes (log $K_f=17.4$) and the Zn-ligand complexes (log $K_f=17.5$) (Martell and Smith, 1974; 1982).

After 10 mmol EDTA kg^{-1} had been added for 21 days, Zn and Pb concentrations in the soil solution were approximately double and 30 times those of the control (Fig. 3b and 3c). Lombi et al. (2001a) found that EDTA can be found in the soil pore water five months after EDTA solution is applied.

The result of this study found that these three heavy metals remained in soil solution for more than 21 days after 10 mmol EDTA kg^{-1} was added, showed that EDTA treatments increased the mobility of metals in the contaminated soils. This finding also implied a potential risk of groundwater contamination when EDTA was used to mobilize metals in contaminated sites.

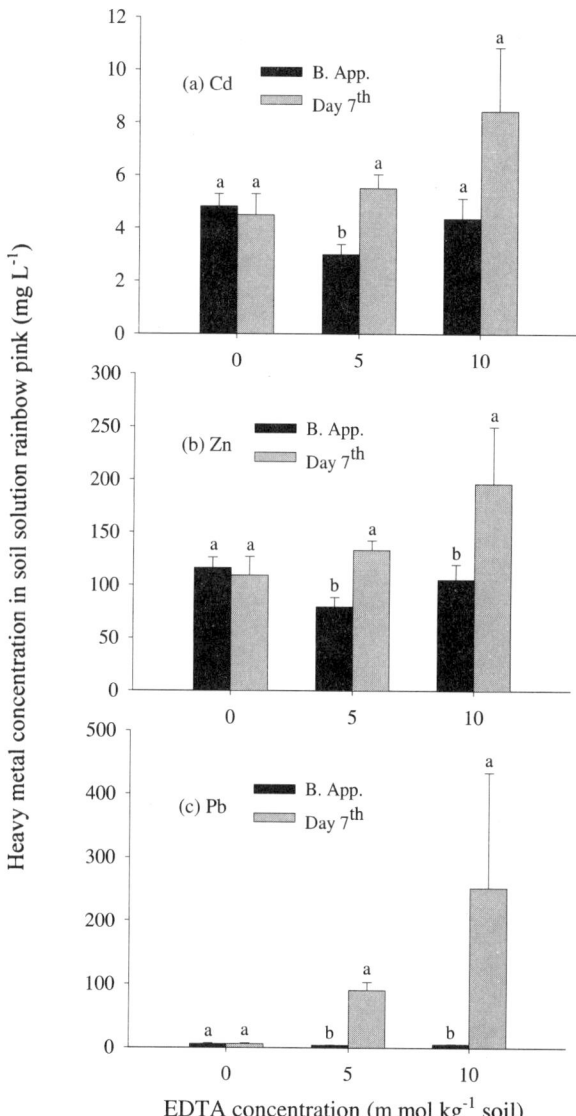

Fig. 2. The concentrations of (a) Cd, (b) Zn, and (c) Pb in the soil solutions of rainbow pink, which was sampled before applying (B. App.) and at 7th day (Day 7th) after applying different concentrations of EDTA. The probability level of significant difference is at p=0.05. Replicates (n)=3.

3.3. Effect of EDTA on the Metal Concentration in the Shoot of Rainbow Pink

The Cd concentration in the shoot of rainbow pink was significantly increased after applying 10 mmol EDTA/kg ($p< 0.05$) (Fig. 4a). The Zn concentration in the shoots of plants also increased from around 700±132 mg kg^{-1} (control) to 1,200±369 mg kg^{-1} (10 mmol EDTA kg^{-1}) (Fig. 4b). The Pb concentration in the shoots of plants was not detectable (ND, below detection limit) when no EDTA treatment was applied. However, at 7th day after 5 or 10 mmol EDTA kg^{-1} was added, the Pb concentration in the shoots of rainbow pink was significantly increased to 243±61.3 or 348±101 mg kg^{-1}, respectively ($p< 0.05$) (Fig. 4c). Lead appears to form the Pb-EDTA complex, which is available for uptake by plants after applying EDTA (Vassil et al., 1998). Barona et al. (2001) also pointed out that Pb remained weakly adsorbed to soil components after EDTA solution was applied. The results of this study indicate that EDTA solution enhances the removal of Pb from contaminated soil and also the accumulation of Pb in the shoots, these results are consistent with those of previous studies (Blaylock et al., 1997; Huang et al., 1997; Wu et al., 1999; Chen and Cutright, 2001; Lombi et al., 2001a; Jarvis and Leung, 2002).

After 5 or 10 mmol EDTA kg^{-1} was applied, no significant influence on the biomass of rainbow pink was observed. The mean dry biomass of rainbow pink following 0, 5, and 10 mmol EDTA kg^{1} was 4.91±0.93, 5.35±0.56, and 4.08±0.70 g plant^{-1}, respectively. With respect to the enhancement of phytoextraction by rainbow pink in a Cd, Zn, and Pb-contaminated soil, adding 5 or 10 mmol EDTA kg^{-1} did not significantly increase the total removal of Cd and Zn from the contaminated soil (Fig. 5a and 5b); however, the increase was significant for the removal of Pb ($p< 0.001$) (Fig. 5c). The total removal of Pb by the rainbow pink obtained by adding 5 mmol EDTA kg^{-1} did not differ significantly from that obtained by adding 10 mmol EDTA kg^{-1} (Fig. 5c).

We propose that rainbow pink is a potential phytoextraction plant for removing Cd or Zn from a combined metals-contaminated soil without applying any soil amendment. The amount removed per harvest is around 0.13±0.05 mg Cd plant^{-1}, 3.49±1.37 mg Zn plant^{-1}, and trace Pb plant^{-1}, respectively (Fig. 5). We also propose that Pb in contaminated soil is not available to rainbow pink, but adding 5 mmol EDTA kg^{-1} significantly increases the bioavailability of Pb in contaminated soil. The treatment of 5 or 10 mmol EDTA kg^{-1} can promote the removal of Pb by rainbow pink, far in excess of that obtained from the control treatment in this study. Therefore, further study must be undertaken to identify the critical concentration of EDTA solution from 0 to 5 mmol EDTA kg^{-1}, to remove a significant amount of Pb from the contaminated soil.

4. THE EFFECT OF EDTA ON THE PHYTOEXTRACTION OF SINGLE OR COMBINED METALS-CONTAMINATED SOIL BY RAINBOW PINK

The result of above study showed that applying 5 or 10 mmol EDTA kg^{-1} was significantly to increase the concentration of Cd, Zn, and Pb in soil solution when planting rainbow pink in a combined metals-contaminated soil. In further study, we used different metals-treated soil to investigate the effect of EDTA on the enhanced phytoextraction by rainbow pink for single or combined metals-contaminated soil. The soil used in this study was collected from a Cd-contaminated site in northern Taiwan the same as described in above study. The surface soil (0-20 cm) of the contaminated site was sampled, air-dried, and stored in large plastic bottles for laboratory analysis and pot experiments. Single or combination of these three solutions, $Cd(NO_3)_2.4H_2O$, ZnO, or $Pb(NO_3)_2$, was added to the air-dried soil, respectively, to control the total concentration of the three metals as following treatments. (1) CK: initial contaminated soil, (2) Cd-treated soil: 10 mg Cd kg^{-1}, (3) Zn-treated soil: 100 mg Zn kg^{-1}, (4) Pb-treated soils: 1,000 mg Pb kg^{-1}, (5) Cd-Zn-treated soil: combined contaminated with 10 mg Cd kg^{-1} and 100 mg Zn kg^{-1}, (6) Cd-Pb-treated soil: combined contaminated with 10 mg Cd and 1,000 mg Pb

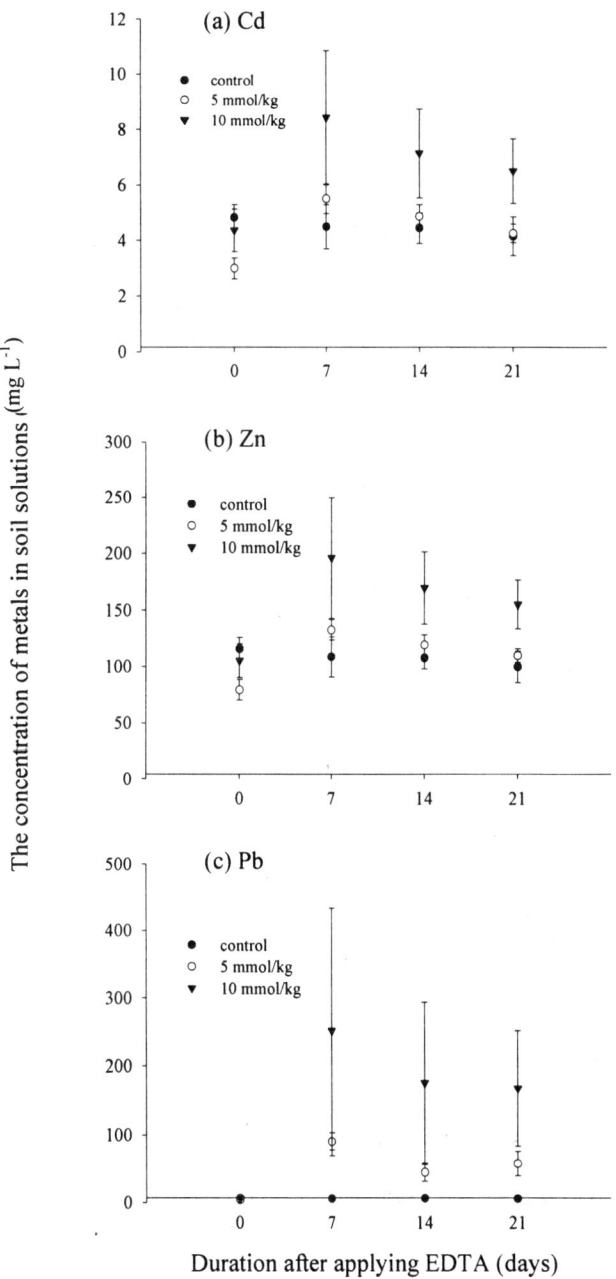

Fig. 3. The concentrations change of (a) Cd, (b) Zn, and (c) Pb in the soil solutions of rainbow pink. Replicates (n)=3.

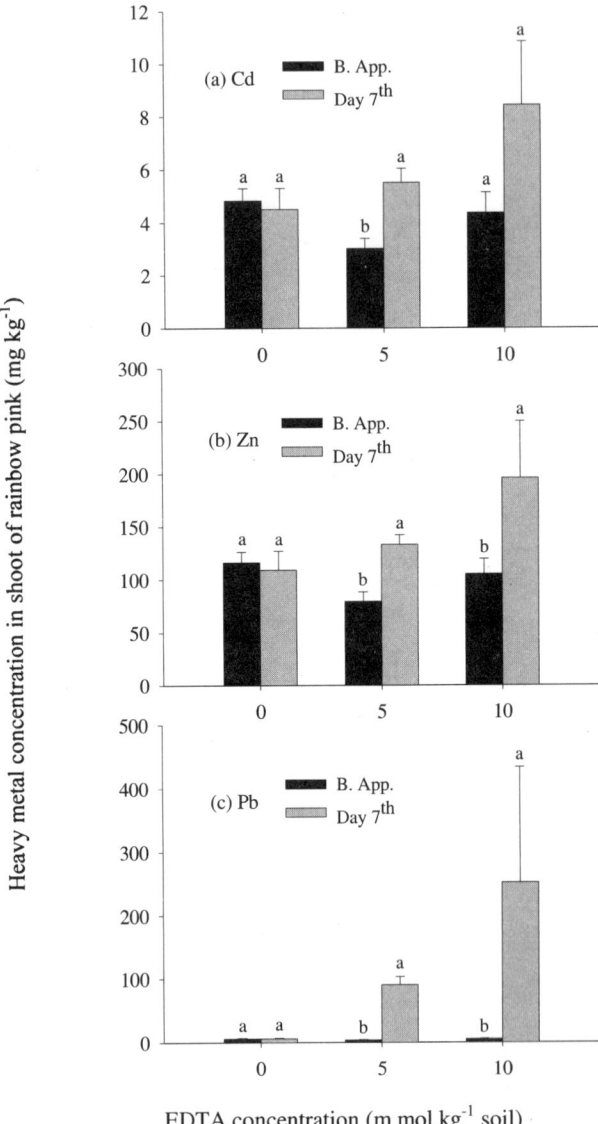

Fig. 4. The concentrations of (a) Cd, (b) Zn, and (c) Pb in the shoots of rainbow pink, which was harvested at 7th day after applying with different concentrations of EDTA. The probability level of significant difference is at p=0.05. Replicates (n)=3.

kg^{-1}, (7) Zn-Pb-treated soil: combined contaminated with 100 mg Zn kg^{-1} and 1,000 mg Pb kg^{-1}, and (8) Cd-Zn-Pb-treated soil: combined contaminated with 10 mg Cd kg^{-1}, 100 mg Zn kg^{-1}, and 1,000 mg Pb kg^{-1}. The different metals-treated soils were subjected to three cycles of wet and dry processes as described above before pot experiments were conducted (Blaylock et al., 1997).

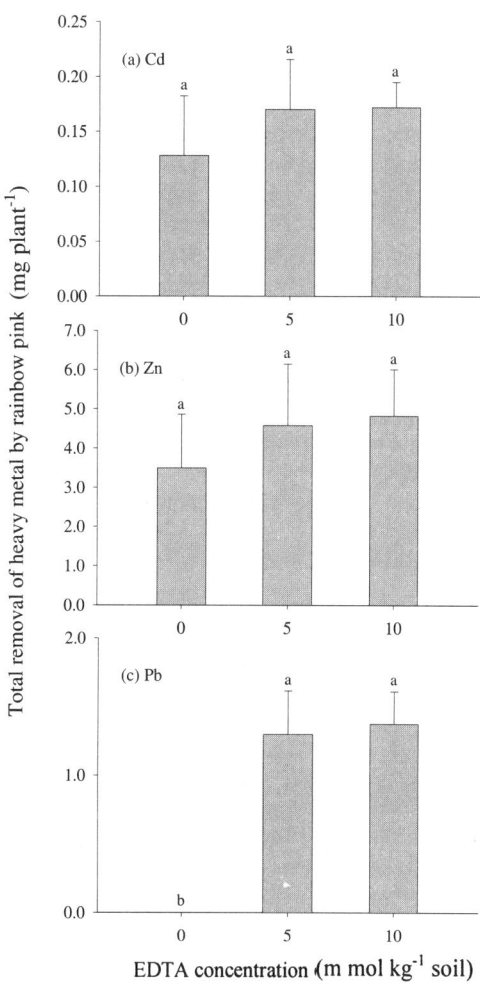

Fig. 5. The total removal of (a) Cd, (b) Zn, and (c) Pb in the shoots of rainbow pink, which was harvested at 7th day after applying with different concentrations of EDTA. The probability level of significant difference is at p=0.05. Replicates (n)=3.

Each pot (16 cm diameter and 19 cm height) contained 3.0 kg (dry weight) of artificially metals-treated soils. Two RSMS were put in each pot at a depth of around 10 cm below the surface of the soil. One seedling of rainbow pink (*Dianthus chinensis*) was planted in each pot. The pot experiments were conducted in a phytotron using a RCBD with a number of treatments and three replicates. The soil moisture content was maintained at 60% WHC, by weighing and adding deionized water every two days.

The application of 5 or 10 mmol EDTA kg^{-1} was significantly to increase the metal concentration in soil solution for a combined metals-contaminated soil used in above study. To decrease the potential of groundwater contamination, three concentrations of 2Na-EDTA solutions (0 (control), 2, and 5 mmol kg^{-1}) were added to the different metals-treated soils after planting for 21 days. The growth conditions and observed symptoms of toxicity, displayed by the plants, were also recorded. The shoots of rainbow pink were harvested at 7th day after the EDTA solution was added to the pot. The harvested plants were rinsed with tap-water to remove soils adhered to the surface of stems and leaves and then oven-dried at 60°C for 72 hr, weighed, and ground to 0.5 mm for analysis. Then, 0.2 g of the plant was digested with the H_2SO_4/H_2O_2 digestion method (Harmon and Lajtha, 1999). The digested solutions were filtered through a Whatman No. 42 filter paper, and then diluted with deionized water to a volume of 50 mL in a flask. The Cd, Zn, and Pb concentrations were determined with ICP-OES (Perkin Elmer 2000 DV).

Soil solutions were also collected directly with a RSMS before adding EDTA solutions and at 7th day after adding EDTA solutions, respectively. The concentration of Cd, Zn, and Pb in the tested soil after harvesting was extracted with deionized water (Mench et al., 1994), 0.05M EDTA (pH 7.0) (Mench et al., 1994), and 0.005M DTPA (pH 5.3) (Lindsay and Norvell, 1978), respectively. The concentrations of Cd, Zn, and Pb in soil solutions or in different extracts were determined with ICP-OES. To understand the concentration change of other elements after applying EDTA solutions, the concentrations of Ca, Mg, Fe, and Mn in soil solutions were also determined with ICP-OES.

4.1. Effect of EDTA on the Element Concentration in Soil Solution

Artificially adding Cd, Zn, or Pb into the original contaminated soil increased the concentration of

Table 3. The concentrations change of elements in soil solutions before applying and at 7th day after applying 2 and 5 mmol kg^{-1} soils of EDTA solutions

Treatment	EDTA concentration (mmol kg^{-1})	Cd	Zn	Pb	Ca	Mg	Fe	Mn
CK	0	Ns	ns	ns	ns	ns	ns	ns
	2	*	*	ns	ns	ns	**	**
	5	**	**	*	ns	ns	**	*
Cd	0	ns	ns	ns	ns	ns	ns	ns
	2	**	***	***	ns	ns	**	***
	5	***	***	***	ns	ns	***	*
Zn	0	ns	ns	ns	ns	ns	ns	ns
	2	ns	ns	ns	ns	ns	*	*
	5	**	***	***	ns	ns	***	*
Pb	0	ns	ns	ns	ns	ns	ns	ns
	2	ns	ns	*	ns	ns	ns	ns
	5	ns	ns	**	ns	ns	*	ns
Cd+Zn	0	ns	ns	ns	ns	ns	ns	ns
	2	***	***	***	ns	ns	***	ns
	5	***	***	***	ns	ns	**	***
Cd+Pb	0	ns	ns	ns	ns	ns	ns	ns
	2	ns	ns	*	ns	ns	**	ns
	5	*	*	**	ns	ns	***	ns
Zn+Pb	0	*	ns	ns	ns	ns	ns	ns
	2	*	**	*	ns	ns	ns	**
	5	***	***	***	ns	ns	***	**
Cd+Zn+Pb	0	ns	ns	ns	ns	ns	ns	ns
	2	*	*	***	ns	ns	*	ns
	5	*	**	**	*	ns	***	***

"ns": No significantly increased after applying EDTA solutions
*: $p < 0.05$, **: $p < 0.01$, ***: $p < 0.001$

specific added metal in soil solution. For instance, adding $Cd(NO_3)_2$ raised the Cd concentration in soil solution from not detectable (ND, below detection limit) in original contaminated soil (CK) to 0.14±0.07 mg L^{-1} in Cd-treated soil, 0.03±0.01 mg L^{-1} in Cd-Zn-treated soil, 1.34±0.47 mg L^{-1} in Cd-Pb-treated soil, and 1.38±0.43 mg L^{-1} in Cd-Zn-Pb-treated soil, respectively. Possessing the same result with that treated with $Cd(NO_3)_2$, artificial added with ZnO or $Pb(NO_3)_2$ also increased the Zn or Pb concentration in soil solution (data not shown).

Table 3 showed the concentrations change of elements in soil solutions before applying and at 7th day after applying 2 and 5 mmol kg^{-1} soils of EDTA solutions. In the different heavy metals-treated soils, applying two concentrations of EDTA solutions (2 and 5 mmol kg^{-1}) significantly increased the concentration of Cd, Zn, or Pb in soil solution ($p< 0.05$) except for applying 2 mmol EDTA kg^{1} in Zn-treated and Cd-Pb-treated soil (Table 3). The soil used in this study was innately contaminated with Cd because of the wastewater discharged by plants, and the total concentration of Cd, Zn, and Pb in original contaminated soil analyzed with aqua regia method was 2.58±0.08, 80.0±5.32, and 31.3±2.09 mg kg^{-1}, respectively. In the original contaminated soil (CK) without artificially adding metals, applying 2 or 5 mmol EDTA kg^{-1} significantly increased the concentrations of Cd, Zn, or Pb in soil solutions ($p< 0.05$) except for Pb after applying 2 mmol EDTA kg^{-1}. However, the increase was slight because of the low concentration of Cd, Zn, and Pb in the studied soil innately. Because of the studied soil was a silty clay soil and also of the low applying rate of EDTA, there was no identical change of Cd, Zn, or Pb concentration in soil solution after applying 2 mmol EDTA kg^{-1} compared with that of 5 mmol EDTA kg^{-1}.

Except for the Ca concentration in the Cd-Zn-Pb-treated soil after applying 5 mmol EDTA kg^{-1}, there was no significant increase on the Ca and Mg concentration in soil solution after applying two EDTA treatments. However, most of the concentration of Fe and Mn in soil solution significantly increased after applying 5 mmol EDTA kg^{-1} ($p< 0.05$), but not for all of those 2 mmol EDTA kg^{-1} treatments. Consistent with the results of Cd, Zn, and Pb concentration in soil solution after applying EDTA, there was no identical result of Fe and Mn concentration in soil solution after applying 2 mmol EDTA kg^{-1}.

The result revealed that applying EDTA increased the concentrations of Cd, Zn, Pb, Fe, and Mn in soil solutions of different metals-treated soils, but not for Ca and Mg. However, because of the low permeability of the tested soil compared with other studies (Huang et al., 1997; Chen and Cutright, 2001; Puschenreiter et al., 2001; Hammer and Keller, 2002) and also because of the low EDTA applying rate (2 mmol kg^{-1} soils) used in this study, the effect of EDTA was only statistic significant for 5 mmol EDTA kg^{-1} ($p< 0.05$).

The used soil was contaminated with metals, which have low concentration of metals innately. Thus the concentration of Cd, Zn, and Pb in soil solution of original contaminated soil was significantly increased after applying 5 mmol EDTA kg^{-1} ($p< 0.05$) (Fig 6). When considering the Cd concentration in soil solutions of different Cd-treated soils (i.e. Cd, Cd-Zn, Cd-Pb, and Cd-Zn-Pb-treated soil), applying 2 or 5 mmol EDTA kg^{-1} significantly increased the Cd concentration in soil solution ($p< 0.05$), except for the Cd-Pb-treated soil (Fig 6a). After applying 5 mmol EDTA kg^{-1}, the Cd concentration in soil solution was significantly increased from 0.14±0.07 to 5.90±1.14 mg L^{-1} (42-fold) in Cd-treated soil, and from 0.03±0.01 to 5.65±0.10 mg L^{-1} (188-fold) in Cd-Zn-treated soil, respectively ($p< 0.05$). Because of the low applying rate of 2 mmol EDTA kg^{-1}, there was no significantly increased in CK, Cd-Pb, and Cd-Zn-Pb-treated soils after applying 2 mmol EDTA kg^{-1}.

The result of Zn concentrations in soil solutions after applying different concentrations of EDTA solutions was similar to the result of Cd (Fig 6b). Applying 5 mmol EDTA kg^{-1} was significantly to increase Zn concentration in soil solution of different Zn-treated soils (i.e. Zn, Cd-Zn, Zn-Pb, and Cd-Zn-Pb-treated soil) ($p< 0.001$). The Zn concentrations in soil solutions of different Zn-treated soils were all below 4.0 mg L^{-1} before adding EDTA treatments. After treating with 5 mmol EDTA kg^{-1}, Zn concentration was significantly increased to the levels of 13-22 mg L^{-1} ($p< 0.001$). In soil only artificially treated with Zn, the Zn concentration in soil solution increased significantly from 3.37±0.66 to 21.3±2.67 mg L^{-1} (6.32-fold) after applying 5 mmol EDTA kg^{-1} treatment ($p< 0.001$), which was the highest concentration after applying 5 mmol EDTA kg^{-1}. Because of the low Zn concentration artificially added into the Zn-treated soils, the Zn concentration in soil solution after applying EDTA treatments is not as high as pervious study (Lai and Chen, 2004).

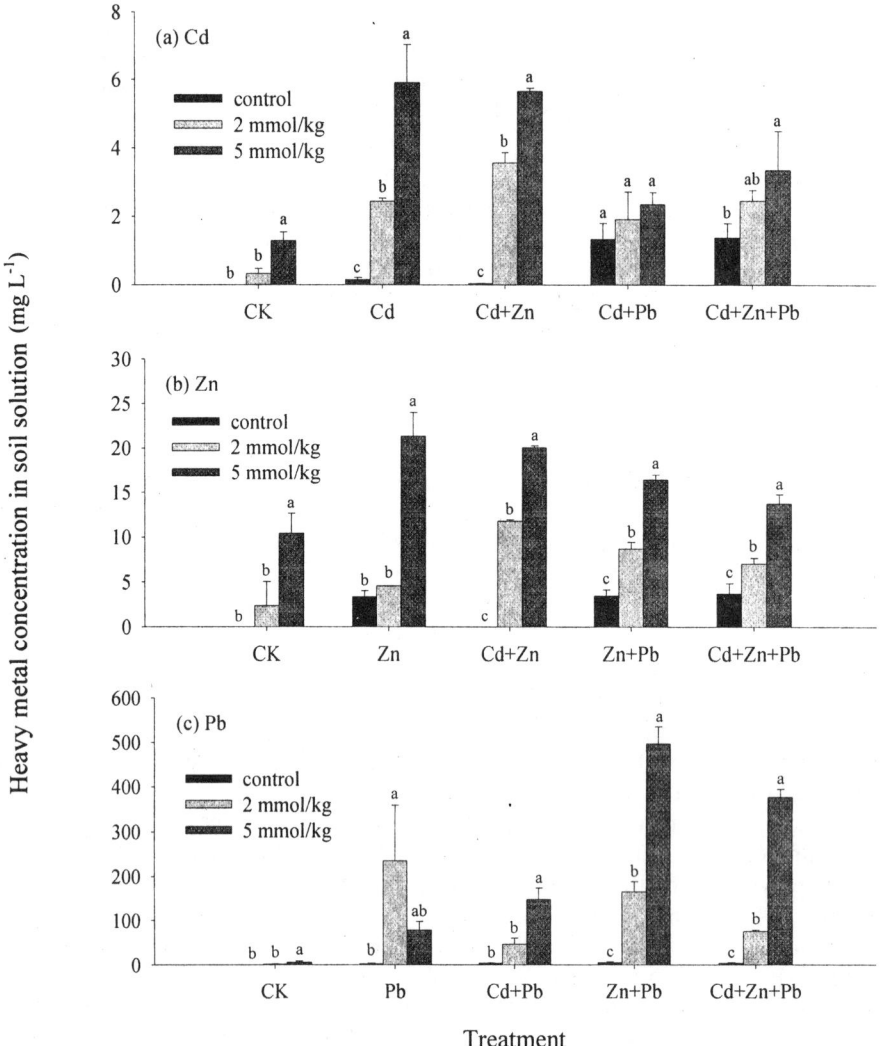

Fig. 6. The effect of applying different concentrations of EDTA on (a) Cd, (b) Zn, and (c) Pb concentration in soil solution of different metals-treated soils. The probability level of significant difference is at $p=0.05$. Replicates (n)=3.

Because of the low solubility of Pb in contaminated soil (Alloway, 1995), the Pb concentrations in the soil solutions of different Pb-treated soils were low without adding EDTA (Fig 6c). They were

2.49±1.38 mg L^{-1} in Pb-treated soil, 3.87±0.88 mg L^{-1} in Cd-Pb-treated soil, 5.29±2.23 mg L^{-1} in Zn-Pb-treated soil, and 4.11±0.99 mg L^{-1} in Cd-Zn-Pb-treated soil, respectively. Except for applying 2 mmol EDTA kg^{-1} in artificially Pb-treated soil, applying both concentrations of EDTA solutions was significantly to increase the Pb concentration in soil solution of different Pb-treated soils (i.e. Pb, Cd-Pb, Zn-Pb, and Cd-Zn-Pb-treated soil) ($p< 0.05$), especially for the 5 mmol EDTA kg^{-1} (Fig 6c). These results indicated that applying EDTA solutions was effectively to increase the concentration of metal, especially for Pb, in the soil solution. The differences of the chelating effects between EDTA and metals are reflected by the formation constant (log K_f =19.0) of the Pb-ligand complexes, which exceeds those of the Cd-ligand complexes (log K_f=17.4) and the Zn-ligand complexes (log K_f=17.5) (Martell and Smith, 1974; 1982). The maximum concentration of Pb in soil solution was 498±38.2 mg L^{-1} in the Zn-Pb-treated soil after applying 5 mmol EDTA kg^{-1} solutions. After applying 2 mmol EDTA kg^{-1}, the Pb concentration in soil solution of artificially Pb-treated soil was higher than that of 5 mmol EDTA kg^{-1} treatment, and significantly increased the Pb concentration from 2.49±1.38 mg L^{-1} (without applying EDTA solutions) to 236±123 mg L^{-1} ($p< 0.05$). However, there was no significantly different between 2 and 5 mmol EDTA kg^{-1} in artificially Pb-treated soil. This might because of the high variation of Pb concentration in soil solution after applying 2 mmol EDTA kg^{-1}.

The result in this study showed that applying 5 mmol EDTA kg^{-1} was efficient to increase Cd, Zn, or Pb concentration in soil solution of different metals-treated soil. This finding is consistent with those obtained following the application of EDTA treatments on contaminated soils in other studies (Blaylock et al., 1997; Huang et al., 1997; Wu et al., 1999; Luo et al., 2000; Chen and Cutright, 2001; Lombi et al., 2001a; Hopgood et al., 2003). Because the soil used in this study was a silty clay soil, which have lower permeability compared with sandy soils, the effect of applying 2 mmol EDTA kg^{-1} on increasing metal concentration in soil solution was not identically. Lai and Chen (2004) reported that applying 5 and 10 mmol EDTA kg^{-1} significantly increased the metal concentration in soil solution of the same soil used in this study. We thus proposed that when applying EDTA in the silty clay soil to increase the effect of phytoextraction, more than 5 mmol EDTA kg^{-1} was necessary to efficiently and identically increase metal concentration in soil solution.

The concentrations of Ca, Mg, Fe, and Mn in soil solutions were also analyzed in the study to investigate the concentration change of other elements after applying EDTA solutions. Results showed that Ca and Mg concentrations in soil solution did not have identical tendency after applying 2 or 5 mmol EDTA kg^{-1} (Fig 7a and Fig 7b), but Fe or Mn concentration in soil solution was significantly increased after applying both concentrations of EDTA solutions ($p< 0.05$), especially for the 5 mmol EDTA kg^{-1} treatment (Fig 7c and Fig 7d). Cooper et al. (1999) reported that applying high concentration of DTPA significantly increased the Fe and Mn concentration in plants, but was not significantly for Ca and Mg. Applying EDTA significantly increased the Fe and Mn concentration in soil solution in this study, which was easier to uptake by plants. The result of this study was similar to the result of Copper et al. (1999). The Fe concentrations in soil solutions without EDTA treatments were not detected in different metal-treated soils. Seven days after 2 mmol kg^{-1} EDTA solutions were added to the soils, Fe concentration in soil solution was increased drastically and significantly to the levels of 70-210 mg L^{-1} ($p< 0.05$). Applying 5 mmol EDTA kg^{-1} was more efficient than that applying 2 mmol EDTA kg^{-1}, and the Fe concentration significantly increased to the levels of 150-590 mg L^{-1} ($p< 0.05$). The concentration of Mn in soil solution after applying EDTA had the same tendency with that of Fe, even some of the increases were not statistic significant.

4.2. Effect of EDTA on the Metal Concentration of Different Extractors

After applying 5 mmol EDTA kg^{-1}, the deionized water extractable Cd, Zn, or Pb concentration of different metals-treated soils was increased significantly ($p< 0.05$) (Fig 8). The concentrations of metals

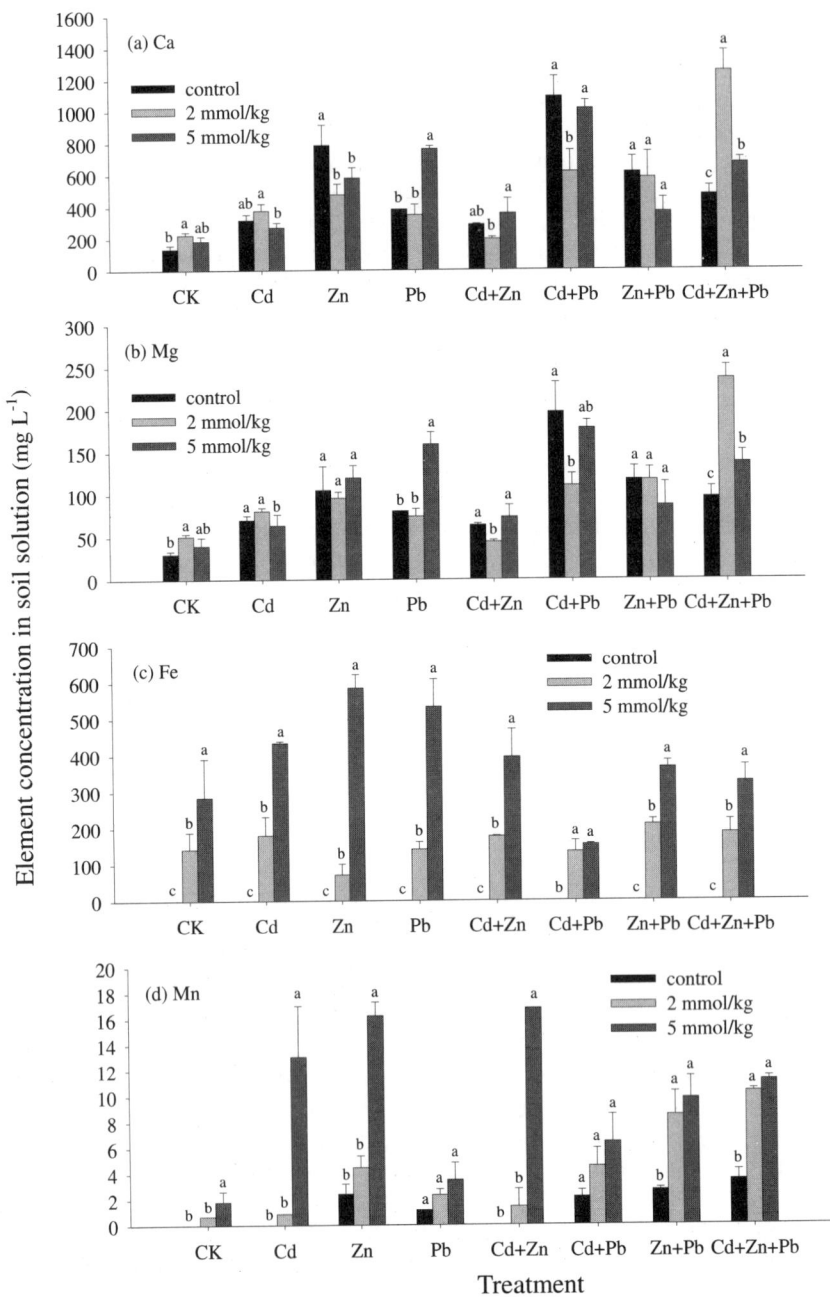

Fig. 7. The effect of applying different concentrations of EDTA on (a) Ca, (b) Mg, (c) Fe, and (d) Mn concentration in soil solution of different metals-treated soils. The probability level of significant difference is at p=0.05. Replicates (n)=3.

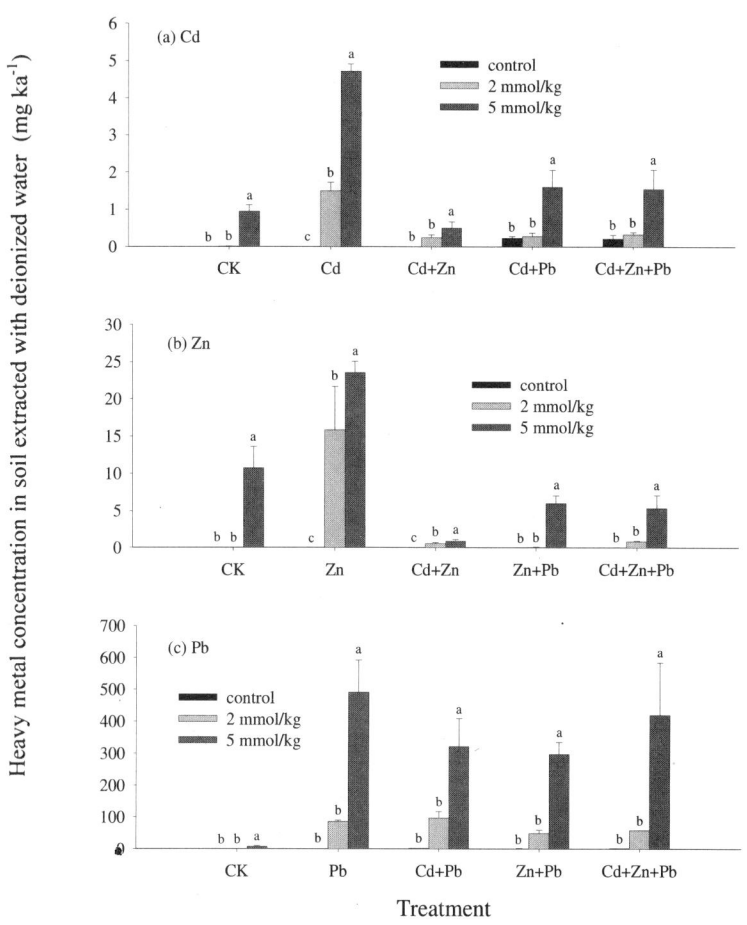

Fig. 8. The effect of applying different concentration of EDTA on deionized water extractable (a) Cd, (b) Zn, and (c) Pb concentration in different metals-treated soils. The probability level of significant difference is at p=0.05. Replicates (n)=3.

extracted with deionized water were equal to the water soluble form in the contaminated soil (Zyrin et al., 1974; Gupta and Chen, 1975; Gatehouse et al., 1976; Mathur and Levesque, 1983; Jones et al., 1984; Miller et al., 1986), which were easily leaching into the groundwater along with rainfall. Because the soil used in this study was contaminated with heavy metal innately, applying 5 mmol EDTA kg^{-1} significantly increased the deionized water extractable concentration of Cd, Zn, and Pb in original contaminated soil. Applying 2 mmol EDTA kg^{-1} was not significantly to increase the water soluble metals in most of the treatments, except for Cd concentration in Cd-treated soil and Zn concentration in Zn and Cd-Zn-treated soil (p< 0.05). Among these three metals, the effect of applying 5 mmol EDTA kg^{-1} was most efficient on increasing the Pb concentration in soil extracted with deionized water. In the different artificially Pb-treated soils (i.e. Pb, Cd-Pb, Zn-Pb, and Cd-Zn-Pb-treated soil) without treating with EDTA solutions, deionized water extractable Pb concentrations were all below 2 mg kg^{-1}. After applying 5 mmol EDTA kg^{-1}, the Pb concentrations were significantly increased (p< 0.05), and the concentration levels were from 300 to 500 mg kg^{-1} (Fig 8c). The results showed that applying EDTA into the different metals-treated soils can significantly increase the water soluble fraction of metals in soils ($p < 0.05$).

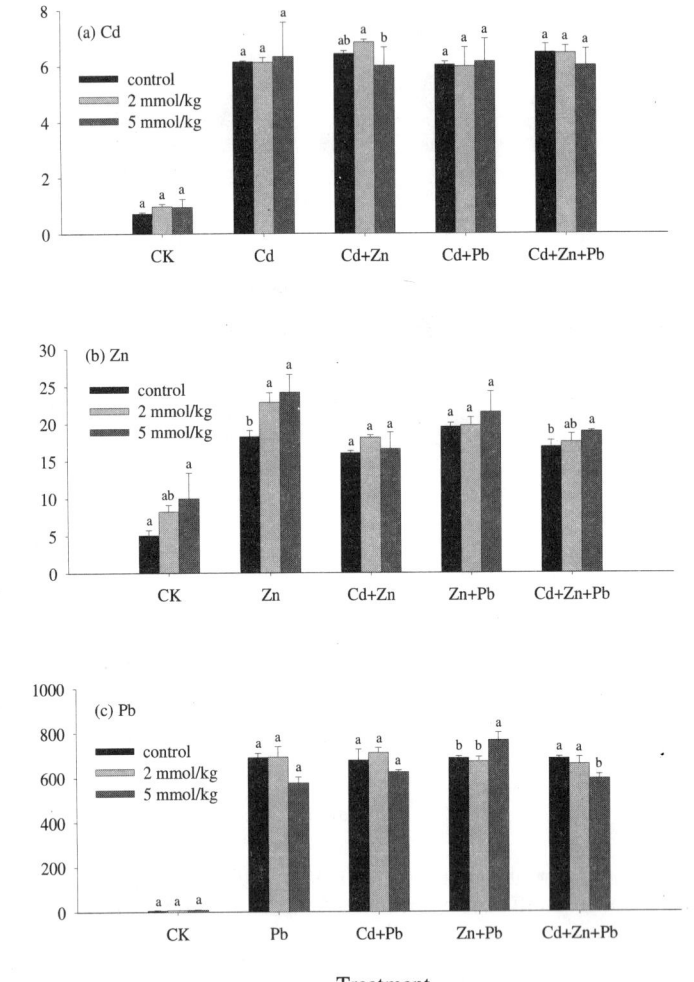

Fig. 9. The effects of applying different concentrations of EDTA on (a) Cd, (b) Zn, and (c) Pb concentration of different metals-treated soils extracted with $0.005M$ DTPA (pH 5.3). The probability level of significant difference is at $p=0.05$. Replicates (n)=3.

Fig. 9 showed the 0.005M DTPA (pH 5.3) extractable concentrations of different metals-treated soils after applying 2 or 5 mmol EDTA kg^{-1} solutions. There was no significantly different on those concentrations extracted with DTPA for most of the metals-treated soils after applying EDTA treatments. This might because of (1) the metals existed in contaminated soils were artificial added with different solutions of meals and only contaminated for a short period (about 3 months), in which the metals were easier to extract with DTPA than the long-term contaminated soils, and (2) the extracting agents (0.005M DTPA) used in the study is too strong to extract metals in different metals-treated soils. The bioavailability of metals decreased with the contamination time (Vig et al., 2003); Soil artificially contaminated with solutions of heavy metals will have high bioavailability compared with long-term contaminated soils. The DTPA can extract higher amounts of metals in artificially contaminated soils than the long-term contaminated soils. Some authors indicated that 0.005M DTPA was able to extract heavy metals which were water soluble, exchangeable, organic matter bound, and even the fractions occluded in oxides (Stover et al., 1976; Sposito et al., 1982; Mathur and Levesque, 1983; Jones et al., 1984; Goldberg and Smith, 1984). We consider that most of the metals in different metals-treated soils of this study were existed in the fractions of water soluble, exchangeable, or bounding by organic matter. Because of the strong extracting capacity of 0.005M DTPA, which can extract most fractions of the metals in this tested soil, the Cd, Zn, and Pb concentrations in extracts after adding different concentrations EDTA were not significantly increased for most of the metals-treated soils.

The result of 0.05M EDTA (pH 7.0) extractable Cd, Zn, and Pb concentrations was similar to that result of 0.005M DTPA (pH 5.3) (Fig 10). There was no significantly different for most of the EDTA extractable metals in different metals-treated soils after adding different concentrations EDTA treatments. Pervious studies indicated that 0.05M EDTA (pH 7.0) can extract the same fractions with 0.005M DTPA (pH 5.3) (Jones et al., 1984; Miller et al., 1986). There were no significantly increased on the concentrations of extracts after applying 2 or 5 mmol kg^{-1} EDTA solutions because of the strong extracting capacity of 0.05M EDTA.

4.3. Effect of EDTA on the Metal Concentration in the Shoot of Rainbow Pink

In the different artificial Cd-treated soils (i.e. Cd, Cd-Zn, Cd-Pb, and Cd-Zn-Pb-treated soil), applying both 2 and 5 mmol EDTA kg^{-1} was significantly to increase the Cd concentration in shoots of rainbow pink ($p< 0.05$) except for the Cd-Pb-treated soil (Fig 11a). The Cd concentrations in the shoots of rainbow pink when growing in the Cd-Pb-treated soils were not statistic significant because of (1) the Cd concentrations in soil solutions were not significantly increased after applying 2 and 5 mmol kg^{-1} EDTA solutions, and (2) high variation of the Cd concentration in Cd-Pb-treated soils after applying 2 or 5 mmol EDTA kg^{-1} solutions. Applying EDTA solutions was significantly to increase the Cd concentrations both in soil solutions and extracted with deionized water in Cd, Cd-Zn, and Cd-Zn-Pb-treated soil ($p< 0.05$), thus significantly increased the Cd concentration in the shoots of rainbow pink ($p< 0.05$). The result is similar to that reported by Lai and Chen (2004).

Even through both the Zn concentrations in soil solutions and deionized water extractable concentrations were significantly increased after applying both concentrations of EDTA solutions ($p< 0.05$), the Zn concentrations in shoots of rainbow pink were not significantly increased except for Cd-Zn-Pb-treated soil ($p< 0.05$) (Fig 11b). The Zn concentrations levels in shoots of rainbow pink were from 50 to 110 mg kg^{-1} in different Zn-treated soils. Zinc is an essential element to maintain the growth of plants, and the mean content in the plants were from 1.2 to 73 mg kg^{-1} (Kabata-Pendias and Pendias, 2001). We consider that because of the low concentration of Zn added into the different artificial metals-treated soil (0.05M EDTA and 0.005M DTPA extracted Zn were all below 30 mg kg^{-1}) and the Zn concentration in shoot is sufficient for the growth of rainbow pink. Even through the Zn concentration both in soil solution or extracted with deionized water were increased significantly after applying EDTA solutions ($p< 0.05$), the Zn concentration in shoots of rainbow pink when growing in different Zn-treated soils was not significantly affected by the application of EDTA solutions. Some interactions of synergistic or antagonistic effects may also contribute to the result described above.

When growing in the different Pb-treated soils without adding EDTA solutions, all the Pb concentration in shoots of rainbow pink were below 30 mg kg^{-1} (Fig 11c). The pattern can be illustrated by the low concentration of Pb both in the soil solution (<6 mg L^{-1}) (Fig 6c) and extracted with deionized water (<2 mg kg^{-1}) (Fig 8c). After applying 2 and 5 mmol EDTA kg^{-1}, Pb concentration in shoots of rainbow pink was significantly increased ($p< 0.05$).

Among the three metals, EDTA was most efficient in increasing the Pb concentration in shoots of rainbow pink when planting in different metals-treated soils. The result is similar to other studies when applying EDTA in the metal-contaminated soil (Blaylock et al., 1997; Lai and Chen, 2004).

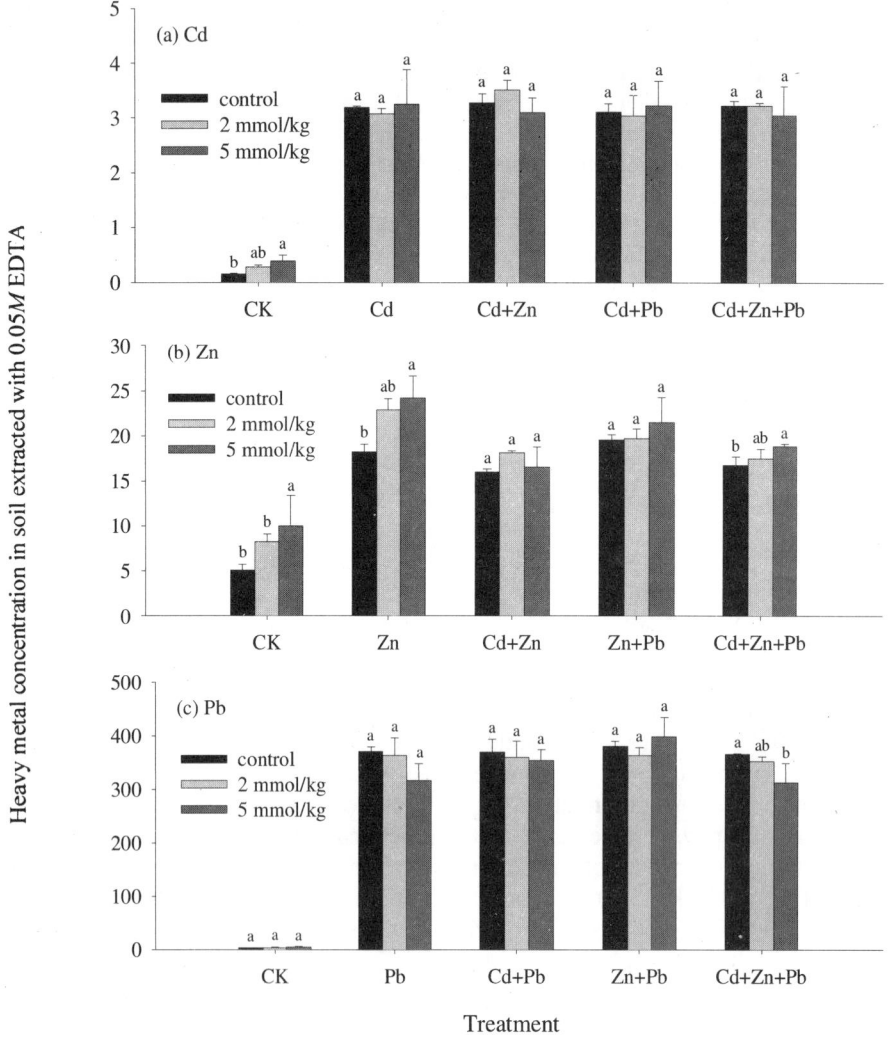

Fig. 10. The effects of applying different concentrations of EDTA on (a) Cd, (b) Zn, and (c) Pb concentration of different metals-treated soils extracted with 0.05M EDTA (pH 7.0). The probability level of significant difference is at p=0.05. Replicates (n)=3.

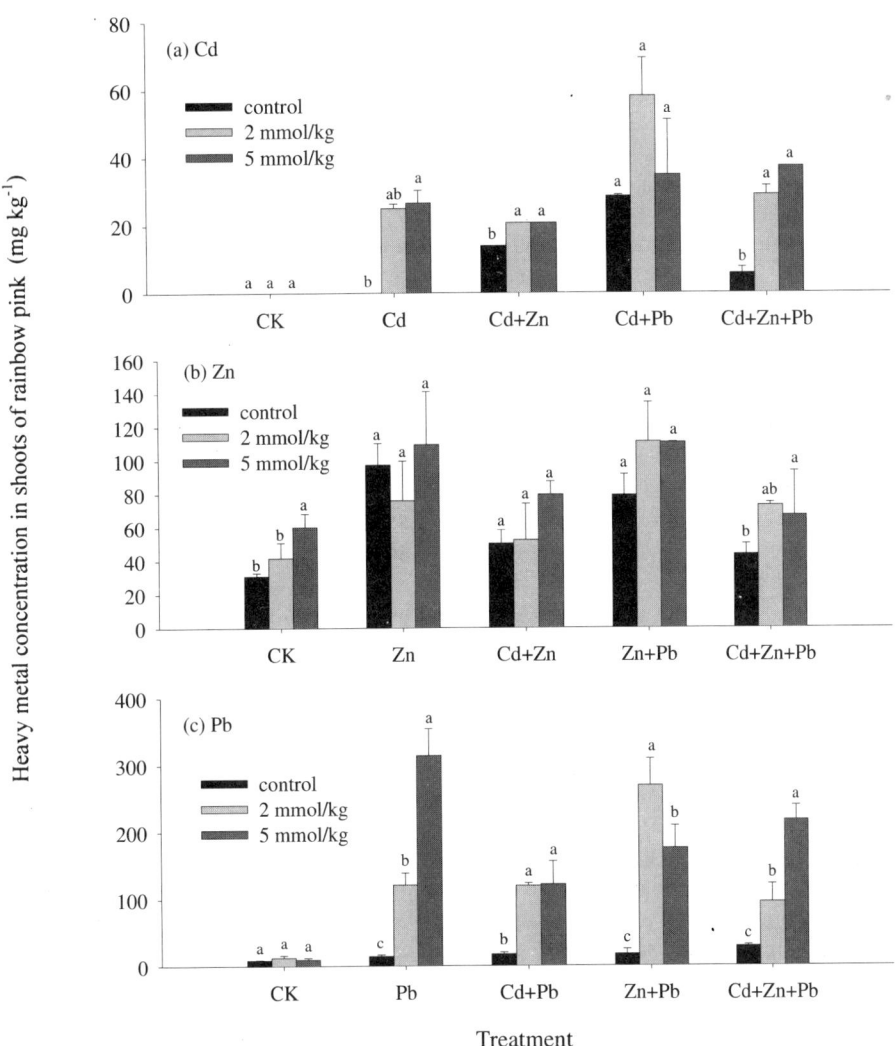

Fig. 11. The concentrations of (a) Cd, (b) Zn, and (c) Pb in the shoots of rainbow pink, which was harvested at 7th day after applying different concentrations of EDTA. The probability level of significant difference is at p=0.05. Replicates (n)=3.

5. POSSIBLE MECHANISMS OF Cd HYPERACCUMULATOR

The mechanisms of plants to tolerate the toxicities of metals included exclusion and detoxification (Baker, 1981). By both to increase the soil pH value in the rhizosphere or to release excludes by roots avoiding plants to uptake metals or to transfer metals from roots to shoots (De Voc et al., 1991). The result of previous study showed that *Triticum aestivum* raised the soil pH value of rhizosphere to decrease the bioavailability of metals and exclude in the root of *Silene cucubalus* was release to avoid the uptake of copper (Jackson et al., 1990). Ross and Kaye (1994) indicated that the ligand in the tip of root form chelate with metals, decrease the bioavailability of metals, and thus to decrease the metals concentrations in the roots of plants. *Agrostis tenuis* can also release excludes into the rhizosphere and formed precipitation with Cd, Cu, Pb, or Zn to avoid the uptake of plants (Dahmani-Muller et al., 2000). These kinds of plants were used to phytostabilization the metals-contaminated soils to avoid the erosion of wind and rain. To avoid the toxicities of metals in the tissues of plants, ions or complexes of metals were released back to the rhizosphere by the mechanism of detoxification (Jackson et al., 1990).

The three responses of plants to uptake metals in the soils were hyperaccumulator, indicator, and excluder, respectively (Baker, 1981; McGrath et al., 2000). Hyperaccumulator, with cut-off point, can accumulate high concentration of metals in the plants even growing in the low concentration of metal-contaminated soils. Those plants of indicator can uptake and transfer metals and there was a good relationship between the metals concentrations in soils and that in the tissues of plants. One can assess or monitor the level of contamination by analyzing the concentration of metals in the tissues of plants. The plant of excluder with special mechanism to avoid the uptake of metals in soils and the transfer of metals from root to shoot was also limited. The metals concentration in the excluder was not increase with the increasing metal concentration in the soil, however, this mechanism disappear when the soil was contaminated with high concentration of metal (Ross and Kaye, 1994; Greger, 1999).

Suitable concentration of essential elements is necessary for the growth of plants, but high energy is needed to uptake high concentrations of metals. Previous studies showed that some plants were able to accumulate high concentration of metals in the leaves or shoots to avoid biting by insects, fungi, or bacteria (Boyd and Martens, 1994; Pollard and Baker, 1997). The mechanisms of plants to avoid the toxicities of accumulated high concentrations of metals including the following three reasons: (1) The excludes released by the roots of plants can increase the bioavailability of metals (Lee et al., 1978; Brown et al., 1994; Krämer et al., 1996; Knight et al., 1997; McGrath et al., 1997; Whiting et al., 1997; Klassen et al., 2000; Zhao et al., 2000; Hammer and Keller, 2002), (2) there are transporter proteins in the cell membrane which can transfer metals from root to shoot of plants efficiently (Lasat et al., 1996; Lasat et al., 2000; Lasat and Kochian, 2000; Lombi et al., 2001a and b), and (3) the accumulated metals were sequestrate in the cell membrane or vacuole to reduce the toxicities of metals (Ortiz et al., 1992; Vázquez et al., 1994; Salt et al., 1995; Küpper et al., 1999).

Klassen et al. (2000) indicated that those excludes released by the roots of some plants can change the soil pH value of the rhizosphere and to increase the bioavailability of metals. Hammer and Keller (2002) also indicated that exclude of a hyperaccumulator *Thlaspi caerulescens* can change the fractions of metals in the rhizosphere. *T. caerulescens* was able to extract the Zn in the soils which with low bioavailability (McGrath et al., 1997). The non-hyperaccumulator, *T. arvense*, can accumulate more concentration of Zn in the shoot of plant when planting with hyperaccumulator *T. caerulescens* (Whiting et al., 1997). These results also prove that excludes of hyperaccumulators can formed complexes with metals and thus increase the bioavailability of metals and also the metals concentrations in the plants. For the hyperaccumulator of Ni, the possible excludes including citrate (Lee et al., 1978) and histidine (Krämer et al., 1996). Histidine played an important role in the balance of Zn in the root of plant. However, many researchers hold different opinions about this and indicated that the excluding of citrate and histidine can decrease the uptake of Ni by non-accumulators (Cumming and Taylor, 1990; Salt et al., 2000). Even through that, excludes of roots is an important mechanism to increase the bioavailability of metals for hyperaccumulators. Besides excludes of roots, soil pH value is also important to control the fractions of metals. The bioavailability of metals decreased with the increasing of soil pH value for most of the metals.

Hyperaccumulator may have transport protein in the cell membrane which was efficient to increase the uptake and transfer of metals from soil to root or from root to shoot. Lasat et al. (1996) and Lasat et al. (2000) indicated that because of the high density of Zn transporters in the roots, the Zn hyperaccumulator *T. caerulescens* can uptake high concentration of Zn. Among those Zn transporters, *ZNT1* was the most important. The high concentration of Cd in the *T. caerulescens* also has relationship with the Cd transporter in the root of plant (Lombi et al., 2001b). However, large amount of unknowns were existed for most of the transporters. Some of them even cannot distinguish the types of metals and transport the wrong metals. Chaney et al. (1994) also indicated that the transporter protein of plant cannot distinguish between Cd and Zn, which resulting in the accumulation of non-essential element Cd.

The third mechanism of plants to accumulate high concentrations of metals was the sequestration of metals in the cell wall or vacuole (Vázquez et al., 1992). For the leaves of *Arabidopsis thaliana*, the main part to the sequestrate Cd was in the trichomes (Ager et al., 2002). The sequestration of metals in the vacuoles to avoid the toxicities of metals have also proved by many researches (Ortiz et al., 1992; Vázquez et al., 1994; Lasat et al., 1996; Lasat et al., 1998; Küpper et al., 1999). Jarvis and Leung (2002) used transmission electron microscopy (TEM) to observe the accumulation of Pb in the plants also showed that the Pb was mainly accumulated in the cell membrane.

The toxicities of metals decreased when metals formed combination with low molecular organic compounds (<10 kD) (Salt et al., 1996; Lee et al., 1978). The increasing of metals concentrations can induce the synthesis of phytochelatins (PCs) (Steffens, 1990; Rauser, 1995; Schmöger et al., 2000). Some metals, like Cd, Cu, Ag, Hg, or Pb can induce the synthesis of PCs, especially for Cd and Cu. The combination of metals with PCs can reduce the toxicities of metals (Niebore and Richardson, 1980). Phytochelatins can also call the cadystin, phytometallothionein, or metallothiopeptide (Robinson, 1990; Tomsett et al., 1988), The PCs content the cysteine (a sulphur-containing amino acid) and is a protein in the cytoplasm with the structure of γ-GluCys)$_n$-Gly (n=2~12). Some researches also focused on the types of complexes sequestrated in the vacuole after complex with PCs. Krämer et al. (2000) showed that the main form in the vacuole for Ni hyperaccumulator *T. goesingense* was Ni-organic acid. Van Stevenink et al. (1990) indicated that the Zn was complex with oxalic acid or formed the precipitation of Zn-phytate. Organic cid also played an important role on the transport in the xylem and sequestration in the shoot of Zn (Salt et al., 1999). Figure 2 showed the possible mechanisms of the uptake and transport of metals (Greger, 1999).

For the hyperaccumulators, these excludes of roots were first release into the rhizosphere to decrease the pH value of soil and also to increase the bioavailability of metals. The metals in the soils then enter the roots of plants, formed complexes with organic acids, and then transfer into the shoots of plants by the transport proteins and sequestrate in the trichomes of leaves (Ager et al., 2002) or the vacuoles (Küpper et al., 1999; Clemens et al., 2002).

References

Adriano DC (1986) Cadmium. In: Adriano DC (ed) Trace elements in the terrestrial environment. Spring-Verlag New York Inc., USA, pp. 106-155.
Ager FJ, Ynsa MD, Domínguez-Solís JR, Gotor C, Respaldiza MA, Romero LC (2002) Cadmium localization and quantification in the plant *Arabidopsis thaliana* using micro-PIXE. Nucl Instr Meth Phys Res B 189: 494-498.
Alloway BJ (1995) Soil processes and the behavior of metals. In: Alloway BJ (ed) Heavy metals in soils. Blackie Academic & Professional, Glasgow, UK, pp. 11-37.
Alloway BJ, Steinnes E (1999) Anthropogenic additions of cadmium to soils. In: Laughlin MJ, Singh BR (eds.) Cadmium in Soils and Plants. Kluwer Academic Publishers, Dordrecht, Netherlands, pp. 97-123.
Baker AJM (1981) Accumulators and excluders: strategies in the response of plants to heavy metals. J Plant Nutr 3: 643-654.
Baker AJM, Reeves RD, Hajar ASM (1994) Heavy metal accumulation and tolerance in British populations of the metallophyte *Thlaspi caerulescens* J.&C. Presl (*Brassicaceae*). New Phytol 127: 61-68.

Baker AJM, McGrath SP, Reeves RD, Smith JAC (2000) Metal hyperaccumulator plants: A review of the ecology and physiology of a biological resource for phytoremediation of metal-polluted soils. In: Terry N, Bañuelos G (eds.) Phytoremediation of Contaminated Soil and Water. CRC Press LLC, USA, pp. 85-107.

Barona A, Aranguiz I, Elías A (2001) Metal associations in soils before and after EDTA extractive decontamination: implications for effectiveness of further cleanup procedures. Environ Pollut 113: 79-85.

Bingham FT, Page AL, Mahler RJ, Ganje TJ (1976) Yield and cadmium accumulation of forage species in relation to cadmium content of sludge amended soils. J Environ Qual 5: 57-60.

Blaylock MJ, Salt DE, Dushenkov S, Zakharova O, Gussman C, Kapulnik Y, Ensley BD, Raskin I (1997) Enhanced accumulation of Pb in Indian mustard by soil-applied chelating agents. Environ Sci Technol 31: 860-865.

Boyd RS, Martens SN (1994) Nickel hyperaccumulated by *Thlaspi caerulescens* var. *montanum* is acutely toxic to an insect herbivore. Oikos 70: 21-25.

Brown KW, Thomas JC (1983) Metal accumulation by bermudagrass grown on four diverse soils amended with secondarily treated swageefflent. Water Air Soil Poll 20: 431-446.

Brown SL, Chaney RL, Angel JS, Baker AJM (1994) Phytoremediation potential of *Thlaspi caerulescens* and bladder campion for zinc- and cadmium-contaminated soil. J Environ Qual 23: 1151-1157.

Brown SL, Chaney RL, Angel JS, Baker AJM (1995) Zinc and cadmium uptake by hyperaccumulator *Thlaspi caerulescens* grown in nutrient solution. Soil Sci Soc Am J 59: 125-133.

Chaney RL, Green CE, Filcheva E, Brown SL (1994) Effect on iron, manganese, and zinc-enriched biosolids compost on uptake of cadmium by lettuce from cadmium-contaminated soils. p.205-207. In: Clapp CE, Larson WE, Dowdy RD (eds.) Sewage Sludge: Land Utlization and the Environment. Amer Soc Agron, Madison, WI, pp. 85-107.

Chaney RL, Malik M, Li YM, Brown SL, Brewer EP, Angle JS, Baker AJM (1997) Phytoremediation of soil metals. Environ Biotechnol 8: 279-284.

Chen H, Cutright T (2001) EDTA and HEDTA effects on Cd, Cr, and Ni uptake by *Helianthus annuus*. Chemosphere 45: 21-28.

Chen ZS, Lee DY (1997) Evaluation of remediation techniques on two cadmium-polluted soils in Taiwan. In: Iskandar IK, Adriano DC (eds.). Remediation of Soils Contaminated with Metals. Science Reviews, Northwood, UK, pp. 209-223.

Clemens S, Palmgren MG, Krämer U (2002) A long way ahead: understanding and engineering plant metal accumulation. Trends Plant Sci 7: 309-315.

Cooper EM, Sims JT, Cunningham SD, Huang JW, Berti WR (1999) Chelate-assisted phytoextraction of lead from contaminated soils. J Environ Qual 28: 1709-1719.

Cumming JR, Taylor GJ (1990) Mechanisms of metal tolerance in plants: physiological adaptations for exclusion of metal ions from the cytoplasm. In: Alscher RG, Cimming JR (eds.) Stress Responds in Plants: Adaptation and Acclimation Mechanisms. Wiley-Liss, pp. 329-359.

Dahmani-Muller H, van Oort F, Gélie B, Balabane M (2000) Strategies of heavy metal uptake by three plant species growing near a metal smelter. Environ Poll 109: 231-238.

Davis RD, Calton-Smith C (1980) Crops as Indicators of the Significance of Contamination of Soil by Heavy Metals. WRC, Stevenage TR140.

De Voc CHR, Schat H, De Waal MAM, Voojs R, WHO Ernst (1991) Increased resistance to cooper-induced damage of root cell plasmalemma in cooper tolerant *Silene cucubalus*. Physiol Plant 82: 523-528.

Ebbs SD, Kochian LV (1998) Phytoextraction of zinc by oat (*Avena sativa*), barley (*Hordeum vulgare*), and Indian mustard (*Brassica juncea*). Environ Sci Technol 32: 802-806.

Farrah H, Pickering WF (1977) Water Air Soil Poll 8: 189-197.

Garbisu C, Alkorta I (2001) Phytoextraction: a cost-effective plant-based technology for the removal of metals from the environment. Biores Technol 77: 229-236.

Gatehouse S, Russell DW, Van Moort JC (1976) Sequential soil analysis in exploration geochemistry. J Geochem Explor 8: 483-494.

Goldberg SP, Smith KA (1984) Soil manganese: E values, distribution of manganese-54 among soil fractions, and effects of drying. Soil Sci Soc Am J 48: 559-564.

Greger M (1999) Metal availability and bioconcentration in plants. In: Prasad MNV, Hagemeyer J (eds.) Heavy metal stress in plants: From molecules to ecosystems. Springer, NY, USA, pp. 1-27.

Gupta SK, Chen KY (1975) Partitioning of trace metals in selective chemical fractions of near shore sediments.

Environ Lett 10: 129-158.
Hammer D, Keller C (2002) Changes in the rhizosphere of metal-accumulating plants evidenced by chemical extractant. J Environ Qual 31: 1561-1569.
Harmon ME, Lajtha K (1999) Analysis of detritus and organic horizons for mineral and organic constitutes. In: Robertson GP, Coleman DC, Bledsoe CS, Sollins P (eds.) Standard Soil Methods for Long-Term Ecological Research. Oxford University Press, Inc., NY, USA, pp. 143-165.
Hopgood MJ, Hodson ME, Mott CJB, Alloway BJ (2003) Effect of soil properties on the ability of EDTA to mobilize lead. In: 7th International Conference on the Biogeochemistry of Trace Elements. 15-19 June 2003. Uppsala, Sweden. pp. 152-153.
Huang JW, Cunningham SD (1996) Lead phytoextraction: species variation in lead uptake and translocation. New Phytol 134: 75-84.
Huang JW, Chen J, Berti WR, Cunningham SD (1997) Phytoremediation of lead-contaminated soils: Role of synthetic chelates in lead phytoextraction. Environ Sci Technol 31: 800-805.
Huebert DB, Shay M (1993) The response of *Lemna trisulca* L. to cadmium. Environ Poll 80: 247-253.
Iimura K, Ito H, Chino M, Morishita T, Hirata H (1977) Behavior of contaminant heavy metals in soil-plant system. In: Proc Inst Sem AEFMIA, Tokyo, pp. 357.
Ito H, Iimura (1975) Adsorption of cadmium by rice plants in respond to change of oxidation reduction condition of soils. J Sci Soil Manure Japan 46: 82-88.
Jackson AP, Alloway BJ (1992) Transfer of cadmium from soils to the human food chain. In: Adriano DC (ed.) Biogeochemistry of Trace Elements. Lewis Publisher, Botan Rouge, FL, USA, pp. 109-158.
Jackson PJ, Unkefer PJ, Delhaize E, Robinson NJ (1990) Mechanisms of trace metal tolerance in plants. In: Katterman F (ed.) Environmental Injury to Plants. Academic Press, San Diego, USA, pp. 231-258.
Jarvis MD, Leung DWM (2002) Chelated lead transport in Pinus radiate: an ultrastructural study. Environ Exp Bot 48: 21-32.
Jones KC, Peterson PJ, Davies BE (1984) Extraction of silver from soils and its determination by atomic absorption spectrometry. Geoderma 33: 157-168.
Kabata-Pendias A, Pendias H (2001). Trace Element in Soils and Plants. 3rd edn. CRC Press, Boca Raton, FL, USA.
Kitagishi K, Yamane I (1981) Heavy Metal Pollution in Soils of Japan. Japan Sci Soc Press, Tokyo, Japan.
Klassen SP, McLean JE, Grossl PR, Sims RC (2000) Fate and Behavior of lead in soils planted with metal-resistant species (River Birch and Smallwing Sedge). J Environ Qual 29: 1826-1834.
Knight B, Zhao FJ, McGrath SP, Shen ZG (1997) Zinc and cadmium uptake by the hyperaccumulator *Thlaspi caerulescens* in contaminated soils and its effects on the concentration and chemical speciation of metals in soils solution. Plant Soil 197: 71-78.
Krämer U, Cotter-Howells JD, Charnock JM, Baker AJM, Smith JAC (1996) Free histidine as a metal chelator in plants that accumulate nickel. Nature 379: 635-638.
Krämer U, Pickering IJ, Prince RC, Raskin I, Salt DE (2000) Subcellular localization and speciation of nickel in hyperaccumulator and non-accumulator *Thlaspi* species. Plant Physiol 122:1343-1353.
Küpper H, Zhao FJ, McGrath SP (1999) Cellular compartmentation of zinc in leaves of the hyperaccumulator *Thlaspi caerulescens*. Plant Physiol 119: 305-311.
Kuleff I, Djingova R (1984) The dandelion (*taraxacum officinale*): A monitor for environmental pollutions? Water Air Soil Poll 21: 77-85.
Kumar PBAN, Dushenkov V, Motto H, Raskin I (1995) Phytoextraction: The use of plants to remove heavy metals from soils. Environ Sci Technol 29: 1232-1238.
Lai HY, Chen ZS (2004) Effects of EDTA on solubility of cadmium, zinc, and lead and their uptake by rainbow pink and vetiver grass. Chemosphere 55: 421-430.
Lasat MM, Baker AJM, Kochian LV (1996) Physiological characterization of root Zn^{2+} absorption and translocation to shoots in Zn hyperaccumulator and nonaccumulator species of *Thlaspi*. Plant Physiol 112: 1715-1722.
Lasat MM, Fuhrmann M, Ebbs SD, Cornish JE, Kochian LV (1998) Phytoremediation of a radiocesium-contaminated soil: Evaluation of cesium-137 bioaccumulation in the shoots of tree plant species. J Environ Qual 7: 165-169
Lasat MM, Kochian L (2000) Physiology of Zn hyperaccumulation in Thlaspi caerulescens. In: Terry N, Bañuelos G (eds.) Phytoremediation of Contaminated Soil and Water. CRC Press LLC, FL, USA, pp.159-169.
Lasat MM, Pence NS, Garvin DF, Ebbs SD, Kochian SD (2000) Cadmium accumulation in populations of *Thlaspi*

caerulescens and *Thlaspi goesingense*. New Phytol 145: 11-20.

Lee J, Reeves RD, Baker RR, Jaffré T (1978) Isolation and identification of a citrato-complex of nickel from nickel-accumulating plants. Phytochemistry 16: 1502-1505.

Lindsay WL, Norvell WA (1978) Development of a DTPA soil test for zinc, iron, manganese, and copper. Soil Sci Soc Am J 42: 421-428.

Lombi E, Wenzel WW, Adriano DC (1998) Soil contamination, risk reduction and remediation. Land Contamin Reclam 6: 183-197.

Lombi E, Zhao FJ, Dunham SJ, McGrath SP (2000) Cadmium accumulation in populations of *Thlaspi caerulescens* and *Thlaspi goesingense*. New Phytol 145: 11-20.

Lombi E, Zhao FJ, Dunham SJ, McGrath SP (2001a) Phytoremediation of heavy metal-contaminated soils: natural hyperaccumulation versus chemically enhanced phytoextraction. J Environ Qual 30: 1919-1926.

Lombi E, Zhao FJ, McGrath SP, Young SD, Sacchi GA (2001b) Physiological evidence for a high-affinity cadmium transporter highly expressed in a *Thlaspi caerulescens* ecotype. New Phytol 149: 53-60.

Luo YM, Wu LH, Jiang XL, Wu SC, Christie P (2000) Chelate-enhanced phytoextraction of metal-contaminated soils and its environmental risk. In: Luo YM, Mcgrath SP, Cao ZH, Zhao FJ, ChenYX, Xu JM (eds.) Proc 2000 Int Conf of Soil Remediation. 15-19 October, 2000. Hangzhou, China.

Martell AE, Smith RM (1974) Peptides. In: Martell AE, Smith RM (eds.) Critical Stability Constants. Vol. 1: Amino acids. Plenum Press, NY, USA, pp. 294-337.

Martell AE, Smith RM (1982) Inorganic ligands. In: Martell AE, Smith RM (eds.) Critical Stability Constants. Vol. 5: First supplement. Plenum Press, NY, USA, pp. 393-424.

Mathur SP, Levesque MP (1983) The effects of using copper for mitigating histosol subsidence on: 2. the distribution of copper, manganese, zinc, and iron in an organic soil, mineral sublayers, and their mixtures in the context of setting a threshold of phytotoxic soil-copper. Soil Sci 135: 166-176.

McGrath SP, Shen ZG, Zhao FJ (1997) Heavy metal uptake and chemical changes in the rhizosphere of *Thlaspi ochroleucum* grown in contaminated soils. Plant Soil 188: 153-159.

McGrath SP, Dunham SJ, Correll RL (2000) Potential for phytoextraction of zinc and cadmium from soils using hyperaccumulator plants. In: Terry N, Bañuelos G. (eds.) Phytoremediation of Contaminated Soil and Water. Lewis Publisher, Boca Raton, FL, USA, pp. 109-128.

Mench MJ, Didier VL, Loffler M, Gomez A, Masson P (1994) A mimicked in-situ remediation study of metal-contaminated soils with emphasis on cadmium and lead. J Environ Qual 23: 58-63.

Miller WP, Martens DC, Zelazny LW (1986) Effect of sequence in extraction of trace metals from soils. Soil Sci Soc Am J 50: 598-601.

Mulligan CN, Yong RN, Gibbs BF (2000) Remediation technologies for metal-contaminated soils and groundwater: an evaluation. Engin Geol 60: 19-207.

Muramoto S, Oki Y (1983) Remove of some heavy metals from polluted water by water hyacinth (*Eichhornia crassipes*). Bull Environ Contam Toxicol 30: 170-177.

Niebore E, Richardson DHS (1980) The replacement of the non-descriptive term "heavy metal" by a biologically and chemically significant classification of metal ions. Environ Poll Ser 1: 3-26.

Ortiz DF, Kreppel L, Speiser DM, Scheel G, McDonald G, Ow DW (1992) Heavy metal tolerance in the fission yeast requires an ATP-binding cassette-type vacuolar membrane transporter. EMBO J 11: 3491-3499.

Page AL, Bingham FT, Chang AC (1981) In: Lepp NW (ed.) Effect of Heavy Metal Pollution on Plant. Applied Science, London, pp. 72-109.

Pollard JA, Baker AJM (1997) Deterrence of herbivory by zinc hyperaccumulation in *Thlaspi caerulescens* (*Brassicacea*). New Phytol 135: 655-658.

Rauser W (1995) Phytochelatins and related peptides. Plant Physiol 109: 1141-1149.

Robinson NJ (1990) Metal binding polypeptides in plants. In: Shaw AJ (ed.) Metal Tolerance in Plants: Evolutionary Aspects. CRC Press, Boca Raton, FL, USA, pp. 195-215.

Robinson BH, Leblanc M, Petit D, Brooks RR, Kirkman JH, Gregg PEH (1998) The potential of *Thlaspi caerulescens* for phytoremediation of contaminated soils. Plant Soil 23: 47-56.

Ross SM, Kaye KJ (1994) The meaning of metal toxicity in soil-plant systems. In: Ross SM (ed.) Toxic Metals in Soil-Plant System. John Wiley & Sons Ltd., England, pp. 27-61.

Salt DE, Blaylock MJ, Kumar NPBA, Dushenkov V, Ensley BD, Chet I, Raskin I (1995) Phytoremediation: a novel strategy for the removal of toxic metals from the environment using plants. Biotechnol 13: 468-474.

Salt, DE, Prince RC, Baker AJM, Raskin I, Pickering IJ (1996) Zinc ligands in the metal hyperaccumulator *Thlaspi caerulescens* as determined using X-ray adsorption spectroscopy. Environ Sci Technol 33: 713-717.

Salt DE, Kato N, Krämer U, Smith RD, Raskin I (2000) The role of root exudates in nickel hyperaccumulation and tolerance in accumulator and nonaccumulator species of *Thlaspi*. In: Terry N, Bañuelos G (eds.) Phytoremediation of Contaminated Soil and Water. CRC Press LLC, FL, USA, pp. 189-200.

Schmöger MEV, Oven M, Grill E (2000) Detoxification of arsenic by phytochelatins in plants. Plant Physiol 122; 793-801.

Sela M, Garty J, Telor E (1989) The accumulation and the effect of heavy metals on the water fern *Azolla filiculoides*. New Phytol 112: 7-12.

Simon L (1998) Cadmium accumulation and distribution in sunflower plant. J Plant Nutr 21: 341-352.

Simon L, Martin HW, Adriano DC (1996) Chicory(*Cichorium intybus* L.) and danelion (*taraxacum officinale* web.) as phytoindicators of cadmium contamination. Water Air Soil Poll 91: 351-362.

Singh BR, McLaughlin MJ (1999) Cadmium in soils and plants. In: Lauglin MJ, Singh BR (eds.) Cadmium in Soils and Plants. Kluwer Academic Publishers, Dordrecht, Netherlands, pp. 257-267.

Sposito G, Lund LJ, Chang AC (1982) Trace metal chemistry in arid-zone field soils amended with sewage sludge: .Fractionation of Ni, Cu, Zn, Cd, and Pb in solid phases. Soil Sci Soc Am J 46: 260-264.

Steffens JC (1990) The heavy metal-binding peptides of plants. Ann Rev Plant Physiol Plant Mol Biol 41: 553-575.

Stover RC, Sommers LE, Silviera DJ (1976) Evaluation of metals in wastewater sludge. J Water Poll Control Fed 48: 2165-2175.

Tomsett AB, Thurman, DA (1988) Molecular biology of metal tolerance in plants. Plant Cell Environ 11: 383-394.

Van Steveninck RFM, Van Steveninck ME, Wells AJ, Fernando DR (1990) Zinc tolerance and the binding of zinc as zinc phytate in Lemma minor. X-ray microanalytical ecidence. J Plant Physiol 137: 140-146.

Vassil AD, Kapulnik Y, Raskin I, Salt DE (1998) The role of EDTA in lead transport and accumulation by Indian Mustard. Plant Physiol 117: 447-453.

Vázquez MD, Barceló J, Poschenrieder CH, Mádico J, Hatton P, Baker AJM, Hatton P, Cope GH (1992) Locilization of zinc and cadmium in *Thlaspi caerulescens* (*Brassicaceae*), a metallophyte that can hyperaccumulate both metals. J Plant Physiol 140: 350-355.

Vázquez MD, Poschenrieder C, Barceló J, Baker AJM, Hatton P, Cope GH (1994) Compartmentation of zinc in root and leaves of the zinc hyperaccumulator *Thlaspi caerulescens* JC Presl Bot Acta 107: 243-250.

Vig K, Megharaj M, Sethunatnan N, Naidu R (2003) Bioavailability and toxicity of cadmium to microorganisms and their activities in soil: a review. Adv Environ Res 8: 121-135.

Wallace A, Romney EM, Alexander GV (1980) Zinc-cadmium interactions on the availability of each to bush bean plants grown in solution culture. J Plant Nutr 2: 51-54.

Wenzel WW, Adriano DC, Salt D, Smith R (1999) Phytoremediation: A plant-microbe-based remediation system. In: Adriano DC, Bollag JM, Frankenberger Jr. WT, Sims RC (eds.) Bioremediation of Contaminated Soils. American Society of Agronomy, Inc., Madison, Wisconsin, USA, pp. 457-508.

Whiting SN, Leake JR, Baker AJM, McGrath SP (1997) Changes in phytoavailability of zinc to plants sharing a rhizosphere with the zinc hyperaccumulator *Thlaspi caerulescens* J&C Presl. In: Proceeding of Extended Abstract from the Fourth International Conference on the Biogeochemistry of Trace Elements. 23-26 June 1997. University of California Berkeley, California, USA, pp. 469-470.

Wu J, Hsu FC, Cunningham SD (1999) Chelate-assisted Pb phytoremediation: Pb availability, uptake, and translocation constraints. Environ Sci Technol 33: 1898-1904.

Zayed AM, Growthaman S, Terry N (1998) Phytoaccumulation of trace elements by wetland plants: I. Duckweed. J Environ Qual 27: 715-721.

Zhao FJ, Lombi E, Breedon T, McGrath SP (2000) Zinc hyperaccumulation and cellular distribution in *Arabidopsis halleri*. Plant Cell Environ 23: 507-514.

Zhu YL, Zayed AM, Qian JH, de Souza M, Terry N (1999a) Phytoaccumulation of trace elements by wetland plants: II. Water hyacinth. J Environ Qual 28: 339-344.

Zyrin WG, Obukhov AI, Motuzova GV (1974) Forms of micronutrients in the soils of the USSR and methods of their investigation. Trans 10th Int Congr Soil Sci II: 350-357.

Cadmium Toxicity and Tolerance in Plants
Editors: Nafees A. Khan and Samiullah
Copyright © 2006, Narosa Publishing House, New Delhi, India

Cadmium Uptake and its Toxicity in Higher Plants

Pallavi Sharma and R. S. Dubey[*]

Department of Biochemistry, Faculty of Science, Banaras Hindu University, Varanasi 221005, India

1. INTRODUCTION

Cadmium (Cd) is a potent heavy metal pollutant of the environment. It is a heavy metal of widespread occurrence and is released into environment by power stations, heating systems, metal working industries, waste incinerators, urban traffic, cement factories and as a byproduct of phosphate fertilizers (di Toppi and Gabbrielli, 1999; Nolan et al., 2003). Cadmium released to the environment enters biogeochemical cycle and tends to accumulate in soils and sediments, where it is potentially available to rooted plants. The degree to which higher plants take Cd depends on its concentration in the soil and its bioavailability which are modulated by the presence of organic matter, pH, redox potential, temperature and concentration of other elements (di Toppi and Gabbrielli, 1999).

Cd in higher concentration often damages root tip, reduces nutrient and water uptake, impairs photosynthesis and inhibits growth of the plants. The basis of these dysfunctions is due to formation of metal-thiolate bonds and alteration of cell wall and membrane permeability by binding to nucleophilic groups. Furthermore, Cd directly or indirectly induces increased formation of reactive oxygen species, which interfere with the redox status of the cell and cause oxidative damage to proteins, lipids and other biomolecules (Stohs et al., 2000; Schutzendubel et al., 2001). Plants growing in heavy metal contaminated soil tend to adapt to prevailing levels of metals for normal functions. Plants have evolved a wide range of Cd-tolerance mechanisms not only in different species of a genus but also among genotypes within a species (Ozturk et al., 2003; Dunbar et al., 2003).

Cd is considered as an extremely potent pollutant affecting all life forms due to its high toxicity and great solubility of its salts in soil and water. The level of Cd in the soil is continuously increasing over time (Jones et al., 1992). Soils contaminated with Cd show a sharp decline in crop productivity and therefore Cd serves as a serious problem for the agriculture (Shah and Dubey, 1997a).

The present chapter focuses on the sources of Cd; its uptake, transport and localization within plants; physical, biochemical and ultrastructural changes due to Cd toxicity; Cd tolerance in plants as well as possible remediation measures for Cd contaminated soils.

2. SOURCES, UPTAKE, TRANSPORT AND LOCALIZATION OF CADMIUM

Cadmium occurs naturally in rocks and soils usually in concentrations less than $1\mu g\ g^{-1}$ (Thornton, 1992). Cd pollution is increasing in the environment due to rock mineralization, mining activities, industrial usage and anthropogenic activities (Astolfi et al., 2003). It is emitted to the atmosphere from coal-fired power plants, steel mills, metal smelting and roasting operations, incineration of wastes and

---

[*] Corresponding Author (rsdbhu@rediffmail.com, rsd@bhu.ac.in)

electroplating processes (Foy et al., 1978). Besides, Cd accumulates in soil polluted by traffic exhaust, whereas gasoline and rubber tyres are additional sources of soil Cd contamination (Mankovska, 1977). Some phosphate fertilizers and phosphorites contain high concentrations of Cd (4.77 µg g^{-1}) and are considered as the potential cause of increasing Cd contamination in rice plants (Muramoto and Aoyama, 1990). Acid soils and acid rains have contributed to a great extent to the solubilization of Cd and its transformation into an available ionic form, which is preferentially absorbed by plant roots (Hagemeyer et al., 1986). In acid soils mobility and availability of Cd is much higher than in non-calcareous, neutral and slightly alkaline soil. Figure 1 projects various sources which contribute to Cd pollution in the environment and its effect on agriculture, animal and human health.

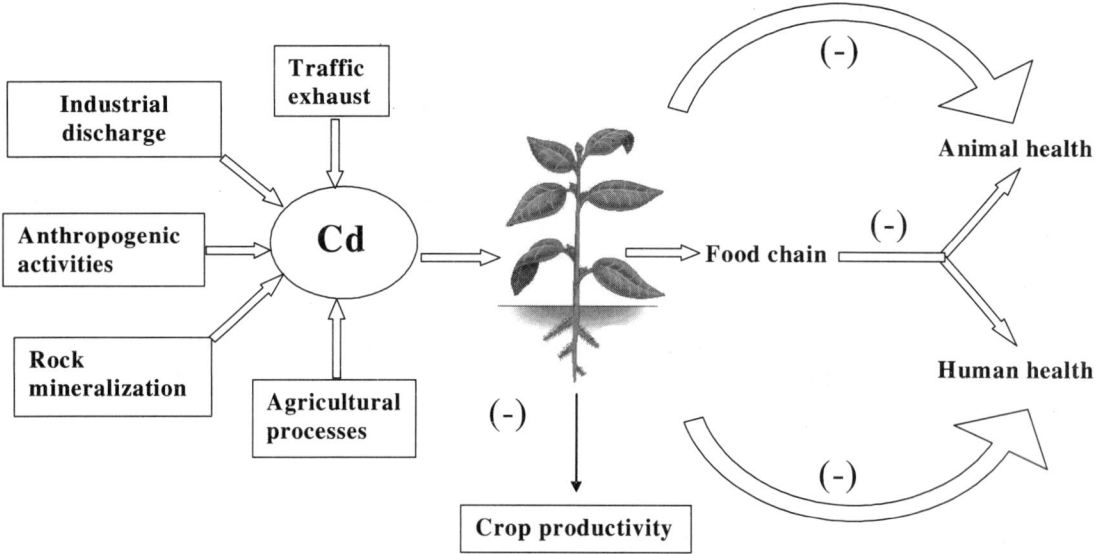

Fig. 1. Sources of Cd input to the soil and effects of Cd on crop productivity as well as animal and human health. The processes related to industrial and anthropogenic, rock mineralization, agricultural processes as well as traffic exhaust add Cd to the soil environment. Cd reduces crop productivity and adversely affects human and animal health.

Primarily Cd is taken up by the plants from soil *via* the root system. To a lesser extent Cd gets access into plant system *via* leaves. The ability of leaves to absorb Cd depends on specific leaf morphology (Godzik, 1993). Forms of Cd in soils can be operationally described on the basis of selective extractions, e.g. water soluble, exchangeable, reducible, diethyltriamine pentaacetic acid extractable and organically bound (insoluble) forms. It was demonstrated that an increase in redox potential (Eh) from –150 to 500 mV led to a decrease in exchangeable Cd and an increase in its reducible form (Khalid et al., 1981). It appears that the availability of Cd to plants is dependent on the pH and ionic strength of the soil medium (Hatch et al., 1988; Ahumada and Schalscha, 1993). Free Cd ion (Cd^{2+}) is considered to be the Cd species which is widely available to the plants (Grant et al., 1998). Chloride levels in the soil have an antagonistic effect on heavy metal toxicity (Bharti and Singh, 1994). The soil organic matter of different vegetational cover preferentially binds to selected heavy metal ions (Krosshavn et al., 1993).

In aquatic plants Cd is taken up not only through roots but also by shoot system. Cd interaction at the shoot or root surface is affected by H$^+$ ions (pH) and Ca^{2+} concentrations; in both cases the availability of Cd is sensitive to the presence of dissolved organic matter in the soil sediment as well as terrestrial or ambient water surrounding the shoots (Ornes and Sajwan, 1993; Campbell et al., 1998). Linear time dependent accumulation of Cd occurs in roots at low concentration of Cd (20-500 nM) in the soil (Homma and Hirata, 1984) whereas at higher Cd concentrations (90 μM) saturable time dependent Cd accumulation occurs. Hart and coworkers (1998) reported that at 90 μM soil Cd concentration, the bulk of Cd accumulated in roots remained in the form of Cd bound to apoplastic components and the cell wall binding capacity could get saturated at such high Cd concentrations. Cd gets transported across the plasma membrane. Low temperature inhibits the uptake of Cd as well as its transport across plasma membrane (Hart et al., 1998).

Studies of Cd uptake by plants have indicated that at lower Cd concentrations (2.5 to 90 nM), its transport across membrane is an active, energy requiring H^+-ATPase mediated process whereas at high concentrations of Cd the uptake is a non metabolic (passive) process, involving diffusion coupled with sequestration (Grant et al., 1998). Although the mechanism of transport of Cd across the plasmalemma of root cells is not well understood, the electrochemical potential across the membrane appears to be an important factor (Grant et al., 1998). It is suggested that the uptake of cationic solutes is likely to be driven largely by the negative membrane potential across the plasma membrane, which is generated in part by metabolically dependent processes such as proton extrusion *via* the plasma membrane H^+-ATPase (Kochian, 1991). According to Hart and coworkers (1998) Cd uptake is a transporter mediated process that exhibits Michaelis-Menten kinetics and is controlled by a transport protein present in the membrane. Similar carrier mediated uptake has been reported for a number of divalent cationic micronutrients (Kochian, 1991). Zn competitively inhibits Cd uptake by plant roots, suggesting that Cd is transported across the plasma membrane *via* a native Zn-transport system. However, the reported kinetic constants of Zn and Cd uptake are quite different. The Km value for Zn uptake by roots ranges between 2 to 6 μM, and for Cd uptake nearly two orders of magnitude lower. Thus if Zn and Cd share a common influx pathway, the affinity of the transporter appears to be considerably higher for Cd than for Zn (Hart et al., 1998).

Table 1. Amount of Cd absorbed and length as well as fresh weight of roots and shoots of rice seedlings grown for 15 days in sand culture. Seedlings of rice cultivar Ratna were raised in sand cultures containing nutrient solutions either without Cd (control) or supplemented with 100 μM and 500 μM $Cd(NO_3)_2$. Absorbed Cd was estimated by atomic absorption spectrophotometer. Data represent mean values ± s.d. based on three independent determinations. From Shah and Dubey, 1995 and 1998a.

Plant material	Concentration of $Cd(NO_3)_2$ (μM)	Amount of Cd absorbed (μ mol Cd g^{-1} dry wt.)	Length (cm)	Fresh weight (mg)
ROOT	0	0	8.10 ± 0.41	40.8 ± 2.11
	100	1.2 ± 0.07	6.98 ± 0.34	35.2 ± 1.70
	500	1.9 ± 0.09	3.16 ± 0.15	10.9 ± 0.54
SHOOT	0	0	16.28 ± 0.81	60.0 ± 3.21
	100	0.52 ± 0.003	11.42 ± 0.57	52.8 ± 2.61
	500	0.63 ± 0.004	3.8 ± 0.19	46.0 ± 2.32a

After its uptake by plant roots, Cd gets accumulated in cytosol, cytosolic organelles and vacuoles. At low level of exposure, Cd forms complexes in the cytosol with glutathione whereas at higher Cd exposure levels, it is transported into the vacuoles, where it forms complexes with organic acids and phytochelatins (Grant et al., 1998). In most of the plant species much of the Cd taken up by plants is retained in the root and its translocation to aerial portion is low. In soybean plants, about 98 percent of the accumulated Cd is strongly retained by roots and only 2 percent is transported to the shoot system (Cataldo et al., 1983). Rice (*Oryza sativa* L.) plants absorb Cd from the rooting medium against the concentration gradient and the localization of absorbed Cd in rice is greater in roots than in shoots (Shah and Dubey, 1998a). When rice seedlings were raised in sand cultures for 20 days in nutrient medium containing 500 µM Cd, localization of absorbed Cd was about 3 times in roots compared to its level in shoots (Table 1) (Shah and Dubey, 1998a). Movement of Cd from roots to shoots is likely to occur *via* the xylem and is driven by the transpiration from the leaves. Evidence for this was provided by the Salt and coworkers (1995) who showed that ABA-induced stomatal closure dramatically reduced Cd accumulation in shoots of Indian mustard. Reduced movement of Cd from roots to shoots in plants is believed to result from barrier function of the root endodermis and mechanism involving sequestration and decreased xylem loading of Cd (Hart et al., 1998). The casparian strips of root endodermis retard the entrance of Cd into the central cylinder (Seregin et al., 2004), vascular compartmentation of Cd tends to limit symplastic movement of Cd. Movement of Cd across the tonoplast occurs *via* Cd^{2+}/H^+-antiport system (Salt and Wagner, 1993) as well as by phytochelatin-Cd transporter (Vogeli-Lange and Wagner, 1990) that appears to be Mg-ATP dependent (Salt and Rauser, 1995). Though little is known about the processes involved in Cd movement into developing fruits, it appears likely that loading into developing seeds occurs *via* the phloem (Popelka et al., 1996). Because Zn and Cd appear to compete for transport at plasma membrane (Costa and Morel, 1993), it is possible that Cd moves into developing fruits *via* the phloem in a manner similar to that of Zn.

Absorption of Cd by green microalgae *Chlorella vulgaris*, *Ankistrodesmus braunii* and *Eremosphera viridis* revealed that Cd is mainly sorbed in the cell wall. The binding sites seem to be the acidic groups in the cell wall structure (Geisweid and Urbach, 1983). In water hyacinth (*Eichhornia crassipes*) Cd was found to accumulate throughout the roots (Hosayama et al., 1994). In *Solidago altissima* Cd was found in most parts of the plant, but located mainly in the cambium, cortex and phloem tissues (Hosayama et al., 1994). Within the cell, most of the Cd accumulates in the vacuole; together with the cell wall vacuole comprises up to 96 percent of the absorbed metal. The fact that Cd is found in golgi apparatus and endoplasmic reticulum is apparently related to metal secretion through the cell surface and into the vacuole. A small quantity of Cd reaches nuclei, chloroplast, mitochondria and exerts toxic effects on these organelles (Miller et al., 1973; Malik et al., 1992). Localization of Cd in different parts of a plant appears to be greatest in roots followed by stem, seeds indicating that in roots accumulation of Cd is much larger than any other organ (Hart et al., 1998).

3. PHYSIOLOGICAL, BIOCHEMICAL AND ULTRASTRUCTURAL CHANGES INDUCED BY CADMIUM

Cd causes inhibition of plant growth and even leads to plant death (Shah and Dubey, 1995; Urwin et al., 1996). Symptoms induced by elevated level of Cd (over 10 ppm) in the soil include growth retardation and root damage in plants, chlorosis of leaves and red brown colouration of leaf margins and veins (Fediuc and Erdei, 2002). Cd toxicity adversely affects germination of seeds, seedling vigour, plant growth and metabolism (Shah and Dubey, 1997a) by limiting water transport to growing tissues and leaves, impairing transpiration rate, causing ultrastructural changes in cell organelles and altering activity behaviour of enzymes of various metabolic pathways (Shah et al., 2001; Fediuc and Erdei, 2002). Inhibition of plant growth due to Cd results from direct effect of Cd on functioning of various enzymes and growth processes (Ubio et al., 1994; Shah and Dubey, 1995) including interaction with the cell wall polysaccharides, decreasing cell wall plasticity, etc. In rice Cd concentration of

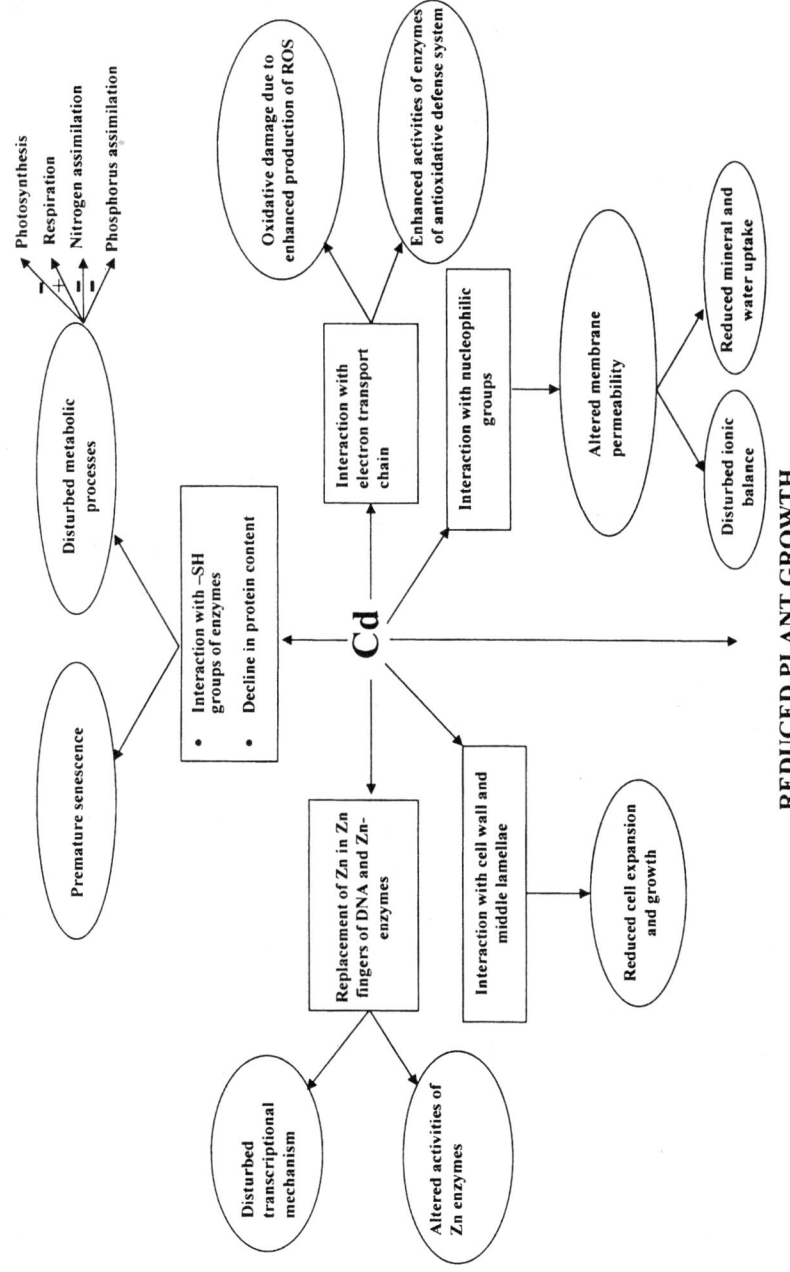

Fig. 2. An overview of the physiological and metabolic processes associated with Cd toxicity in plants. All events ultimately result into reduced growth of the plant. (+) and (-) signs represent increase or decrease in the metabolic processes respectively. For details see text.

500 µM decreased the germination of seeds by about 10 to 20% (Shah and Dubey, 1998b). A drastic reduction in seed setting occurred in sesamum (*Sesamum indicum*) due to Cd treatment (Bora, 1981). Increasing Cd supply markedly reduced root and shoot length as well as fresh and dry weight in *Glycine max*, *Oryza sativa* and *Triticum durum* (Huang et al., 1974; Shah and Dubey, 1995; Ozturk et al., 2003). A significant reduction in plant height, fresh and dry weight of various organs, leaf area, relative growth rate, root/shoot ratio occurred in sesamum plants following Cd treatment (Bora, 1981). Cd concentration of 500 µM in sand cultures reduced the growth of rice seedlings by 61 to 77% and fresh weight by 23 to 73% at 15th day of growth (Shah and Dubey, 1995) (Table 1).

Inhibition of cell growth due to Cd appears to be as a result of increased cross-linking of pectins in the middle lamellae. This cross-linking might be responsible for inhibition of cell expansion and its further growth (Poschenrieder et al., 1989). The inhibition of cell expansion and its growth might also be due to direct or indirect effects of Cd on auxin metabolism and auxin carriers (Barcelo and Poschenreider, 1990). Even at moderate concentration levels Cd adversely affects the growth, survival and yield of several crops (Bingham et al., 1976; Shah and Dubey, 1995). An overview of Cd induced various physiological and biochemical changes leading to altered metabolism and reduced growth of the plant has been presented in Fig. 2.

4. EFFECTS OF CADMIUM ON BIOMOLECULES AND ENZYMES

Cellular protein patterns change both qualitatively and quantitatively with changes in environmental conditions (Muntz, 1996). Altered levels of proteins and amino acids and inhibition in the activity of protease and alteration in the behaviour of peptidases occur in rice seedlings growing under 100 µM Cd in the growth medium (Shah and Dubey, 1997b). An apparent increase in protein level occurred in Cd exposed rice seedlings when data recorded as mg protein g^{-1} dry weight of plant parts, which appeared to be due to the synthesis of new proteins as well as due to decreased proteolysis under Cd treatment (Shah and Dubey, 1997b). However, a decline in total protein content with a substantial increase in protease activity was observed in Cd treated plants by Hortensteiner and Feller (2002). These changes which appear to be similar to those occurring during natural senescence of plants suggest that Cd may induce premature senescence. Cd induced proteolysis and accumulation of amino acids occur in plants just prior to Cd toxicity (chlorophyll degradation) symptoms, which suggests that proteolysis or accumulation of amino acids might play some role in regulating Cd toxicity (Hsu and Kao, 2003). A common response of plants to many abiotic stresses includes accumulation of free amino acids. The amino acid proline preferentially accumulates in large amounts (Aspinall and Paleg, 1981) in response to environmental stresses followed by other amino acids such as those derived from asparatic acid, including asparagine, isoleucine, leucine, methionine and valine (Munns et al., 1979). Cd-exposed plants accumulate proline in substantial amounts and it has been suggested that the accumulation of proline in response to Cd is induced by a Cd imposed increase in water deficit rather than by a Cd-accumulation *per se* (Schat et al., 1997). However, in Cd treated leaf discs proline accumulation has been reported to be due to Cd uptake *per se* (Kastori et al., 1992). Significance of proline in Cd stressed plants lies in its contribution to water balance maintenance, protection of enzymes and biomolecules as well as detoxification of reactive oxygen species. Proline mediated alleviation of water deficit could substantially contribute to the Cd tolerance of plants.

Proteins perform numerous crucial functions in the cell, primarily in the form of enzymes that mediate biochemical reactions required for cellular functions. Cd treatment influences the activity behaviour of key enzymes of various metabolic pathways. In most of the cases inhibition exerted by Cd on enzyme activity results due to the interaction of Cd with enzyme –SH groups (Seregin and Ivanov, 2001). Cd interacts with free –SH group that are present at active site of enzyme and are essential for enzyme reaction as well as with –SH groups that are necessary for the stabilization of enzyme tertiary structure. Cd forms mercaptide with –SH group of cysteine. Moreover, Cd may displace Zn bound to –SH groups as the negative logarithms of the binding of the former (27.2) exceed that of the later (25.2). Table 2 presents an overview of the effect of Cd treatment on the activities of various enzymes reported from different plant species. Chlorophyll synthesis is inhibited by Cd at the level of the rate limiting enzymes of porphyrin biosynthesis, *viz.* ALA synthase and ALA dehydratase which are inactivated by Cd due to its reaction with –SH groups present at the active site of the enzymes (Padmaja et al., 1990). A decline in the activity of carbonic anhydrase, a Zn-enzyme, is observed in soybean (Lee et al., 1976), following Cd treatment which is possibly due to the inhibitory effect of Cd on Zn uptake leading to Zn deficiency in plants and thus causing decreased activity of the enzyme.

Table 2. Effect of Cd on activities of enzymes of different metabolic processes.

Metabolic processes	Enzymes	Plant species	Effect of Cd	References
Chlorophyll Synthesis	ALA synthase	*Phaseolus vulgaris*	-	Padmaja et al., 1990
	ALA dehydratase	*Phaseolus vulgaris*	-	Padmaja et al., 1990
Protein metabolism	Protease	*Oryza sativa*	-	Shah and Dubey, 1997b
RNA metabolism	RNase	*Oryza sativa*	-	Shah and Dubey, 1995
Sugar metabolism	Acid invertase	*Oryza sativa*	+	Verma and Dubey, 2001
	Sucrose synthase	*Oryza sativa*	+	Verma and Dubey, 2001
	Sucrose phosphate synthase	*Oryza sativa*	-	Verma and Dubey, 2001
N metabolism	Nitrate reductase	*Sesamum indicum*	-	Singh et al., 1994
	Glutamine synthatase	*Pisum sativum*	-	Chugh et al., 1992
	Glutamate dehydrogenase	*Phaseolus vulgaris*	+	Gouia et al., 2003
Phosphorus metabolism	Acid phosphatase	*Oryza sativa*	-	Shah and Dubey, 1998c
	Alkaline phosphatase	*Oryza sativa*	-	Shah and Dubey, 1998c
	Inorganic pyrophosphatase	*Oryza sativa*	-	Shah and Dubey, 1998c
Photosynthesis	RUBP Carboxylase	*Phaseolus vulgaris*	-	Siedlecka et al., 1997
	PEP Carboxylase	*Phaseolus vulgaris*	-	Siedlecka et al., 1997
Ion-transport	H^+-ATPase	*Avena sativa*	-	Astolfi et al., 2003
	K^+-ATPase	*Cucumis sativus*	-	Burzynski, 1987
Antioxidative metabolism	Catalase	*Helianthus annus*	+	Gallego et al., 2002
	Guaiacol peroxidase	*Oryza sativa*	+	Shah et al., 2001
	Ascorbate peroxidase	*Helianthus annus*	+	Gallego et al., 2002
	Glutathione reductase	*Oryza sativa*	+	Shah et al., 2001
	Superoxide dismutase	*Oryza sativa*	+	Shah et al., 2001

'+' and '-' signs denote stimulatory and inhibitory effects of Cd on enzymes respectively.

Cadmium induces oxidative stress in several plant species and increases the activity of enzymes of antioxidative defense system (Shah et al., 2001; Metwally et al., 2003). Enhancement in the activities of superoxide dismutase, catalase, peroxidase and glutathione reductase is observed in Cd-stressed rice seedlings (Shah et al., 2001). Barley seedlings exposed to 25 µM Cd showed increased activity of H_2O_2-metabolising enzymes catalase and ascorbate peroxidase (Metwally et al., 2003). Apparently it is the oxidative stress induced by Cd that enhances the activities of stress related enzymes by increasing the levels of free radicals and peroxides. Lipids that contain phosphate groups are essential components of membrane that surround the cell as well as other cellular structures, such as the chloroplast, mitochondria, nucleus. Cd induced oxidative damage involves peroxidation of polyunsaturated fatty acids of membrane lipids by reactive oxygen species (ROS) generated by Cd. Lipid peroxidation eventually increases membrane fluidity and membrane permeability. Elevated levels of lipid peroxides were observed in Cd stressed rice seedlings (Shah et al., 2001). Under 500 µM Cd treatment about 1.4 to 1.6 times increase in malondialdehyde (MDA) content was observed indicating enhanced peroxidation of lipid due to Cd (Shah et al., 2001).

The level of thylakoid total lipids, total glycolipids, total phospholipids and total neutral lipids decreases under Cd treatment (Malik et al., 1992). Cd application caused decrease in the concentration of phosphatidyl glycerol and phosphatidyl choline. On the other hand phosphatidic acid and free fatty acid contents showed an increase. These compositional changes in thylakoid membrane might be responsible for reduced PS II activity and decreased rate of photosynthesis observed under cadmium treatment (Malik et al., 1992).

In root-tip cells of plant, Cd damages nucleoli (Liu et al., 1995) and in growing plants, it alters the synthesis of RNA and inhibits ribonuclease activity (Shah and Dubey, 1995). Rice seedlings raised under 500 µM Cd showed decreased RNase activity as well as decreased number of its isoforms compared to control grown plants (Shah and Dubey, 1995). Under *in vitro* conditions $Cd(NO_3)_2$ upto 100 µM in the reaction medium stimulated RNase activity whereas with concentration beyond this level a gradual inhibition in enzyme activity was observed (Shah and Dubey, 1995). The affinity of RNase towards its substrates increased slightly in rice seedlings grown at moderately toxic Cd concentration (100 µM), whereas significant decrease in enzyme affinity towards its substrates was observed in seedlings grown at higher concentration of Cd (500 µM). An increase in Michaelis constant (K_m) value of RNase, extracted from 500 µM Cd-grown rice seedlings, suggest possible alteration in conformation or inactivation of RNase due to Cd treatment (Shah and Dubey, 1997c). Increased production of ROS under Cd toxicity serves as major source of DNA damage leading to strand breakage, removal of nucleotides and variety of modifications in organic bases of nucleotides. Besides, Cd replaces Zn ions in the Zn fingers and consequently interferes with the transcriptional mechanism (Sarkar, 1995). Inhibition of DNA repair is an important genotoxicity mechanism of Cd. It has been suggested that this may occur when Cd competes with Zn for a common binding site on enzymes involved in DNA synthesis (Rossman et al., 1992).

In addition to the role of sugars in osmoregulation under abiotic stresses, the soluble sugars allow growing plants to maximize their carbohydrate storage reserves to support basal metabolism under stressful environments (Dubey and Singh, 1999). Altered level of soluble sugars as well as alteration in activity behaviour of sugar metabolizing enzymes have been observed in rice plants growing under increasing (100-500 µM) Cd levels (Verma and Dubey, 2001). As Cd decreases water transport to the leaves (Chardonnens et al., 1998), the accumulating sugars in rice plants growing in presence of Cd could possibly provide adaptational significance to the plants in maintaining a favourable osmotic potential under adverse condition of Cd toxicity. When the role of acid invertase and sucrose synthase was investigated in Cd-exposed rice seedlings, a decline in the level of non-reducing sugars accompanied with increased activities of sucrose degrading enzymes acid invertase and sucrose synthase was noticed under Cd treatment (Verma and Dubey, 2001). Increase in acid invertase and sucrose synthase activities accompanied with decreased level of non-reducing sugars in Cd-grown rice seedlings suggests that Cd-toxicity in rice limits availability of sucrose in the cells by favouring its enhanced degradation due to both invertase and sucrose synthase (Verma and Dubey, 2001).

4.1. Nitrogen Metabolism

Cadmium adversely affects the process of nitrate assimilation (Boussama et al., 1999). The nitrate assimilation process consumes about 25% of energy produced by photosynthesis (Soloman and Barber, 1990). In most of the plants nitrate reduction takes place in the leaves where major part of the reducing power arises directly from light *via* ferredoxin. A high rate of CO_2 assimilation favours an efficient N assimilation and *vice versa* (Ferrario et al., 1998). Cd being potent inhibitor of photosynthetic process, considerably inhibits nitrate assimilation. Enzymes of N metabolism are differentially affected by Cd (Chugh et al., 1992; Singh et al., 1994). The enzymes nitrate reductase (NR) and glutamine synthetase (GS) are sensitive to Cd whereas glutamate dehydrogenase shows a substantial rise under influence of Cd with dramatic build-up of ammonium pool (Gouia et al., 2003). The induction of GDH activity by Cd results from *de novo* synthesis and/or activation of specific isoenzymes that removes excess ammonium (Syntichaki et al., 1996). GDH isoenzymes appear to remove in part the excess of ammonium under Cd toxicity conditions. Under physiological conditions the incorporation of ammonium into organic compounds occurs mainly *via* the GS/GOGAT cycle (Gouia et al., 2000). The most striking change in Cd treated plants appears to be a rapid decay of GS/Fd-GOGAT and NADH-GOGAT activities and accumulation of ammonia (Boussama et al., 1999). This implies a reduced capacity of GS/GOGAT cycle and induced activity of the alternative mode of ammonium assimilation by NADH-GDH pathway under Cd treatment.

4.2. Phosphorus Metabolism

Cd impairs phosphorus metabolism in plants (Shah and Dubey, 1998c). Pi not only plays an important role in energy transfer reactions and other metabolic functions but it also serves as an important structural constituent of many biomolecules. Consequently Pi assimilation, storage and metabolism are of critical importance for plant growth and development (Vincet et al., 1992). Tissue specific inhibition of the activities of phosphatases both under *in situ* and *in vitro* conditions have been observed due to Cd treatment in growing plants (Shah and Dubey, 1997a). Decrease in activity as well as synthesis of acid phosphatase isoforms in embryo axes of Cd-treated germinating rice seeds limits the energy need of germinating seeds thereby decreasing the vigour of establishing seedlings (Shah and Dubey, 1997a). A decline in the level of total phosphate pool along with inhibition in the activities of phosphorolytic enzymes acid phosphatase, alkaline phosphatase and inorganic pyrophosphatase appears to be one of the key reasons for decreased metabolic activity and inhibited growth of rice plants grown under high Cd levels (Shah and Dubey, 1998c).

4.3. Photosynthesis

The process of photosynthesis in higher plants is particularly susceptible to Cd. Plants such as clover, lucern and soybean when grown in presence of Cd show decreased transpiration due to closure of stomata and this parallels with a decrease in net photosynthesis (Huang et al., 1974). The decline in net photosynthetic rate in Cd-exposed plants results from the distorted chloroplast ultrastructure, restrained synthesis of chlorophylls, plastoquinone and carotenoids, disturbed electron transport, inhibited activities of Calvin cycle enzymes and CO_2 deficiency in the cells (Seregin and Ivanov, 2001).

Cd alters lipid composition of thylakoid membranes and causes its swelling (Barcelo et al., 1988). The fine structures of chloroplasts in Cd-treated plants degenerate and the affected plants are characterized by the occurrence of large plastoglobuli and disorganized lamellar structures (Krupa et al., 1993). Cd alters the content of trans Δ^3 hexadecanoic acid in plants, which is widely accepted as a component responsible for the oligomerization of the chlorophyll protein complex (Krupa et al., 1993). Carotenoids are generally less affected than chlorophylls due to Cd resulting in a decreased chlorophyll/carotenoid ratio in plants (Krupa, 1988). Cd damages PS II and affects the water splitting system at the level of protein (Tukendorf, 1993). At low concentrations, Cd acts as a phosphorylation inhibitor in spinach chloroplasts (Tukendorf, 1993). According to Gouia and coworkers (2003) inhibition of photosynthesis by Cd appears to be primarily as a result of inhibition of different reaction steps of the Calvin cycle rather than due to interaction of Cd with photochemical reactions in the thylakoid membrane system. The inhibition of the Calvin cycle may slow down the reactions of the light phase of photosynthesis, either by down-regulation or feedback inhibition due to inefficient consumption of ATP and NADPH. Thus, a high proton gradient across the thylakoid membrane is maintained (Krupa, 1999). Chlorophyll fluorescence decrease ratio (Rfd) was proposed as an indirect indicator of the functional status of the dark phase of the photosynthesis (Lichtenthaler and Rinderle, 1998). Rfd decreases due to Cd which can be interpreted as a result of the toxic effect of Cd both on phosphoenolpyruvate carboxylase (PEPC) and RUBISCO where PEPC is more sensitive to the Cd (Stiborova et al., 1986; Siedlecka et al., 1997).

4.4. Respiration and ATP Contents

The effects of Cd on respiration and ATP content have not been studied in much detail. Cd in 1 mM concentration is reported to reduce oxygen consumption by roots (Keck, 1978) and cells in suspension cultures (Reese and Roberts, 1985). Dithiothreitol, a –SH agent, alleviates the Cd-led inhibition of the mitochondrial respiration and also restores its swelling (Miller et al., 1973). Presumably Cd inhibits the transport of electrons and protons in mitochondria and disorganizes the electron transport chain (Seregin and Ivanov, 2001). Reese and Roberts (1985) demonstrated that Cd did not affect the glycolysis and pentose phosphate pathway but considerably inhibited succinate oxidation *via* the Kreb's cycle. These investigators concluded that the succinate dehydrogenase complex in the mitochondria is the primary target affected by Cd. However, Chugh and Sawhney (1999) demonstrated that Cd under both *in vivo* and *in vitro* conditions inhibited enzymes of glycolysis and pentose phosphate pathway.

4.5. Water Status

Excess of Cd induces changes in water status of plants, *viz.* total water content, specific water content, water saturation deficit and transpiration (Singh and Tewari, 2003). Cd in the soil has negative effects on growth and performance of roots (Mitchell and Fretz, 1977) and influences the ability of roots to absorb water (Carlson et al., 1975). In sunflower leaves Cd caused closure of stomata (Bazzaz et al., 1974). Inhibition of transpiration in *Acer saccharinum* due to Cd has also been explained in terms of its effect on stomatal closure (Lamoreaux and Chaney, 1978). With increase in Cd level the sum total of stomatal aperture per unit leaf area decreases. An increase in number of undeveloped and defective stomata was found in Cd treated plants as compared to untreated plants (Greger and Johansson, 1992). Guard cells are reported to be smaller in Cd-treated plants compared to control plants (Breckle, 1991).

Besides, Cd lowers the content of compounds maintaining cell turgor and cell wall plasticity and in turn lowers the water potential (Barcelo and Poschenrieder, 1990). Increase in the content of ABA is also reported due to Cd treatment, which induces stomatal closure (Hollenbach et al., 1997). Disordered respiration and oxidative phosphorylation in Cd-stressed plants would also cause disarray in plant water regime (Seregin and Ivanov, 2001).

4.6. Mineral Nutrition

Cd checks the uptake of both cations (K^+, Ca^{2+}, Mg^{2+}, Mn^{2+}, Zn^{2+}, Fe^{3+}) and anions like NO_3^-. There are atleast two distinct mechanisms for decreased uptake of macro and micronutrients due to Cd. The first mechanism termed as physico-chemical mechanism depends on the size of metal ion radii. Cd^{2+} (1.03 A°) decreases the uptake of Zn^{2+} (0.83 A°) and Ca^{2+} (1.06 A°) (Yang et al., 1996), whereas the second mechanism relies on the metal induced disorder in the cell metabolism leading to the changes in membrane enzyme activities and membrane structure (Keck, 1978; Burzynski, 1987). The efflux of K^+ occurs from the roots, apparently due to the extreme sensitivity of K^+-ATPase and -SH groups of cell membrane proteins to Cd (Burzynski, 1987). Cd drastically alters the lipid composition of membranes. The resulting changes in the membrane permeation together with membrane enzymes could shift the ionic balance in the cytoplasm. Nitrate uptake declines in Cd-stressed plants which could also be due to moisture stress induced by Cd. Decreased nitrate uptake in turn results in lower nitrate reductase activity and disturbed nitrogen metabolism (Hernandez et al., 1997). Under exposure to Cd, content of K^+ decreases in root and cotyledon, whereas Ca content declines in cotyledons and Fe content declines in roots (Burzynski, 1987). In clover and cabbage plants Cd decreases both uptake and transport of Zn^{2+}, Fe^{3+}, Mn^{2+}, Ca^{2+} and Mg^{2+} (Yang et al., 1996).

4.7. Oxidative Metabolism

Cd is a non-redox metal, unable to perform single electron reactions, however it can cause oxidative stress by reducing the antioxidant glutathione (GSH) pool, activating calcium-dependent system and affecting iron-mediated processes. Cd can disrupt the photosynthetic electron chain, leading to increased production of O_2^- and O_2 ($^1\Delta g$) (Asada and Takahashi, 1987). Cd induced production of ROS within plants depends on the intensity of the stress, repeated stress periods and age of the plants (Shah et al., 2001; Singh and Tewari, 2003; Milone et al., 2003). Rice plants grown for 20 days in presence of 500 µM Cd showed about 0.8-1.7 times increase in superoxide anion generation and about 1.4-1.6 times increase in lipid peroxidation products as measured in terms of malondialdehyde (MDA) levels indicating thereby that Cd induces oxidative stress in rice plants (Shah et al., 2001). Lipid peroxidation is regarded as an indicator of oxidative damage involving oxidative degradation of polyunsaturated fatty acyl residues of membranes (Girotti, 1990). On Cd exposure the level of lipid peroxidation was elevated in *Phaseolus aureus* (Shaw, 1995), *Phaseolus vulgaris* (Chaoui et al., 1997) and *Pisum sativum* (Lozano-Rodriguez et al., 1997). Like all aerobic organisms plants possess the antioxidative mechanism comprising of antioxidant molecules and enzymes to protect themselves from the oxidative damage caused due to harmful oxygen species. Effect of Cd on generation of reactive oxygen species as well as the response of antioxidative enzymes in Cd-exposed plants has been shown in Fig. 3. Increased activity of the antioxidative enzyme superoxide dismutase and peroxidase is observed when plants are exposed to Cd (Shah et al., 2001). The increased activity of antioxidative enzymes in metal exposed plants appears to serve as an important component of antioxidant defense mechanism of plants to combat metal-induced oxidative injury (Shah et al., 2001). The activity of another antioxidative enzyme catalase increased in rice seedlings grown at moderately toxic Cd (100 µM) level whereas with highly toxic Cd (500 μM) level a marked inhibition in catalase activity was noted (Shah et al., 2001). Decline in catalase activity in plants growing under higher levels of Cd appears to be supposedly due to inhibition of enzyme synthesis or a change in assembly of enzyme subunits (Shah et al., 2001).

Glutathione together with ascorbic acid affects plant tolerance to reactive oxygen species by participation in the detoxification of these species in plant cells (Noctor and Foyer, 1998). The involvement of glutathione and ascorbic acid in the tolerance of plants to Cd phytotoxicity has been shown in different plant species (El-Naggar and El-Sheekh, 1998; Wu and Zhang, 2002; Mendoza-Cozatl et al., 2002; Ozturk et al., 2003). The adaptation of *Helianthus annus* plants growing in presence of Cd has been shown to be due to increased activity of key antioxidative enzymes superoxide dismutase, catalase, ascorbate peroxidase, glutathione reductase together with unaltered values of reduced to oxidized glutathione (GSH/GSSG) and reduced to oxidized ascorbic acid (AsA/DHA) ratios (Gallego et al., 2002).

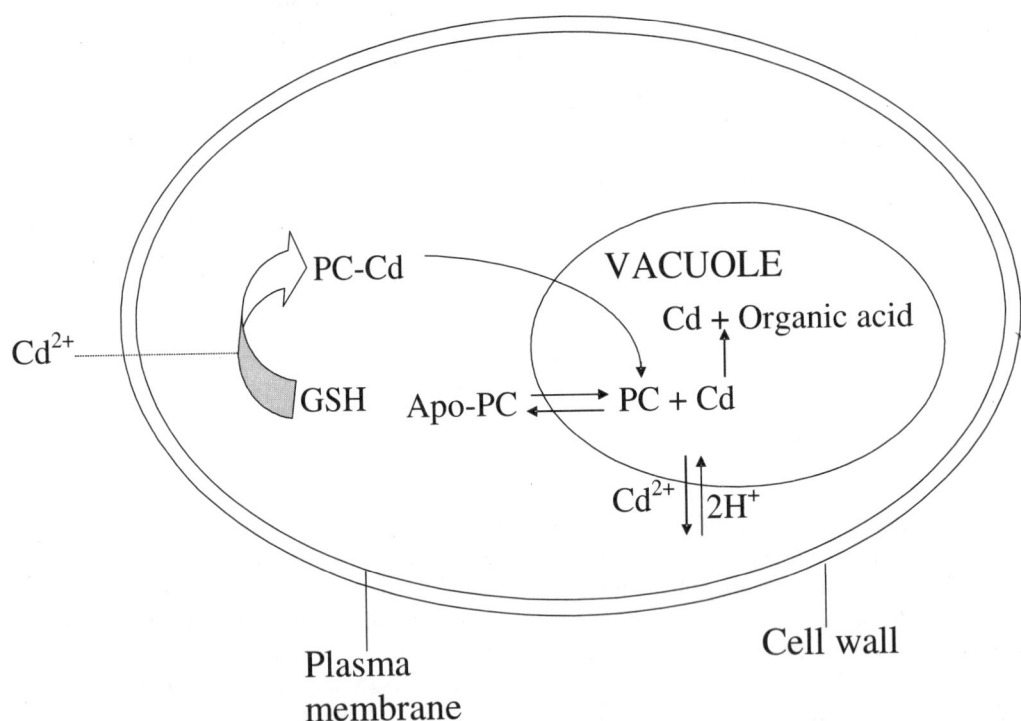

Fig. 3. Effect of Cd on generation of reactive oxygen species (ROS) and activities of antioxidative enzymes. Cd induces increased formation of ROS (H_2O_2, O_2^- and $\cdot OH$), increases the activities of antioxidative enzymes superoxide dismutase (SOD), catalase (CAT), guaiacol peroxidase (GPX), ascorbate peroxidase (APX), monodehydroascorbate reductase (MDHAR), dehydroascorbate reductase (DHAR) and glutathione reductase (GR). The compounds ascorbic acid (AsA) and glutathione (GSH) are important non-enzymic antioxidants within the cell. Their oxidized forms are dehydroascorbic acid (DHA) and GSSG. Haber-Weiss cycle and Fenton mechanism generate hydroxyl radical ($\cdot OH$) from superoxide anion (O_2^-) and H_2O_2. (+) sign denotes induction due to Cd.

5. CADMIUM TOLERANCE IN PLANTS

Primarily two types of mechanisms are recognized that confer tolerance to the toxicity of metal ions in plants. These are (i) avoidance involving various ways of preventing toxic ions to reach their target sites and (ii) tolerance to metal ions in symplasm by complexation (Rengel, 1997). Seed coat presents the first barrier for Cd absorption by germinating seeds. Cd does not enter the embryos even at lethal concentrations (Seregin and Ivanov, 2001). At the root level, the immediate barriers against Cd uptake include immobilization of Cd by extracellular carbohydrates like mucilage, callose etc. (Wagner, 1993) as well as by the constituents of cell wall (Nishizono et al., 1987). In roots and leaves of bush bean Cd ions seem to get bound mostly on pectic sites and histidyl groups of the cell wall (Leita et al., 1996). After absorption by roots Cd ions accumulate primarily in the rhizodermis and cortex. The multilayer cortex seems to reduce the toxic effects of Cd on other tissues by binding most of the Cd ions in the cell wall and in this way serves as the second barrier defending plants from the toxic effects of Cd (Seregin and Ivanov, 2001). Further, casparian strips present in endodermis serve as a barrier for Cd entrance into the central cylinder (Seregin et al., 2004).

On entering the cell, Cd ions get bound to low molecular weight proteins and peptides. Two major groups of Cd binding complexes have been isolated from different plant species. The first group includes 8-14 KDa complexes similar to those of metallothioneins (MTs) of animals and the second group represents 1.5-4.0 KDa complexes, the phytochelatins with a common structure $(\gamma\text{-Glu-Cys})_n$ Gly. The former group of metallothionein like complexes possibly appears to be the aggregates of the later group of complexes and contain the amino acids glutamic acid, cysteine and glycine as the major constituents (Rauser, 1990; Rauser, 1993). Besides these complexes, certain novel proteins with molecular weight greater than 14 KDa have been isolated and well characterized from plants, the synthesis of these proteins is induced when plants are exposed to cadmium (Choi et al., 1995; Shah and Dubey, 1998a). Isolation of a c-DNA was reported by Choi and coworkers (1995), which was expressed by 150 mM Cd in *Arabidopsis* plants and encoded for a 18.3 KDa protein. Similarly, a Cd binding protein complex with an apparent molecular weight of 18 KDa was purified by Shah and Dubey (1998a) from rice plants. This complex had specific Cd content of 3.7 $\mu M\ mg^{-1}$ peptide and had 4 –SH groups per protein molecule. It is suggested that these protein complexes bind Cd with the help of –SH groups of the peptide in mercaptide bonds and help in sequestration of excess Cd ions in plants (Shah and Dubey, 1998a).

Cd exposure has been shown to induce the synthesis of considerable number of stress proteins presumably heat shock proteins with molecular weight ranging between 10 KDa to 70 KDa (di Toppi and Gabbrielli, 1999). In cell cultures of *Lycopersicon peruvianum*, exposed to 1 mM Cd, significant amounts of hsp 70 appeared bound to plasmalemma, mitochondrial membranes and endoplasmic reticulum (Neumann et al., 1994). Hsp 70 has a strong affinity for misfolded proteins and helps them to find their native conformation by reintegrating them into the proper membrane complex (di Toppi and Gabbrielli, 1999). In presence of ATP and various auxiliary factors, Cd-denatured proteins provide a substrate for ubiquitin. The ubiquinated substrate is then degraded by the proteasome, a multi subunit protease complex (Bachmair et al., 1990).

Synthesis of phytochelatins has been reported in many plant species when exposed to Cd. Phytochelatins (PC) represent a group of short non-protein heavy metal binding peptides $(\gamma\text{-glutamyl cysteinyl})_n\text{-X}$, where n = 2-11 and X is glycine, serine, β-alanine, glutamate or glutamine (Grill et al., 1985; Rauser, 1995). Cadmium binds primarily to the thiol group of the cysteine residues in the PC-peptide and the Cd-PC complex is about 1000 times less toxic to the plant enzymes as compared to free Cd ions (Kneer and Jenk, 1992). The use of Cd-sensitive mutants of *Arabidopsis thaliana* was extremely useful in determination of the crucial role of phytochelatins in Cd detoxification (Howden et al., 1995). In particular, the *cd1* mutant deficient in phytochelatin synthase activity does not form any Cd- phytochelatin complex (although it has glutathione level comparable with the wild type plants) and is consequently sensitive to Cd. The production of classic phytochelatins is a wide spread mechanism of Cd detoxification in higher plants (Gekeler et al., 1989). The use of phytochelatins as biomarkers for Cd toxicity was proposed by Keltjens and Van Beusichem (1998). PC-Cd complexes have been shown to get transported across the tonoplast (Salt and Rauser, 1995). In tobacco leaves and many other plant species PC-heavy metal complexes have shown to accumulate in the vacuoles (Vogeli-Lange and Wagner, 1990). Vacuolar compartmentalization of PC-Cd complexes prevents the free circulation of Cd ions as well as PC-Cd complexes in the cytosol and forces them to localize within a limited area (Hart et al., 1998). PC-SH to PC-Cd ratio depends on the level of Cd to which plant is exposed (De Knecht et al, 1994). Vacuolar transport of PC-Cd complexes in the root cells may delay the radial transport of Cd to xylem and its further transport to shoot (De Knecht et al., 1994). In the vacuole due to the acidic pH, PC-Cd complexes dissociate and Cd may get complexed with vacuolar organic acids like citrate, oxalate, malate, etc. (Krotz et. al., 1989). Apo-phytochelatin may get degraded by vacuolar hydrolases and inturn phytochelatins may return to cytosol where they could continue to carry out their shuttle role. Fig. 4 describes the sequence of events associated with PC chelation and compartmentalization of Cd in vacuole.

Cd induces increased production of ethylene in roots compared to shoots. The molecular mechanisms associated with Cd toxicity and ethylene biosynthesis are not properly understood, however it appears that Cd-induced ethylene synthesis might represent a signal capable of accelerating the lignification process (Ecker and Davis, 1987) by increasing the activities of enzymes phenylalanine ammonia-lyase and peroxidase (Ecker and Davis, 1987). Fuhrer (1982) reported that Cd-induced stress-ethylene restricted water and Cd flux into bean leaves, by the induction of cell wall alterations in the vascular system. Ethylene has also been shown to induce the activity of ascorbate peroxidase in order to detoxify hydrogen peroxide (Mehlhorn, 1990) and to regulate the expression of genes encoding metallothioneins and defense proteins (Whitelaw et al., 1997). Ethylene is thought to bind its receptor through a transition metal cofactor, possibly Zn or Cu. It is likely that due to the presence of pi (π) bond between carbon atoms ethylene may interact directly with Cd and influence glutathione metabolism as well as phytochelatin synthesis (Ecker, 1995; Bleeker and Schaller, 1996).

6. REMEDIATION OF CADMIUM CONTAMINATED SOILS

The background cadmium level in agricultural soil is less than 1 mg kg^{-1}. However higher level of Cd is observed in many agricultural soils due to long term use of phosphatic fertilizers and sewage sludge application (Chaney, 1980). Increased Cd levels have been found in surface soils near metal processing industries all over the world. High mobility of this metal in soil-plant system allows its easy entry into food network (Ryan et al., 1982) where it may provoke a variety of human diseases (Nogawa et al., 1987), as well as it can cause toxicity effects in animals, microorganisms and plants.

Metal phytoextraction is a promising approach applicable to slightly or moderately contaminated soils as an alternative to the *ex situ* decontamination techniques, which are very expensive and unacceptable from ecological viewpoint (Mc Grath et al., 2001). Two groups of plant species are considered for metal phytoextraction purposes: (i) hyperaccumulator species which are able to accumulate and tolerate extraordinary high levels of metals as well as (ii) high biomass producing species which compensate lower level of metal accumulation with a higher biomass yield. Metal phytoextraction is a branch of modern biology associated with phytoremediation. Phytoextraction of Cd is a rare phenomenon in higher plants. So far, only *Thlaspi caerulescens* plant has been identified as Cd hyperaccumulator, which is able to meet Cd hyperaccumulation criteria of 100 mg kg^{-1} shoot dry

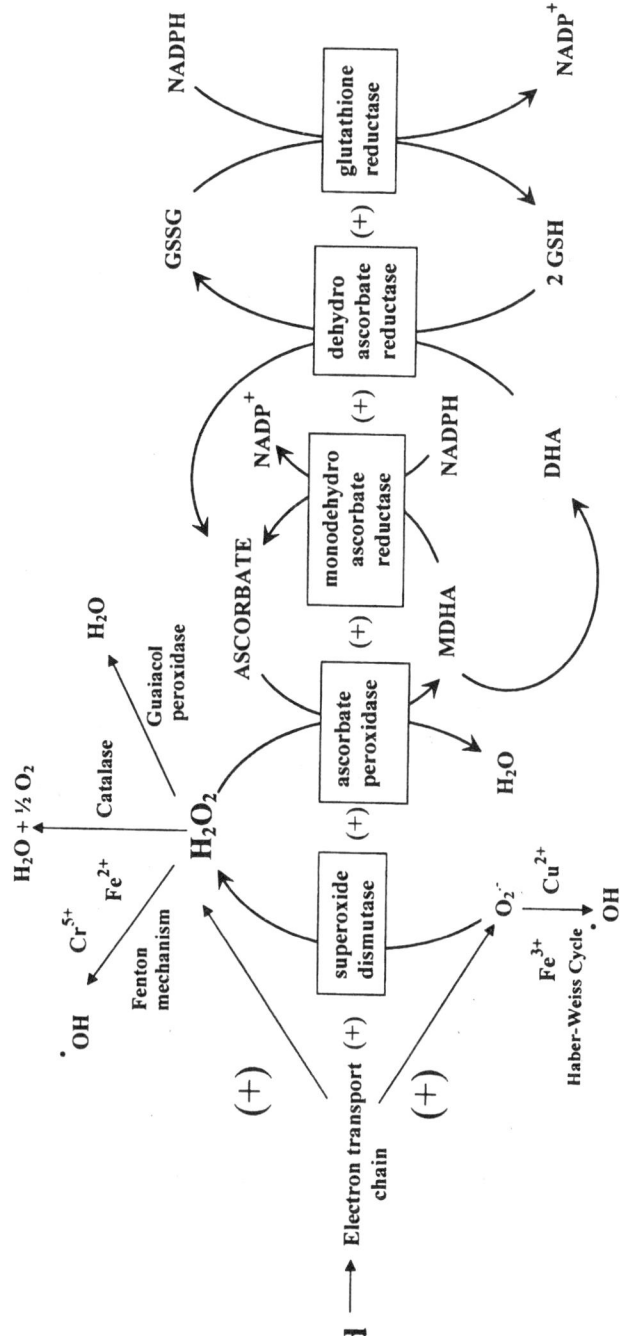

Fig. 4. Mechanism involved in phytochelatin (PC) chelation and compartmentalization of Cd in vacuole. For detail, see text.

weight (Baker et al., 2000). Another possible Cd hyperaccumulator plant could be *Arabidopsis halleri*, which was found to hyperaccumulate Cd in hydroponics (Kupper et al., 2000). Robinson and coworkers (1998) reported accumulation of Cd in the leaves of *T. caerulescens* upto 1600 mg Cd kg^{-1} dry wt, whereas in hydroponics experiments leaves of these plants accumulated Cd over 10,000 mg kg^{-1} dry wt without any reduction in biomass (Lombi et al., 2000).

An ideal plant suitable for metal phytoextraction should be highly productive in biomass and should have capacity to assimilate and translocate significant amount of absorbed metals to aerial organs like shoots, leaves, etc. Additional desirable traits include fast growth, easy propagation and a deep root system. Some tree species like willow and poplar (*Populus*) exhibit these traits and are used in phytoremediation programmes. Short rotation of crops using willow has yielded encouraging results, as willow is regarded as renewable energy crop having ability to hyperaccumulate Cd (Rulford et al., 2002). The use of non-woody biomass plants has also been in practice for phytoextraction in order to overcome certain limitations, such as low dry biomass of hyperaccumulators as well as low phytoavailability of certain metals. Chemically assisted metal phytoextraction is primarily based on EDTA application to the soil, however EDTA has some toxic properties and there is possibility of leaching of metals to ground water (Geebelen et al., 2002). EDTA applied at a rate of 2 g kg^{-1} soil caused a transient increase in uptake of Cd by poplar but also resulted in significant reduction in its growth as well as abscission of leaves (Robinson et al., 2000). Attempts are being made to identify certain potential genotypes of crops like *Brassica napus*, *Nicotiana tabacum*, *Linium usitatissimum*, *Mentha piperita*, *Gossipium hirsutum*, *Zea mays*, *Helianthus annus*, *Brassica juncea*, etc. which could serve as potential phytoaccumulators in order to provide remedy to Cd contaminated soils (Zhelijazkov and Nielsen, 1996; Vassilev and Zaprianova, 1999; Stoinova et al., 1998/1999; Yankov et al., 2000; Griga et al., 2002; Schmidt 2003).

7. CONCLUSIONS

Cadmium contamination poses a serious hazard to human health and uptake of Cd into plants is the primary avenue through which it enters the food chain. Infact Cd is not essential for plants but it is readily taken up by plants from the soil. Soil pH, chloride level, presence of organic matter, concentration of cadmium and its bioavailability as well as plant factors affect the uptake of Cd. Cd is transported across the plasmamembrane *via* a native Zn transporter system. After its uptake Cd is transported to different parts of the plant but gets localized predominantly in cambium cortex and phloem tissues. Within the cell major part of Cd accumulates in vacuole. Diminished translocation of Cd from roots to shoots and other organs of the plant is due to the barrier function of the root endodermis and mechanisms involving sequestration or decreased xylem loading of cadmium. Visual non-specific symptoms of Cd toxicity include leaf roll, chlorosis, growth inhibition, browning of root tips and finally death of the plant. Cd phytotoxicity adversely affects germination of seeds, seedling vigour, decreases nutrient and water uptake, induces ultra structural changes in cell organelles, decreases rate of photosynthesis, respiration as well as transpiration and impairs a wide array of metabolic processes by altering the activities of many key enzymes. In addition Cd induces changes in redox status of the cell, interferes with the uptake and transport of essential metals, results in increased production of free radicals and reactive oxygen species leading to non-specific damage to proteins, lipids and other biomolecules. Cd greatly alters the level of bimolecules of the cell like proteins, lipids, nucleic acids and enzymes of metabolic pathways. Cd interacts with –SH groups present at the active site of enzyme or –SH groups that are essential for tertiary structure of the enzyme and hence it inhibits enzyme activity. Moreover Cd may disrupt metal thiolate bonds leading to decreased activity of many enzymes.

Cd exclusion capacity of plant is related to efficiency to immobilize Cd in rhizosphere as well as binding capacity of the cell wall. Mechanisms of Cd detoxification within the cell include binding of Cd to low molecular weight proteins and peptides including metallothioneins like 8-14 KDa complexes and phytochelatins. These complexes are eventually degraded in the vacuole where Cd can get complexed with vacuolar organic acids. Cd exposure induces the synthesis of considerable number of stress proteins, which have a strong affinity for misfolded proteins and help them to find their native conformation. Cd induces ethylene synthesis, which accelerates lignification process, restricts water and Cd flux in leaves and induces activity of antioxidative enzymes.

Cd phytoextraction is a promising and environment friendly approach for soil decontamination. Plant potential for Cd phytoextraction generally depends on shoot Cd concentration and shoot biomass yield. Among the ecotypes of Cd hyperaccumulator species *Thlaspi caerulescens* possesses an extraordinary Cd accumulation capacity and tolerance, however it has low biomass yield. Genotypes of certain fast growing trees like willow and poplar as well as high biomass producing crops like *Brassica*, *Helianthus*, *Nicotiana* appear to have potential use in phytoextraction and soil decontamination of Cd.

References

Ahumada IT, Schalscha EB (1993) Fractionation of cadmium and copper in soils-effect of redox potential. Agrochimica 37: 281-289.

Asada K, Takahashi M (1987) Production and scavenging of active oxygen in photosynthesis. In: Kyle DJ, Osmond CB, Arntzen CJ (eds.) Photoinhibition. Elsevier Science Publishers, The Netherlands, pp. 227-287.

Aspinall D, Paleg LG (1981) Proline accumulation: physiological aspects. In: Aspinall D, Paleg, LG (eds.) Physiology and Biochemistry of Drought Resistance in Plants. Academic Press, Australia, pp. 206-241.

Astolfi S, Zuchi S, Chiani A, Passera C (2003) *In vivo* and *in vitro* effects of cadmium on H^+-ATPase activity of plasma membrane vesicles from oat (*Avena sativa* L.) roots. J. Plant Physiol. 160: 387-393.

Bachmair A, Becker F, Masterson RV, Schell J (1990) Perturbation of the ubiquitin system causes leaf curling, vascular tissue alterations and necrotic lesions in a higher plant. Euro Mol Biol Organiz. J 9: 4543-4549.

Baker AJM, McGrath SP, Reeves RD, Smith JAC (2000) Metal hyperaccumulator plants: A review of the ecology and physiology of a biochemical resource for phytoremediation of metal-polluted soils. In: Terry N, Banuelos G (eds.) Phytoremediation of Contaminated Soil and Water. Lewis Publishers, United States of America, pp. 85-107.

Barcelo J, Poschenreider C (1990) Plant water relation as affected by heavy metal stress: A review. J Plant Nutr 13: 1-37.

Barcelo J, Vazquez MD, Poschenrieder C (1988) Structural and ultrastructural disorders in cadmium treated bush bean plants (*Phaseolus vulgaris* L.). New Phytol 108: 37-49.

Bazzaz FA, Rolfe GL, Carlson RW (1974) Effect of cadmium on photosynthesis and transpiration of excised leaves of corn and sunflower. Physiol Plant 32: 373-376.

Bharti N, Singh RP (1994) Antagonistic effect of sodium chloride to differential heavy metal toxicity regarding biomass accumulation and nitrate assimilation in *Sesamum indicum* seedlings. Phytochemistry 35: 1157-1161.

Bingham FT, Page AL, Mahler RJ, Ganje TJ (1976) Cadmium availability to rice in sludge amended soil under flood and non-flood culture. Soil Sci Soc Amer J 40: 715-719.

Bleeker AB, Schaller GE (1996) The mechanism of ethylene perception. Plant Physiol 111: 653-660.
Bora KK (1981) Studies on the role of bio-regulants and pollutants in growth and productivity of desert plants. Ph.D. Thesis, University of Jodhpur, Rajasthan, India.
Boussama N, Ouariti O, Suzuki A, Ghorbal MH (1999) Cd-stress on nitrogen assimilation. J Plant Physiol 155: 310-317.
Breckle S (1991) Growth under stress: heavy metals. In: Waisel Y, Eshel A, Kafkafi U (eds.) Plant Roots: The Hidden Half. Marcel Dekker, New York, pp. 351-373.
Burzynski M (1987) The influence of lead and cadmium on the absorption and distribution of potassium, calcium, magnesium and iron in cucumber seedlings. Acta Physiol Plant 9: 229-238.
Campbell PGC, Lewis AG, Chapman PM, Crowder AA, Fletcher WK, Imber B, Luomas N, Stokes PM, Winfrey M (1998) Biologically active metals in sediments. National Research Council. Canada, No. 27694. Association of Common Scientific Criteria for Environmental Quality.
Carlson RW, Bazzaz FA, Rolfe GL (1975) The effects of heavy metals on plants. II. Net photosynthesis and transpiration of whole corn and sunflower plants treated with Pb, Cd, Ni and Ti. Environ Res 10: 113-120.
Cataldo CD, Garland TR, Wildung RE (1983) Cadmium uptake, kinetics in intact soybean plants. Plant Physiol 73: 844-848.
Chaney R (1980) Health risks associated with toxic metals in municipal sludge. In: Bitton G, Damron, B, Edds, G, Davidson, M (eds.) Sludge-Health Risks of Land Application. Ann Arbor Scientific Publishers, Ann Arbor, pp. 59-83.
Chaoui A, Mazhoudi S, Ghorbal MH, El Ferjani E (1997) Cadmium and zinc induction of lipid peroxidation and effects on antioxidant enzymes activities in bean (*Phaseolus vulgaris* L.). Plant Sci 127: 139-147.
Chardonnens AN, Bookum WM, Kuijper LDJ, Verkleij JAC, Ernst WHO (1998) Distribution of cadmium in leaves of cadmium tolerant and sensitive ecotypes of *Silene vulgaris*. Physiol Plant 104: 75-80.
Choi SY, Beak EM, Lee SY (1995) A cDNA differentially expressed by cadmium stress in *Arabidopsis*. Plant Physiol 108: 849-859.
Chugh LK, Sawhney SK (1999) Photosynthetic activities of *Pisum sativum* seedlings grown in presence of cadmium. Plant Physiol Biochem 37: 297-303.
Chugh LK, Gupta VK, Sawhney SK (1992) Effect of cadmium on enzymes of nitrogen metabolism in pea seedlings. Phytochemistry 31: 395-400.
Costa G, Morel JL (1993) Cadmium uptake by *Lupinus albus* (L.): Cadmium excretion, a possible mechanism of cadmium tolerance. J Plant Nutr 16: 1921-1929.
De Knecht JA, Van Dillen M, Koevoets PLM, Schat H, Verkleij JAC, Ernst WHO (1994) Phytochelatins in cadmium-sensitive and cadmium-tolerant *Silene vulgaris*. Chain length distribution and sulphide incorporation. Plant Physiol 104: 255-261.
di Toppi SL, Gabbrielli R (1999) Response to cadmium in higher plants. Environ Exp Bot 41: 105-130.
Dubey RS, Singh AK (1999) Salinity induces accumulation of soluble sugars and alters the activity of sugar metabolizing enzymes in rice plants. Biol Plant 42: 233-239.
Dunbar KR, Laughlin MJ, Reid RJ (2003) The uptake and partitioning of cadmium in two cultivars of potato (*Solanum tuberosum* L.). J Exp Bot 54:349-354.
Ecker JR, Davis RW (1987) Plant defense genes are regulated by ethylene. Proc Natl Acad Sci USA 84: 5202-5206.
Ecker JR (1995) Ethylene signal transduction pathway in plants. Science 268: 667-675.

El-Naggar AH, El-Sheekh MM (1998) Abolishing cadmium toxicity in *Chlorella vulgaris* by ascorbic acid, calcium, glucose and reduced glutathione. Environ Poll 101: 169-174.

Fediuc E, Erdei L (2002) Physiological and molecular aspects of cadmium toxicity and protective mechanisms induced in *Phragmites australis* and *Typha latifolia*. J Plant Physiol 159: 265-271.

Ferrario S, Valadier MH, Foyer CH (1998) Overexpression of nitrate reductase in tobacco delays drought-induced decreases in nitrate reductase activity and mRNA. Plant Physiol 117: 239-302.

Foy CD, RL Chaney, White MC (1978) The physiology of metal toxicity in plants. Ann Rev Plant Physiol 29: 511-566.

Fuhrer J (1982) Ethylene biosynthesis and cadmium toxicity in leaf tissues of bean (*Phaseolus vulgaris* L.). Plant Physiol 70: 162-167.

Gallego S, Benavides M, Tomaro M (2002) Involvement of an antioxidant defense system in the adaptive response to heavy metal ions in *Helianthus annus* L. cells. Plant Growth Regul 36: 267-273.

Geebelen W, Vangronsveld J, Adriano DC, Van Poucke LC, Clijsters H (2002) Effects of Pb-EDTA and EDTA on oxidative stress reactions and mineral uptake in *Phaseolus vulgaris*. Physiol Plant 115: 377-384.

Geisweid HJ, Urbach W (1983) Sorbtion of cadmium by the green microalgae *Chlorella vulgaris*, *Ankistrodesmus braunii* and *Eremosphaera viridis*. Zeit fur Pflanzenphysiol 109: 127-142.

Gekeler W, Grill E, Winnacker E-L, Zenk MH (1989) Survey of the plant kingdom for the ability to bind heavy metals through phytochelatins. Zeit für Naturfor 44: 361-369.

Girotti AW (1990) Photodynamic lipid peroxidation in biological systems. J Photochem Photobiol 51: 497-509.

Godzik B (1993) Heavy metal contents in plants from zinc dumps and reference area. Polish Bot Studies 5:113-132.

Gouia H, Ghorbal MH, Meyer C (2000) Effects of cadmium on activity of nitrate reductase and on other enzymes of the nitrate assimilation pathway in bean. Plant Physiol Biochem 38: 629-638.

Gouia H, Suzuki A, Brulfert J, Ghorbal MH (2003) Effects of cadmium on the co-ordination of nitrogen and carbon metabolism in bean seedlings. J Plant Physiol 160: 367-376.

Grant CA, Buckley WT, Bailey LD, Selles F (1998) Cadmium accumulation in crops. Can J Plant Sci 78: 1-17.

Greger M, Johansson M (1992) Cadmium effects on leaf transpiration of sugar beet (*Beta vulgaris*). Physiol Plant 86: 465-473.

Griga M, Bjelkova M, Tejklova E (2002) Potential of flax (*Linum usitatissimum*) for heavy metal extraction and industrial processing of contaminated biomass- a review. Proceedings of the 4th workshop of COST action 837, working group 2 held in Bordeaux, April 25-26th, 2002.

Grill E, Winnacker EL, Zenk MH (1985) Phytochelatins: the principal heavy-metal complexing peptides of higher plants. Science 230: 674-676.

Hagemeyer J, Kahle H, Breckle SW, Waisel Y (1986) Cadmium in *Fagus sylvatica* L. trees and seedlings: leaching, uptake and interconnection with transpiration. Water Air Soil Poll 29: 347-359.

Hart JJ, Welch RM, Norvell WA, Sullivan LA, Kochian LV (1998) Characterization of cadmium binding, uptake and translocation in intact seedlings of bread and durum wheat cultivars. Plant Physiol 116: 1413-1420.

Hatch DJ, Jones LHP, Burau RG (1988) The effect of pH on the uptake of cadmium by four plant species grown in flowing solution culture. Plant Soil 105: 121-126.

Hernandez E, Olguín E, Trujillo SY, Vivanco J (1997) Recycling and treatment of anaerobic effluents from pig waste using *Lemma* sp. under temperate climatic conditions. In: Wise DL (ed.) Global Environmental Biotechnology. The Netherlands, pp. 293-304.

Hollenbach B, Schreiber L, Hartung W, Dietz KJ (1997) Cadmium leads to stimulated expression of the lipid transfer protein genes in barley: Implications for the involvement of lipid transfer proteins in wax assembly. Planta 203: 9-19.

Homma Y, Hirata H (1984) Kinetics of cadmium and zinc absorption by rice seedling roots. Soil Science Plant Nutr 30: 527-532.

Hortensteiner S, Feller U (2002) Nitrogen metabolism and remobilization during senescence. J. Exp Bot 53: 927-937.

Hosayama Y, Daiten S, Sakai MI (1994) Histochemical demonstration of cadmium in plant tissues. Kenkyu Kiyo-Nihon daigaku Bunrigakabu shizen Kagaku Kenkyusho 29: 107-110.

Howden R, Goldsbrough PB, Andersen CR, Cobbett CS (1995) Cadmium-sensitive, *cad1*, mutants of *Arabidopsis thaliana* are phytochelatin deficient. Plant Physiol 107: 1059-1066.

Hsu YT, Kao CH (2003) Role of abscisic acid in cadmium tolerance of rice (*Oryza sativa*) seedlings. Plant Cell Environ 26: 867-874.

Huang GY, Bazzaz FA, Vanderhoef LN (1974) The inhibition of soybean metabolism by cadmium and lead. Plant Physiol 54: 122-124.

Jones KC, Jackson A, Johnston AE (1992) Evidence for an increase in the Cd content of herbage since 1860's. Environ Sci Technol 26: 834-836.

Kastori R, Petrovic M, Pertovic N (1992) Effect of excess lead, cadmium, copper and zinc on water relations in sunflower. J Plant Nutr 15: 2427-2439.

Keck RW (1978) Cadmium alteration of root physiology and potassium ion fluxes. Plant Physiol 62: 94-96.

Keltjens WG, Van Beusichem ML (1998) Phytochelatins as biomarkers for heavy metal toxicity in maize: single metal effects of copper and cadmium. J Plant Nutr 21: 635-648.

Khalid RA, Gambrell RP, Patrick WH Jr. (1981) Chemical availability of cadmium in Mississippi river sediment. J Environ Qual 10: 523-528.

Kneer R, Jenk MH (1992) Phytochelatins protect plant enzymes from heavy metal poisoning. Phytochemistry 31: 2663-2667.

Kochian LV (1991) Mechanisms of micronutrient uptake and translocation in plants. In: Mortvedt JJ (ed.) Micronutrients in Agriculture. Soil Science Society of America, Madison, Wisconsin, pp. 251-270.

Krosshavn M, Steinnes E, Varskog P (1993) Binding of Cd, Cu, Pb and Zn in soil organic matter with different vegetational background. Water Air Soil Poll 71: 185-193.

Krotz RM, Evangelou BP, Wagner GJ (1989) Relationships between cadmium, zinc, Cd-peptide, and organic acid in tobacco suspension cells. Plant Physiol 91: 780-787.

Krupa Z, Oquist G, Huner NPA (1993) The effects of cadmium on photosynthesis of *Phaseolus vulgaris*: a fluorescence analysis. Plant Physiol 88: 626-630.

Krupa Z (1988) Cadmium induced changes in the composition and structure of the light-harvesting chlorophyll a/b protein complex II in radish cotyledons. Physiol Plant 73: 518-524.

Krupa Z (1999) Cadmium against higher plants photosynthesis: a variety of effects and where do they possibly come from? Zeit für Naturfor 54: 723-729.

Küpper HE, Lombi E, Zhao FJ, McGrath SP (2000) Cellular compartmentation of cadmium and zinc in relation to other elements in the hyperaccumulator *Arabidopsis halleri*. Planta 212: 75-84.

Lamoreaux RJ, Chaney WR (1978) The effect of cadmium on net photosynthesis, transpiration and dark respiration of excised maple leaves. Physiol Plant 36: 231-236.

Lee KC, Cunningham BA, Poulsen GM, Liang JM, Moore RB (1976) Effects of cadmium on respiration rate and activities of several enzymes in soybean seedlings. Physiol Plant 36: 4-6.

Leita L, Denobili M, Cesco S, Mondini C (1996) Analysis of intercellular cadmium forms in roots and leaves of bush plants. J Plant Nutr 19: 527-533.

Lichtenthaler HK, Rinderle U (1998) The role of chlorophyll fluorescence in the detection of stress conditions in plants. Critical Rev Anal Chem 19: S29-S58.

Liu D, Jiang W, Wang W, Zhai L (1995) Evaluation of metal ion toxicity on root tip cells by the *Allium* test. Israel J Plant Sci 43: 125-133.

Lombi, E, Zhao FJ, Dunham S, McGrath SP (2000) Cadmium accumulation in populations of *Thlaspi caerulescens* and *Thlaspi goesingense*. New Phytol 145: 11-20.

Lozano-Rodriguez E, Hernández LE, Bonay P, Carpena-Ruiz RO (1997) Distribution of Cd in shoot and root tissues of maize and pea plants: physiological disturbances. J Exp Bot 48: 123-128.

Malik D, Sheoran IS, Singh R (1992) Carbon metabolism in leaves of cadmium-treated wheat seedlings. Plant Physiol Biochem 30: 223-229.

Mankovska B (1977) The content of Pb, Cd and Cl in forest trees caused by the traffic of motor vehicles. Biologia 32: 477-490.

McGrath SP, Zhao FJ, Lombi E (2001) Plant and rhizosphere processes involved in phyto-remediation of metal-contaminated soils. Plant Soil 232: 207-214.

Mehlhorn H (1990) Ethylene-promoted ascorbate peroxidase activity protects plants against hydrogen peroxide, ozone and paraquat. Plant Cell Environ 13: 971-976.

Mendoza-Cozatl D, Devars S, Loza-Tavera H, Moreno-Sanchez R (2002) Cadmium accumulation in the chloroplast of *Euglena gracilis*. Plant Physiol 115: 276-283.

Metwally A, Finkerneier I, Georgi M, Dietz KJ (2003) Salicylic acid alleviates the cadmium toxicity in barley seedlings. Plant Physiol 132: 272-281.

Miller RJ, Biuell JE, Koeppe DE (1973) The effect of cadmium on electron and energy transfer reactions in corn mitochondria. Physiol Plant 28: 166-171.

Milone MT, Sgherri C, Cljisters H, Navari-Izzo F (2003) Antioxidative responses of wheat treated with realistic concentration of cadmium. Environ Exp Bot 50: 265-276.

Mitchell CD, Fretz TA (1977) Cadmium and zinc toxicity in seedling white pine, red maple, and Norway spruce. Ohio Agric Res Develop Centre Res Circular 226: 21-25.

Munns R, Brady CJ, Barlow EER (1979) Solute accumulation in the apex and leaves of wheat during water stress. Australian J Plant Physiol 6: 379-389.

Muntz K (1996) Proteases and proteolytic cleavage of storage proteins in developing and germinating dicotyledonous seeds. J Exp Bot 47: 605-622.

Muramoto S, Aoyama I (1990) Effects of fertilizers on the vicissitude of cadmium in rice plant. J Environ Sci Health 25: 629-636.

Neumann D, Lichtenberger O, Gunther D, Tschiersch K, Nover L (1994) Heat-shock proteins induce heavy-metal tolerance in higher plants. Planta 194: 360-367.

Nishizono H, Ichikawa II, Suzuki S, Ishii F (1987) The role of the root cell wall in the heavy metal tolerance of *Athyrium yokoscence*. Plant Soil 101: 15-20.

Noctor G, Foyer CH (1998) Ascorbate and glutathione: keeping active oxygen under control. Ann Rev Plant Physiol Plant Mol Biol 49: 249-279.

Nogawa K, Honda R, Kido T, Tsuritani I, Yamada Y (1987) Limits to protect people eating cadmium in rice, based on epidemiological studies. Trace Substances Environ Health 21: 431-439.

Nolan AL, McLaughlin MJ, Mason SD (2003) Chemical speciation of Zn, Cd, Cu, and Pb in pore waters of agricultural and contaminated soils using Donnan dialysis. Environ Sci Technol 37: 90-98.

Ornes WH, Sajwan KS (1993) Cadmium accumulation and bioavailability in coontail (*Ceratophyllum demersum* L.) plants. Water Air Soil Poll 69: 291-300.

Ozturk L, Karanlik S, Özkutlu F, Çakmak I, Kochian LV (2003) Shoot biomass and zinc/cadmium uptake for hyperaccumulator and non-accumulator *Thlaspi* species in response to growth on a zinc-deficient calcareous soil. Plant Sci 164: 1095-1101.

Padmaja K, Prasad DDK, Prasad ARK (1990) Inhibition of chlorophyll synthesis in *Phaseolus vulgaris* L. seedlings by cadmium acetate. Photosynthetica 24: 399-405.

Popelka JC, Schubert S, Schulz R, Hansen AP (1996) Cadmium uptake and translocation during reproductive development of peanut (*Arachis hypogaea* L.). Angew Bot 70: 140-143.

Poschenrieder C, Gunse G, Barcelo J (1989) Influence of cadmium on water relations, stomatal resistance and abscisic acid content in expanding bean leaves. Plant Physiol 90: 1365-1371.

Rauser WE (1990) Phytochelatins. Ann Rev Biochem 59: 61-86.

Rauser WE (1993) Metal binding peptides in plants. In: Dekok LJ (ed.) Sulfur Nutrition and Assimilation in Higher Plants. SPB academic Publishing, The Hague, The Netherlands, pp. 239-251.

Rauser WE (1995) Phytochelatins and related peptides: structure, biosynthesis, and function. Plant Physiol 109: 1141-1149.

Reese RN, Roberts LW (1985) Effect of cadmium on whole cell and mitochondrial respiration in tobacco cell suspension cultures. J Plant Physiol 120: 123-130.

Rengel Z (1997) Mechanisms of plant resistance to aluminium and heavy metals. In: Basra AS, Basra RK (eds.) Mechanisms of Environmental Stress Resistance in Plants. Harwood Academic Publishers, Amsterdam, pp. 241-276.

Robinson BH, Meblanc L, Petit D, Brooks RR, Kirkman JH, Gregg PEH (1998) The potential of *Thlaspi caerulescens* for phytoremediation of contaminated soils. Plant Soil 203: 47-56.

Robinson B, Mills T, Petit D, Fung L, Green S, Clothier B (2000) Natural and induced cadmium-accumulation in poplar and willow: implications for phytoremediation. Plant Soil 227: 301-306.

Rossman TG, Roy NK, Lin WC (1992) Is cadmium genotoxic? IARC Scientific Publications 118: 367-375.

Rulford ID, Riddell-Black D, Stewart C (2002) Heavy metal uptake by willow clones from sewage sludge-treated soil: the potential for phytoremediation. Int J Phytorem 4: 59-72.

Ryan J, Pahren H, Lucas J (1982) Controlling cadmium in the human food chain: A review and rationale based on health effects. Environ Res 27: 251-302.

Salt DE, Rauser WE (1995) Mg-ATP-dependent transport of phytochelatins across the tonoplast of oat roots. Plant Physiol 107: 1293-1301.

Salt DE, Wagner GJ (1993) Cadmium transport across tonoplast of vesicles from oat roots: Evidence for a Cd^{2+}/H^+ antiport activity. J Biol Chem 268: 12297-12302.

Salt DE, Prince RC, Pickering IJ, Raskin I (1995) Mechanisms of cadmium mobility and accumulation in Indian mustard. Plant Physiol 109: 1427-1433.

Sarkar B (1995) Metal replacement in DNA-binding zinc finger proteins and its relevance to mutagenicity and carcinogenicity through free radical generation. Nutrition 11: 646-649.

Schat H, Sharma SS, Vooijs R (1997) Heavy metal-induced accumulation of free proline in a metal-tolerant and a nontolerant ecotype of *Silene vulgaris*. Physiol Plant 101: 477-482.

Schmidt U (2003) Enhancing phytoextraction: the effect of chemical soil manipulation on mobility, plant accumulation and leaching of heavy metals: A review. J Environ Qual 32: 1939-1954.

Schützendübel A, Schwanz P, Teichmann T, Gross K, Langenfeld-Heyser R, Godbold DL, Polle A (2001) Cadmium-induced changes in antioxidative systems, hydrogen peroxide content, and differentiation in Scots pine roots. Plant Physiol 127: 887-898.

Seregin IV, Ivaniov VB (2001) Physiological aspects of cadmium and lead toxic effects on higher plants. Russian J Plant Physiol 48: 606-630.

Seregin IV, Shpigun LK, Ivaniov VB (2004) Distribution and toxic effects of cadmium and lead on maize roots. Russian J Plant Physiol 51: 525-533.

Shah K, Dubey RS (1995) Effect of cadmium on RNA level as well as activity and molecular forms of ribonuclease in growing rice seedlings. Plant Physiol Biochem 33: 577-584.

Shah K, Dubey RS (1997a) Cadmium alters phosphate level and suppresses activity of phosphorolytic enzymes in germinating rice seeds. J Agron Crop Sci 179: 35-45.

Shah K, Dubey RS (1997b) Effect of cadmium on proteins, amino acids and protease, aminopeptidase and carboxypeptidase in rice seedlings. Plant Physiol Biochem 24: 89-95.

Shah K, Dubey RS (1997c) Effect of cadmium on proline accumulation and ribonuclease activity in rice seedlings: role of proline as a possible enzyme protectant. Biol Plant 40: 121-130.

Shah K, Dubey RS (1998a) A 18 kDa Cd inducible protein complex: Its isolation and characterization from rice (*Oryza sativa* L.) seedlings. J Plant Physiol 152: 448-454.

Shah K, Dubey RS (1998b) Cadmium elevates level of protein, amino acids and alters the activity of proteolytic enzymes in germinating rice seeds. Acta Physiol Plant 20: 189-196.

Shah K, Dubey RS (1998c) Cadmium suppresses phosphate level and inhibits the activity of phosphatases in growing rice seedlings. J Agron Crop Sci 180: 223-231.

Shah K, Kumar RG, Verma S, Dubey RS (2001) Effect of cadmium on lipid peroxidation, superoxide anion generation and activities of antioxidant enzymes in growing rice seedlings. Plant Sci 161: 1135-1144.

Shaw BP (1995) Effects of mercury and cadmium on the activities of antioxidative enzymes in the seedlings of *Phaseolus aureus*. Biol Plant 37: 587-596.

Siedlecka A, Krupa Z, Samuelson G, Oquist G, Garderstrom P (1997) Primary carbon metabolism in *Phaseolus vulgaris* plants under Cd/Fe interaction. Plant Physiol Biochem 35: 951-957.

Singh PK, Tewari RK (2003) Cadmium toxicity induced changes in plant water relations and oxidative metabolism of *Brassica juncea* L. plants. J Environ Biol 24: 107-112.

Singh RP, Bharti N, Kumar G (1994) Differential toxicity of heavy metals to growth and nitrate reductase activity of *Sesamum indicum* seedlings. Phytochemistry 35: 1153-1156.

Soloman L, Barber MJ (1990) Assimilatory nitrate reductase: Functional properties and regulation. Ann Rev Plant Physiol Plant Mol Biol 41: 225-253.

Stiborova M, Doubravova M, Leblova S (1986) A comparative study of the effect of heavy metal ions on ribulose-1,5-bisphosphate carboxylase and phosphoenolpyruvate carboxylase. Biochem Physiol Pflanzen 181: 373-379.

Stohs SJ, Bagchi D, Hassoun E, Bagchi M (2000) Oxidative mechanisms in the toxicity of chromium and cadmium ions. J Environ Pathol Toxicol Oncol 19: 201-213.

Stoinova J, Merakchiiska M, Sabva Z, Phileva S, Paunova S (1998/1999) Productivity and sensibility to heavy metals of new hexaploid lines of triticale. Gen Breed 29: 25-31.

Syntichaki KM, Loulakakis KA, Loulakakis-Roubelakis KA (1996) The amino-acid sequence similarity of plant glutamate dehydrogenase to the extremophilic archaeal enzyme conforms to its stress-related function. Gene 68: 87-92.

Thornton I (1992) Sources and pathways of cadmium in the environment. IARC Scientific publications 118: 149-162.

Tukendorf A (1993) The response of spinach plants to excess of copper and cadmium. Photosynthetica 28: 573-575.

Ubio MI, Escrig I, Martinezcortina C, Lopezbenet FJ, Sanz A (1994) Nickel and cadmium in rice plants: Effects on mineral nutrition and possible interactions of abscisic and gibberellic acids. Plant Growth Regul 14: 151-157.

Urwin PE, Groom QJ, Robin NJ (1996) Characterization of two cDNAs and identification of two proteins that accumulate in response to cadmium in cadmium tolerant *Datura innoxia* (Mill.) cells. J Exp Bot 47: 1019-1024.

Vassilev A, Zaprianova P (1999) Removal of Cd by winter barley (*H. vulgare* L.) grown in soils with Cd pollution. Bulgarian J Agric Sci 5: 131-136.

Verma S, Dubey RS (2001) Effect of cadmium on soluble sugars and enzymes of their metabolism in rice. Biol Plant 44: 117-123.

Vincet JB, Crowder MW, Averill BA (1992) Hydrolysis of phosphate monoesters: a biological problem with multiple chemical solutions. Trends Biochem Sci 17: 105-110.

Vögeli-Lange R, Wagner GJ (1990) Subcellular localization of cadmium and cadmium-binding peptides in tobacco leaves. Plant Physiol 92: 1086-1093.

Wagner GJ (1993) Accumulation of cadmium in crop plants and its consequences to human health. Adv Agron 51: 173-212.

Whitelaw CA, Le Huquet JA, Thurman DA, Tomsett AB (1997) The isolation and characterization of type II metallothionein-like genes from tomato (*Lycopersicum esculentum* L.). Plant Mol Biol 33: 503-511.

Wu FB, Zhang GP (2002) Alleviation of cadmium-toxicity by application of zinc and ascorbic acid in barley. J Plant Nutr 25: 2745-2761.

Yang XE, Baligar VC, Martens DC, Clark RB (1996) Cadmium effects on influx and transport of mineral nutrients in plant species. J Plant Nutr 19: 643-656.

Yankov B, Delibaltova V, Bojinov M (2000) Content of Cu, Zn, Cd and Pb in the vegetative organs of cotton cultivars grown in industrially polluted regions. Plant Sci 37: 525-531.

Zhelijazkov V, Nielsen N (1996) Effect of heavy metals on peppermint and corn mint. Plant Soil 178: 59-66.

Cadmium Toxicity and Tolerance in Plants
Editors: Nafees A. Khan and Samiullah
Copyright © 2006, Narosa Publishing House, New Delhi, India

Cadmium Stress in Higher Plants

Guoping Zhang, Feibo Wu, Kang Wei, Qin Dong, Fei Dai, Fei Chen and Jung Yang*

Agronomy Department, Huajia Chi Campus, Zhejiang University, Hangzhou, China

1. INTRODUCTION

Cadmium (Cd) is one of the most deleterious heavy metals to both plants and animals and has no beneficial biological function in aquatic or terrestrial organisms. It has become one of the most harmful and widespread pollutants in soil-plant-environment system mainly due to industrial emission, application of sewage sludge and phosphate fertilizers and municipal waste disposal containing Cd (Davis, 1984; Guo, 1994). At present, Cd contamination has posed a serious threat to safe food production. Cd is mainly accumulated in human kidney and may cause pulmonary emphysema and renal tubular damage (Ryan et al., 1982). Extreme cases of chronic Cd toxicity can result in osteomalacia and bone fractures, as characterized by the disease called Itai-Itai in Japan in the 1950s and 1960s, where local population were exposed to Cd-contaminated food crops, principally rice. According to recent soil survey done in China, nearly 20 M ha of farmland, being about 1/5 of the total, has been polluted to different extent by toxic heavy metals, including Cd, As, Cr, and Pb etc, and at least 13330 ha of farmland involved in 11 provinces was contaminated by Cd with varying degrees (Zhang and Huang, 2000), which resulted in a sharp reduction of food production by about 10 million ton per year, with about 12 million ton of polluted food, and at least 2.5 billion US dollars of economic loss. On the other hand, Cd is not decomposed by microorganism in soil, with biological half life up to about 20 years. Therefore, Cd pollution is a non-reversible accumulation process. Furthermore, Cd, being biologically easily movable, can be absorbed and accumulated easily by plants, while high accumulation of Cd in plants not only affects crop yield and quality badly, but also gives rise to a threat on human health via a food chain. Thus, Cd contamination in soil has posed a serious issue to the sustainable agriculture and human health worldwide (Davis, 1984; Arthur et al, 2000). Accordingly, increasingly attention has been focused on the toxicity effect of Cd on plants, and in the past 20 years researches were mainly done on Cd toxicity to plants and tolerance mechanisms of plants to Cd toxicity.

2. CADMIUM CONTENT IN SOIL

Cadmium in unpolluted soils is mainly originated from soil parent ores. Globally, Cd content in soil is about 0.01~2 mg kg^{-1}, with an average of 0.35 mg kg^{-1} (Xu and Yang, 1995). In China, the mean Cd content in soil is 0.097 mg kg^{-1}, ranging from 0.017 to 0.333 mg kg^{-1} (Meng et al., 2000).

In general, the natural form of Cd in soil would not do any harm to human, while the harmful Cd in soil is mainly brought about artificially. In America, the amount of Cd entered the environment

---

*Corresponding Author (zhanggp@zju.edu.cn)

was 7613 t within 10 years, and among these, 13%, 5.5%, and 81.8%, respectively, entered the atmosphere, water and soil. Similarly, in Europe Union (EU) countries among 6118 t of Cd, which entered the environment, 94% existed in soil.

3. THE SOURCE OF Cd CONTAMINATION IN SOIL
3.1. The Atmospheric Deposition

The atmospheric deposition of Cd is affected by many factors, and the main source for Cd contamination in atmosphere is industrial exhaust gas. For example, usually atmospheric sedimentation rate of Cd is 0.06-44.9 g ha^{-1} a^{-1}, but the rate could reach to as high as 135.6 g ha^{-1} a^{-1} around some mining/smelting factories. Cd can enter soil from atmosphere via rainfall and deposition, and some of them could be absorbed directly by plant leaves. Therefore, the soil and plant leaves would have high Cd content in the industrially contaminated regions (Ivonin and Shumakova, 1991). Arthur et al. (2000) reported that among the Cd contamination source in agricultural lands in Britain, 50% came from atmospheric deposition, 35% and 11% from the application of sewage sludge and fertilizers (especially phosphorus fertilizer containing Cd) and 4% from manure, respectively.

3.2. The Application of Sewage Sludge and Municipal Waste Disposal

The application of sewage sludge is also one of the important factors for Cd contamination in soil. Usually, 60-80% of sewage is from industrial waste, and the dominating sources for Cd-contaminated sewage are mining, zincification, and some factories relevant to dye and plastic stabilized matter, paint colorant, and tyre production (Guo, 1994). In China, there was about 1.4 M ha farm land irrigated by sewage, and 30% of which was contaminated by heavy metals, to different extent, especially Cd.

Application of solid garbage, such as sludge, in farmland would result in heavy metal contamination. In America, the mean Cd concentration in sludge was 10-16 mg kg^{-1} (Guo, 1994), thus long-term use of sludge results in Cd enrichment of soil.

3.3. The Application of Phosphorus Fertilizer and Others Fertilizers Containing Cd

Cd content in phosphorus fertilizer varies greatly with raw materials and processing methods. In general, Cd content in phosphorite ranges from 1 to 100 mg kg^{-1}, while Cd content in phosphorus fertilizer ranges from 5-50 mg kg^{-1}. However, it could reach to as high as 200 mg kg^{-1} occasionally. The Cd content in phosphorus fertilizer has been reported to nearly 18-91 mg kg^{-1} in Australia, 2.1-9.3 mg kg^{-1} in Canada, 7.4-159 mg kg^{-1} in America, 9-60 mg kg^{-1} in Holland and 2-30 mg kg^{-1} in Sweden. Most of the Cd in phosphorous fertilizer exists in the form of $Cd(H2PO4)_2$ and $CdHPO_4$.

4. THE CRITICAL Cd CONCENTRATION IN SOIL FOR SAFE CROP PRODUCTION

To guarantee a safe agricultural production, many countries put forward a maximal critical content of toxic elements in soil, i.e. below the critical level, the toxic elements neither have effects on crop yield and quality nor cause the contamination of surface/ground-water. However, the criterion varies significantly in different countries. Thus, in terms of the maximal critical content of Cd in soil, a threshold value of Cd has been suggested as 1~3 mg kg^{-1} in EU, 3.56 mg kg^{-1} in USA, 2 mg kg^{-1} in France, 3 mg kg^{-1} in Germany and Italy, 1.6 mg kg^{-1} in Scotland, 3.5 mg kg^{-1} in UK and 1.6 mg kg^{-1} in Canada. In China, different critical Cd content has also been proposed for different kinds of soils. For example, the threshold for Sierozems soil is 2.3 mg kg^{-1}, Cinnamon soil is 0.157 mg kg^{-1}, Black soil is

1.3 mg kg^{-1}, Brown soil is 1.31 mg kg^{-1}, Yellow-brown soil is 0.3 mg kg^{-1}, Purplish soil is 0.5 mg kg^{-1}, Red soil is 0.6 mg kg^{-1} and Yellowish red soil is 0.6 mg kg^{-1} (Xia, 1991). The variation in the critical Cd concentration in soil is understandable because the availability of Cd affects physical and chemical properties of soil.

5. ABSORPTION OF Cd IN PLANTS

Cd content in soil is one of the main factors affecting Cd uptake by plants. For most plants, Cd absorption increases with increasing soil Cd concentration, and a positive association between Cd concentration in plants and available Cd content in soil is commonly found (Baker et al., 1994). However, saturation of Cd uptake can be observed in quite a few plant species (Hagemeyer and Waisel, 1989). Obviously, it results in Cd toxicity. On the other hand, plants can also absorb Cd directly from the atmosphere.

5.1. The Mechanism of Cd Absorption

The mechanisms of Cd absorption by plant roots include non-metabolic (passive absorption) and metabolic (active absorption) processes. The former consists of ion diffusion and cation exchange, and Cd ion enters cells via cell walls through diffusion. Cation exchange is a reversible procedure. Metabolic absorption is a procedure to absorb nutrition from low concentration to high concentration depending on energy metabolism. In low Cd condition, plant absorbs Cd mainly by metabolic absorption, whereas in higher Cd condition, non-metabolic absorption may be dominant (Cataldo et al., 1983).

5.2. The Effect of Soil Properties on Cd Absorption

Cadmium absorption and accumulation in plants is affected by many soil factors, including pH, organic matter, cation exchange capacity (CEC) and Eh. Generally, Cd absorbed by plants reduces with a decreased Eh or an increased pH or CEC in soil. Cadmium in soil can be stabilized by organic matter. Most stabilized Cd is non-exchangeable. Thus available Cd in soil is low in the soils with high organic matter content. Moreover, other ions also affect Cd absorption with different degrees. Ca^{2+} shows a dramatic effect on Cd absorption because it has the similar ion radius with Cd^{2+}, and thus competes with Cd on its combination site. The effect of Fe^{2+} on Cd absorption in plants is relative little (Hagemeyer and Waisel, 1989). The repression of EDTA to Cd absorption is slightly less than that of Ca^{2+}, and greater than those of K^+, Na^+, and Mg^{2+}. Humic acid can decrease Cd availability in plant by chelation, whereas it does not affect the transport of Cd in crop plants (Cabrera et al., 1988).

The absorption and accumulation of Cd in plants is also dependent on growth stage. Generally, the greatest rate of Cd absorption occurs when plants have the most active metabolism. For instance, rice absorbs Cd most rapidly during anthesis, and followed by tillering and spike differentiation stages.

6. CADMIUM TRANSPORT, ACCUMULATION AND DISTRIBUTION IN PLANTS
6.1. Cadmium Transport in Plants

Cadmium absorbed by roots reaches shoots via xylem transport (Salt et al., 1995), but the mechanism is not well known up to now. Pertjt and van de Geijn (1978) reported that more than 50% of Cd ions in xylem flux are in inorganic form with a concentration gradient: decreasing along from the shoot base to top. Meanwhile, Cataldo et al. (1981) and White et al. (1981) observed that Cd existed and was mobilized in xylems of soybean and tomato in the form of anion complex. Chino and Baba (1981) also found an organic complex of Cd in tomato xylem.

The long-distance transport velocity of Cd in plants is about 0.35~0.60 m h^{-1} (Pertjt and Geijn, 1978), speeding up with increasing transpiration rate (Hardiman and Jacoby, 1984). Cd transport rate increased with elevated temperature in the range of 10~30, and was inhibited by some cations, such as Ca, Mg and Zn, but Mn and K had no effect on it (Pawlik and Skowionski, 1994). Other chemicals in xylem sap, such as citric acid, dihydric anions, and inorganic cations also affect Cd transport. Citric acid accelerates Cd transport in xylem, but reduces pumping of Cd out of vascular system and its adhering to the cell walls (Senden and Wolterbeek, 1990). Under Cd stress, the amount of Cd transferred from roots to shoots in tomato plants, exposed to 250 µM citric acid was increased 6~8 times compared with the control (Senden and Wolterbeek, 1990).

Cd accumulation in cereal grains is dependent on its redistribution in plants. Hart et al. (1998) reported that Cd may enter grains via phloem. The xylem-to-phloem transfer in ears is important for Cd accumulation in grains. They used ^{109}Cd to investigate Cd uptake and transport in durum and bread wheat. The results showed that less Cd was found in xylem sap of durum compared with that of bread wheat, but Cd content in mature grains was opposite because durum showed a larger capacity in xylem-to-phloem transfer. Cadmium enters peanut seeds also via phloem (Popelka et al., 1996). The pattern of Cd accumulation in wheat grains is similar to that of Zinc, the latter was firstly transferred into ear phloem, then entered developing grains (Herren and Feller, 1994; Pearson and Rengel, 1995).

Other cations also affect Cd re-transferring into grains. A reduction in Cd transferring into grains was found when Zn was added to nutrient solution. It was presumed that Cd loading and transport in xylem could be inhibited by Zn (Grant et al., 1998).

6.2. Cadmium Accumulation and Distribution in Plants

Both genetic and environmental factors affect Cd uptake and accumulation in plants. It is well documented that there is a great difference among plant species or cultivars within a crop in Cd uptake and accumulation (Florijn et al., 1991; Florijn and Van Beusichem, 1993; Zeng and Hemmasi, 1992; Zhang et al., 2000; Wu and Zhang 2002a, b). The environmental factors include agronomic practices, such as fertilizer application and crop rotation, which affect soil pH and salinity, organic matter content, while the latter may change Cd availability in soil. Arthur et al. (2000) classified the crops into three groups in term of their Cd content: low: soybean and pea; Moderate: rice, wheat, barley, maize, sorghum, onion, leak cucumber, pumpkin carrot and parsley; High: rape, radish, turnip swiss chard, sugar beet, tomato, pepper, eggplant and lettuce. On the other hand, Baker (1981) defined three types of plants based on the difference in absorption, transport and accumulation of heavy metals: Accumulator (hyper-accumulated type), Indicator (sensitive type) and Excluder. *Thlaspi arcense* L. was considered as a Cd hyper-accumulating plant, and its Cd concentration in shoot could be up to 1800 µg g^{-1} DW (Baker 1981; Brown et al., 1994, 1995).

In general, Cd concentration in plant tissues is in order as follows: root > stem > leaf > grain. Cd concentration in grains is much lower than that in roots (Wang and Wu 1995, 1997). Cadmium enters cortex cells of root and then accumulated in the form of macromolecular complexes with protein, polypeptides, amylose, ribose and nucleic acid or insoluble organic macromolecules. But in tobacco and some vegetables, such as carrot, Cd accumulation in leaves was significantly higher than that in roots.

Subcellular distribution of Cd is much concerned as it is possibly associated with Cd tolerance in plants. Cell vacuole is considered to be the organ with greatest amount of Cd accumulation (Heuillet et al., 1986; Krotz et al., 1989; Vögeli-Lange and Wanger, 1990). Rauser and Ackerley (1987) reported that Cd was associated with electron dense granules in the cytoplasm, vacuoles and nuclei of *Agrostis* and maize roots. Khan et al. (1984) reported the presence of Cd only in the walls of sifter element and middle lamella separating the endodermis from the pericycle in maize roots. Lozano-Rodríguez et al. (1997) reported that the total Cd concentration was similar in shoot and root tissues of maize and pea, when exposed to Cd stress, but pea plants showed more severe toxic symptoms. High Cd levels were found in the cell-wall fraction (FI) and in fraction IV (soluble) of maize plants, whereas Cd-treated pea plants accumulated more Cd in the soluble fraction. Wu et al. (2005) found that increased Cd level in the medium caused a significant increase of Cd concentration in all fractions of roots/shoots with most accumulation in FI (cell wall) and FIV (soluble). There was a distinct difference among genotypes in Cd concentration in subcellular and chemical forms. The Cd-sensitive genotype had higher Cd concentration in chloroplast-shoot/trophoplast-root (FII), membrane and organelle (FIII) and in inorganic and water-soluble Cd of roots, while lower in FI, FIV and pectates/protein integrated Cd.

7. THE EFFECT OF Cd TOXICITY
7.1. The Effect of Cd Toxicity on the Growth and Development of Plants

Plant morphogenesis and development are integral physiological processes controlled by a large number of genes, the expression of each gene is dependent on the environment where plants grow. Accordingly, both morphogenesis and development are especially susceptible to all kinds of stresses, including those caused by heavy metals. The root system of plants acts usually as the first barrier or acceptor to heavy metals in the soil. In spite of the different mobility of metals in plants, the root system accumulates them to a significantly higher extent than do the aboveground organs and as a result it is one of the main targets of their toxic effect (Erns et al., 1992; Meharg, 1994). Plants will show visual Cd toxicity symptoms when they are exposed to Cd stress. The most common symptoms are characterized by brown and short roots, chlorosis, fewer tillers, and reduced biomass and senescence (Wu and Zhang, 2002b; Wu et al., 2003a, b). However, it was found that growth would be slightly enhanced when the plants are in the solution with very low Cd level. Greger and Lindberg (1986) reported that sugar beet biomass treated with 50 nM or lower Cd concentration was increased by more than 30%. Wu et al. (2003b) observed that barley(*Hordeum vulgare* L.) plants exposed to 0.1 µM Cd showed a slight increase, compared with the control, in growth parameters and biomass accumulation during ontogenesis, especially in tolerant cultivars. Exposure to 1 µM Cd induced a significant decrease and the deleterious effect became diminished with extended exposure of time, which might be attributed to an adaptation to Cd toxicity occurring during ontogenesis. Increasing Cd concentration in the medium to 5 µM caused a sharp decline ($p<0.05$) in all measurements and the deleterious effect of Cd became more obvious with extended exposure of time. Cotton pot experiment showed that root length, plant height, fruiting branch number, and chlorophyll content decreased with increasing Cd concentration (Bachir et al., 2004).

7.2. Physiological and Biochemical Effects of Cd on Plants

It has been demonstrated that Cd toxicity would reduce root absorption of water and mineral nutrition (Kahle, 1993; Bernal and McGrath, 1994), inhibit nitrogen fixation (Wlckllff et al., 1980), and increase the permeability of cell membrane, and exosmosis of soluble substance in cells, and even disturb distribution and metabolism of enzyme system in cells (Sun et al, 1985; Li et al., 1992).

A variety of abiotic stresses including heavy metal cause molecular damage to plant cells either directly or indirectly through the formation of active oxygen species (AOS) (Greger et al., 1995; Lin and Kao, 2000). Protonation of O^{2*}- can produce hydroperoxyl radical ($*OH$, H_2O_2), which can convert fatty acids to toxic lipid peroxides, destroying biological membranes. Measurement of malondialdehyde (MDA) level is routinely used as an index of lipid peroxidation under stress condition. Lozano-Rodriguez et al. (1997), Chaoui et al. (1997) and Wu et al. (2003c) reported excessive accumulation of MDA when plants were subjected to Cd stress. This suggests that Cd stress indirectly leads to production of superoxide radicals, resulting in increased lipid peroxidation, and oxidative stress.

The activities of superoxide dismutase (SOD), peroxidase (POD) and catalase (CAT) are influenced when plants are exposed to Cd stress. Sandalio et al. (2001) reported that Cd toxicity reduced SOD activity in pea plants. Yang et al. (1995) observed increasing activities of SOD, POD and CAT in tolerant cultivars of wheat, maize, cucumber and soybean exposed to Cd toxicity. The investigation done by Wu et al. (2003c) showed that Cd stress induced a concentration- and genotype-dependent oxidative stress response in barley leaves. A highly significant increase in MDA content, and a stimulation of SOD, POD and CAT activities were recorded in plants subjected to 1 and 5 µM Cd. The effects increased with both Cd concentration in the medium and with the time of exposure in 5 µM Cd treatment. There was a highly significant difference in the alterations of all these parameters among barley genotypes. A sensitive genotype accumulated much more MDA when exposed to 5 µM Cd than relatively tolerant genotypes. By contrast, the tolerant genotypes maintained higher SOD and POD activities than the sensitive one over the whole duration of Cd exposure. These results suggest that the activities of SOD, POD and CAT play a certain role on Cd tolerance in the plants. Meanwhile, the changing tendency of enzyme activity has a significant relationship with Cd level.

Cd toxicity may disturb many metabolic procedures due to its influence on relevant enzymes. The activity of nitrogenase was only 29% of the control when soybean was cultivated in the solution of 18 µM Cd (Huang and Bazzaz, 1974). Cd could inhibit amylase (Hong and Pu, 1991), DNAase and RNAase (Duan et al., 1992 1992), nitrate reductase (Yang et al., 1995, Burzynski, 1990), polyphenoloxidase (Yang et al., 1995) and lactate dehydrogenase (Duan and Wang, 1998). It significant decrease in DNA and RNA, and the activities of DNAase, RNAase was observed in plants subjected to 0.5 mg L^{-1} or higher concentration of Cd (Duan et al., 1992).

Cadmium toxicity inhibited chlorophyll biosynthesis, resulting in decreased chlorophyll content (Stobart et al., 1985; Padmaja et al., 1990). The reduction of biomass by Cd toxicity was considered as a direct consequence of inhibited chlorophyll synthesis (Padmaja et al., 1990) and photosynthesis (Bazzaz et al., 1974, 1975; Baszynski et al., 1980). Bazzaz et al. (1974, 1975) and Baszynski et al. (1980) observed significant reduction in photosynthetic rate in maize, sunflower and tomato, and as a result led to a reduction of biomass accumulation under Cd stress. Larsson et al. (1998) found that seedlings of *Brassica napus* exposed to Cd under high light intensity showed a significant decrease in chlorophyll content and photosynthetic rate. It was reported that light and dark reactions of photosynthesis were inhibited under Cd stress at different target sites (Krupa and Baszynski, 1995), photosystem II being particularly affected (van Assche and Clijsters, 1985; Krupa and Baszynski, 1995). Cadmium is thought to affect PSII on both oxidizing (donor) and reducing (acceptor) sides. Moreover, PSII reaction centers and PSII electron transport are affected by interaction with Cd, metal impairing enzyme activity and/or protein structure (van Assche and Clijsters, 1985). In contrast, Haag-Kerwer et al. (1999) reported that photosynthesis in *Brassica juncea* was not affected by exposure to 25 µM $CdNO_3$, while transpiration showed a significant decline, in particular, under lower light conditions (<300 µmol photons $m^{-2} s^{-1}$). Cd inhibits synthesis of protochlorophyllide reductase resulting in low chlorophyll synthesis (Truong and Claridge, 1996). The studies done by Wu et al. (2003a) on barley demonstrated that negative effect of Cd on net photosynthesis (Pn) which was attributed to a complex of physiological disturbances, mainly inhibition of the chlorophyll biosynthesis, reduction in Fv/Fm (the ratio of variable fluorescence to maximal fluorescence) and ΦPSII (actual photochemical efficiency of PSII in the light), disordered stomata behavior and other photosynthetic processes in barley and tomato plants. Barley plants exposed to 0.1 µM Cd showed a slight increase (p>0.05) in chlorophyll contents, SPAD values (chlorophyll meter readings), Pn and biomass relative to control, whereas exposure to 1 µM Cd induced a slight decrease. Increasing Cd concentration in medium to 5 µM caused a sharp decline (p<0.05) in these measurements and the deleterious effect of Cd became more notable with extended exposure of time. Chlorophyll a showed more sensitivity to Cd toxicity than chlorophyll b. There was a highly significant difference in the reduction of these parameters among four genotypes. The relative tolerant genotype Zhenong 1 was the least affected, while sensitive genotype Wumaoliuling was the most affected. Cd exposure caused a decrease in Fv/Fm ratio and ΦPSII.

Cd toxicity could disturb nutrient metabolism in plants. A hydroponic experiment conducted by Wu *et al* (2003d) showed that Cd addition to the medium not only significantly decreased Zn concentrations in all plant tissues, but also inhibited its translocation from roots to shoots, leading to higher root/shoot Zn ratio in Cd-treated plants. Cd addition also reduced Mn and Cu concentrations in kernels, roots and shoots and of Fe concentration in kernels and shoots. Significantly negative correlation was found between Zn, Cu or Mn concentration and Cd concentration in different plant organs, suggesting the possibility of alleviating Cd accumulation in barley plants through application of these microelements on the Cd-contaminated soils. Wu and Zhang (2002c) observed that the physiological changes caused by Cd toxicity could be alleviated to different extent by the application of 300 µmol L^{-1} Zn in Cd-stressed plants, and the most pronounced effects of adding Zn in Cd-stressed medium were expressed in the decreased MAD and increased biomass accumulation. Hernandez et al. (1998) reported that Mn absorption was inhibited by Cd stress in pea seedlings. Commonly, Cd stress inhibits absorption of N, K, Mg and Mn, but its effect on absorption of P, S, Ca and Zn and Fe seems quite complex, and the studies reported to date provided contradicting results. For example, the effect of Cd on the uptake of Zn can be synergistic (Smith and Brennan, 1983) or antagonistic (Root et al., 1975; Wu et al., 2004). Cataldo et al. (1983) evaluated Cd uptake at concentrations of 0.0025 to 5.0 µM and concluded that Cd transport across root cell membranes to symplast and from root to shoot was inhibited by competing nutrient cations. In these studies, Zn acted as a competitive inhibitor of Cd. Jalil et al. (1994) found that Cd stress decreased concentration of K, Zn, and Mn in roots and shoots of durum wheat, while Fe and Cu concentrations in shoots and roots were not affected. Yang et al. (1998) reported that addition of Cd to growth medium decreased growth rate, dry matter yield as well as accumulation of Fe, Mn, Cu, Ca and Mg in cabbage, ryegrass, maize and white clover, but increased P accumulation. Accordingly, excessive Cd accumulation would affect the rate of uptake and distribution of certain essential nutrients in plants, and consequently would be responsible for mineral deficiencies/imbalance and depression of the plant growth. On the other hand, it may indicate that Zn uptake inhibit Cd uptake and distribution in plants, and be assumed that Zn application to Cd contaminated soil would avert Cd toxicity. Hassen et al. (2005) found that addition of Zn to the medium solution had distinct effect of reducing Cd toxicity of two rice varieties with different Cd tolerance, which was reflected in a significant increase in plant height, biomass, chlorophyll concentration and photosynthetic rate, and a marked decrease in MDA concentration and the activities of antioxidative enzymes. However, there were also contradicting reports on the relationship between Cd and Zn, e.g. Smith and Brennan (1983) reported that uptake of Zn and Cd was synergistic. Wallace et al. (1977) reported that in bush bean, Fe concentration decreased at low Cd level, but increased at high Cd level. In contrast, Zn concentration in *Brassica chinensis* increased at low Cd level but decreased at higher Cd level (Wong et al., 1984). These conflicting results were presumably due to the differences in experimental methods, species, and conditions such as concentration in medium, growth period and temperature.

8. THE MECHANISMS OF Cd TOLERANCE IN PLANTS

Possible mechanisms that govern heavy metal tolerance in plant cells are: (1) binding metal to cell wall (2) reducing metal transport across cell membrane (3) active efflux (4) compartmentalization and (5) chelation (phytochelatin). Phytochelatin (PC), however, is considered as the most important in Cd tolerance (Zenk, 1996; Zhu et al., 1999a). Here the effect of root excretion and PCs on Cd resistance in plants is briefly discussed.

8.1. The Relationship Between Root Excretion and Plant Tolerance to Cd Toxicity

One of the main Cd tolerance mechanisms is involved in depressing Cd availability in soils, thus reducing the amount of Cd uptake. Rhizosphere is an important environmental interface connecting plant roots and soil. Roots excrete some organic substances to rhizosphere during its growth, and the rhizosphere controls the entrance of nutrients, water and other chemicals, beneficial or harmful to plants. Therefore, it is understandable that study on rhizosphere has been one of the important issues in pollution ecology.

Tu et al. (1989) observed that Cd has an inhibitory action on H+ excretion, and proton pump activity decreased by 60% when plants grew in the solution with 50 µmol L^{-1} Cd. Proton pump is important for providing energy in the uptake and movement of ions through plasma membrane. Solubility of most heavy metals is influenced by soil pH. In general, with the decrease in pH, solubility and activity of heavy metals increase. Xian (1989) observed a marked increase of exchangeable Cd in soil, when soil pH decreased from 7.0 to 4.5. Meanwhile, it is well proved that Cd uptake and accumulation are closely related to the amount of exchangeable Cd rather than the total Cd (Wu and Zhang, 2003a; Cheng et al., 2004). Helmisaari et al. (1999) reported that application of lime into pine rhizosphere could significantly reduce solubility of Cd ion in soil. It has been confirmed that liming in acid soil may reduce Cd content in wheat grains wheat by 50%, and Cd absorption in cabbage by 43% (Yang and Yang, 1996). Hence, it may be feasible to avert Cd toxicity in Cd-contaminated acid soil via adjusting soil pH.

Chemical reactions of heavy metals, especially redox reactions may be strongly affected by rhizosphere Eh. Under low Eh in soil, H_2S is produced and then Cd combines with S^{2-} forming insoluble CdS, thus Cd is not easily absorbed by crops. Meanwhile, the presence of plentiful Fe^{3+} and Mn^{4+} is competitive with Cd^{2+} and thereby reducing plant absorption. For instance, Cd absorption and accumulation was significantly reduced with diminishing Eh in reductive conditions formed by flooding rice fields. Application of $(NH_4)_2SO_4$ and sulphate fertilizers would induce vulcanization in soil to produce much H_2S, which form CdS with Cd.

The composition and quantity of root secretion may affect the present form of heavy metals. Liu et al. (1998) reported that organic acid with low molecular weight secreted by roots played an important role in solubility and availability of heavy metals, and Cd^{2+} availability would be reduced if Cd^{2+} turned into Cd-chelatin complex with root secretion. The roots of some plants, such as wheat and buckwheat excrete organic acid (e.g. oxalic acid, malic acid and citric acid), which can chelate with Cd^{2+} to prevent its entrance into roots. Tang (1998) observed that amino acid could also reduce the toxicity of metal ions. Moreover, the combination of organic phosphate acids and Cd ions would produce complex unavailable to plants.

Microorganisms can not degrade Cd but can affect its present form; dispersion and accumulation. Microorganisms participate in various chemical reactions of heavy metals by (1) changing soil pH (2) producing H_2S (3) producing organic substances. In addition, various microorganism metabolites may precipitate and chelate heavy metal ions. Chanmugathas and Bollag (1987) proved sequestration and activation effects of microorganism on soil Cd availability.

The symbiosis of Mycorrhizal fungus and plant roots produces mycorrhiza, which can effectively decrease metal toxicity. Meanwhile, some cell walls in microorganism have the capability of combining with pollutants. The capability may be attributed to chemical components and structure of its cell walls. For example, bacillus can sequester metal ions because of a thick layer of netty peptidoglycan structure in cell walls. In addition, on the surface there are teichoic and teichuronic acids linked to netty peptidoglycan, and it is due to carboxyl group of teichoic acid, the cell walls carry negative charges and have the function of ion sequestration. Li et al. (1998) demonstrated the capacity of yeasts in absorbing Cu^{2+}, Cd^{2+} and Ni^{2+}.

Under the stress of environmental contamination, some microorganisms could excrete organic substances with the ability of chelating and decomposing pollutants, and these organic substances assist with mucous plant root excretion (e.g. amylose). These can form pectin layer covering over root surface and the expansion of the layer would sequester metal ions outside root (Zhang and Chen, 1995). Mucus excreted by epiphyte cells and polyphosphate and organic acid in epiphyte tissues can chelate heavy metal ions including Cd, thus reduce the uptake and transportation of heavy metal ions to shoots.

Rhizosphere environment may be adjusted artificially by (1) applying different fertilizers so as to change rhizosphere pH (2) water management to change Eh status in soil (3) applying organic fertilizer and (4) using fungicides to regulate and control microbial community in rhizosphere. Increasing soil pH through applying physiologically alkaloid fertilizer can reduce Cd bioavailability, thereby decreasing Cd absorption by plants. Kuo and McNeal (1984) reported that liming had a remarkable effect in reducing Cd uptake in the Cd-contaminated soil. Cd content in grains of rice, wheat and maize reduced by 55.9%, 34.6% and 21.0%, respectively when Cd-contaminated soils were amended by liming together with application of Ca, Mg and P fertilizers (Wang and Wu, 1995). The change of cropping system from upland crops to paddy fields will reduce soil Eh and Cd availability. Organic fertilizers not only improve soil fertility, but also provide bioactive substance and energy to soil microorganisms. Accordingly infect soil Cd availability indirectly to decrease Cd biotoxicity. Some organic fertilizers such as waste compost, containing a certain amount of heavy metals may cause contamination in soil. In addition, heavy metal toxicity may be alleviated by forming insoluble compounds. For example, the application of phosphate can increase soil P content and precipitate Cd as $Cd_3(PO_4)_2$. In conclusion, it offers certain realistic significance on safety crop production by artificially regulating and controlling rhizosphere environment to depress Cd absorption and accumulation in plants.

8.2 Phytochelatin and its Function in Cd Tolerance of Higher Plants

Under heavy metal stress, tolerant species and genotypes in plant kingdom could reduce the heavy metal actively to alleviate or eliminate its toxicity through regulating the physiological and biochemical metabolism (Punz and Sieghardt, 1993). Phytochelatin (PC) has recently been considered as a complex closely correlated with the mechanism of heavy metal tolerance in plant (Zenk, 1996; Zhu et al., 1999a).

Grill et al. (1985) first separated Cd-chelatin by gel filtration from *Rauwolfia serpentine*. Later, it was also detected in fungi exposed to Cd (Kondo et al., 1985; Zeng and Hemmasi, 1992). Up to date, more than 200 species of plants have been demonstrated to have their capacity of detoxifying the toxicity of heavy metals by chelating them to some peptides with the different chain length, named as phytochelatin (PC). PC is a kind of thiolate peptides consisting of cysteine, glutamic acid and glycin, with molecular weight of 1-4 kD (Zenk, 1996). The chemical structure of PC suggests close similarity to glutathione (GSH). As in the case of tri-peptide, PC is not a primary gene product that is synthesized on ribosome, because γ-glutamyl linkage is not produced during translation. Grill et al. (1989) found that PC enzyme-catalyzed synthesis process existed in many plant species and GSH was a substrate for PC synthesis. Under heavy metal stress, the enzyme PC synthetase catalyzes the primary reaction: γ-Glu-Cys-Gly + (γ-Glu-Cys)n-Gly =(γ-Glu-Cys)n+1-Gly + Gly. This reaction is strictly dependent on the presence of heavy metal ions. The formation of stable metal-PC complex certainly modifies the reaction and thus prevents the release of toxic metals. Meanwhile, the enzyme-catalyzed synthesis reaction is regulated by reaction products, and GSH, PC and metal-PC.

The vacuole is most likely the ultimate storage compartment for those heavy metal ions that enters cytosol of a plant cell. Heavy metal ions such as Cd^{2+} enters plant cell via the permeable cell wall and cell membrane, and immediately activates the latent constitutive PC synthase that synthesizes at the expense of GSH (Kneer and Zenk, 1992). The inactive toxic metal ions of metal - PC chelatins are subsequently transported from cytosol to vacuole before they cause damage to the enzymes, and inherently stored in vacuole. Thus heavy metal concentration in cytosol is reduced (Salt and Wagner, 1993). Under acidic pH condition in vacuole, it is likely that metals are liberated from PC complex and the metal-free PC molecules are subsequently degraded into amino acids, then they were transferred into cytoplasm for new PC synthesis, and free-metal ions in vacuole were simultaneously combined with organic acid and accumulated.

The synthesis of PC is considered as a kind of "stress adaptive reaction" in plants, and played a major role in heavy metal detoxification (Grill et al., 1985; Steffens, 1990). Grill et al. (1988) found that there was a large amount of PC in plant roots grown in heavy metal polluted soil, but little PC was detected in the plants grown in unpolluted soil. Kneer and Zenk (1992) reported Cd^{2+} -PC content in Cd tolerant species was 10 to 1000 times higher than the control. The most convincing results demonstrating the importance of PC for Cd tolerance in plants were published by Howden et al. (1995a, b). These authors succeeded in isolating Cd-sensitive *Arabidopsis thaliana* mutant (Cad1-1 and Cad2-1). The mutant Cad1-1 was sensitive to Cd ions and deficient in its ability to form Cd-PC complexes while GSH synthesis proceeded at the same rate as did in the wild type. Cad2-1 was a GSH-deficient mutant. Under Cd stress, Cad1-1 and Cad2-1 were sensitive to Cd toxicity, due to their inability to synthesize PC. The enzyme assay for PC synthesis showed deficient in this enzyme in Cad1-1 as compared with Cd tolerant wild type of *Arabidopsis thaliana*. Further study showed that PC synthesis in *Arabidopsis thaliana* was based on GSH, the reaction catalyzed by PC synthetase. However, the deficiency of PC synthetase in Cad1-1 and GSH in Cad2-1 resulted in their failure of PC synthesis. Cobbett et al. (1998) showed that deficiency of γ-glutamylcysteine synthetase (GCS) in Cad2-1 caused accumulation of cysteine, suggesting that PC synthesis would be blocked without the supply of GSH.

A gene coding PC synthetase has been isolated and cloned from plants and yeast (Clemens et al., 1999; Ha et al., 1999; Vatamaniuk et al., 1999). In addition, Clemens et al. (1999) identified and cloned the TaPCS1 gene in *Triticum aestivum* and phytochelatin synthesis from wheat cDNA. The expression of the gene in *Saccharomyces cerevisae* showed a high tolerance to Cd toxicity. The plants with homologous genes AtPCS1 and SpPCS, identified and cloned from *Arabidopsis thaliana* and *Schizosaccharomyces pombe* also showed high tolerance to Cd stress. Moreover, it was demonstrated that addition of an inhibitor limiting GSH synthesis would lead to loss of Cd tolerance. It is clear that PCS gene encodes PC synthetase which positively affects the heavy metal detoxification in eukaryotes. Meanwhile, Arisi et al. (1997) and Noctor et al. (1998) found that high expression of γ-ECS gene (Escherichia coli gshI), which encodes PC synthetase, increased GSH content significantly in the leaves of white poplar. Zhu et al. (1999a, b) reported that GSH, PC, free γ-Glu-Cys (NPT) contents in γ-ECS transgenic seedlings of mustard were significantly higher than those in wild sensitive type under Cd stress. Meanwhile, more Cd was accumulated in transgenic seedlings than in wild sensitive type. It indicated that high expression of γ-ECS gene under Cd stress stimulated synthesis of GSH and PC, which in turn enhanced Cd tolerance.

Reference

Arisi ACM, Noctor G., Foyer CH, Jouanin L (1997) Modification of thiol contents in poplars overexpressing enzymes involved in glutathione synthesis. Planta 203: 362-372.

Arthur E, Crews H, Morgan C (2000) Optimizing plant genetic strategies for minimizing environmental contamination in the food chain. Int J Phytorem 2: 1-21.

Bachir DML, Wu FB, Zhang GP, Wu HX (2004) Genotypic difference in effect of cadmium on the development and mineral concentrations of cotton. Comm Soil Sci Plant Anal 35: 285–299

Baker AJM (1981) Accumulators and excluders- strategies in the response of plants to heavy metals. J Plant Nutr 3: 643-654.

Baker AJM, Reeves RD, Hajar ASM (1994) Heavy metal accumulation and tolerance in British population of the metallophyte *Thlaspi caerulescens* J. & C. Presl (Brassicacceae). New Phytol 127: 61-67.

Baszynski T, Wajda L, Krol M, Wolinska D, Krupa Z, Tuken-dorf A (1980) Photosynthetic activities of cadmium-treated tomato plants. Physiol Plant 48: 365-370.

Bazzaz FA, Rolfe GL, Carlson RW (1974) Effect of Cd on photosynthesis and trnspiration in excised leaves of corn and sunflower. Plant Physiol 32: 373-376.

Bazzaz FA, Carlson RW, Rolfe GL (1975) Inhibition of corn and sunflower photosynthesis by lead. Physiol Plant 34: 326-329.

Bernal MP, McGrath SP (1994) Effects of pH and heavy metal concentrations in solution culture on the proton release, growth and elemental composition. Plant Soil 166: 83-92.

Brown SL, Chaney RL, Angle JS, Baker AJM (1994) Phytoremediation potential of *Thlaspi caerulescens* and bladder campion for zinc- and cadmium-contaminated soil. J Environ Qual 23: 1151-1157.

Brown SL, Chaney RL, Angle JS, Baker AJM (1995) Zinc and cadmium uptake by hyperaccumulator Thlaspi caerulescens and metal tolerant Silene vulgaris grown on sludge-amended soil. Environ Sci Technol 29: 1581-1585.

Burzynski M (1990) Activity of some enzymes involved in NO_3^- assimilation in cucumber seedlings treated with lead or cadmium. Acta Physiol Plant 12: 105-116.

Cabrera D, Young SD, Rowell DL (1988) The toxicity of cadmium to barley plant as affected by complex formations with humic acid. Plant Soil 105: 195-204.

Cataldo DA, Garland TR, Wildung RE (1983) Cadmium uptake kinetics in intact soybean plants. Plant Physiol 73: 844-848.

Cataldo DA, Garland TR, Wildung RE (1981) Cadmium distribution and chemical fate in soybean plant. Plant Physiol 68: 835-839.

Chanmugathas P, Bollag JM (1987) Microbial mobilization of cadmium in soil under aerobic and anaerobic conditions. J Environ Qual 16: 161-167.

Chaoui A, Mazhoudi S, Ghorbal MH, El Ferjani E (1997) Cadmium and zinc induction of lipid peroxidation and effects on antioxidant enzyme activities in bean (*Phaseolus vulgaris* L). Plant Sci 127: 139-147.

Chino M, Baba A (1981) The effect of some environmental factors on the partitioning of zinc and cadmium between roots and top of rice plants. J Plant Nutr 3: 203-214.

Clemens S, Kim EJ, Neumann D, Schroeder J (1999) Tolerance to toxic metals by a gene family of phytochelatin synthases from plants and yeast. EMBO J 18: 3325-3333.

Cobbett CS, May MJ, Howden R, Rolls B (1998) The glutathione-deficient, cadmium-sensitive mutant, cad2-1, of Arabidopsis thaliana is deficient in γ-glutamylcysteine synthetase. Plant J 16: 73-78.

Davis RD (1984) Cadmium- a complex environmental problem: Cadmium in sludge used as fertilizer. Experientia 40: 117-126.

Duan CQ, Wang HX, Qu ZX (1992) Studies on the effects of heavy metals on the contents of nucleic acids and activities of nucleases in root tips of *Vicia faba* L. Environ Sci 13: 31-35.

Duan CQ, Wang HX (1998) The effects of lead, cadmium and mercury ions on LDH in *Vicia faba* L. Acta Ecol Sinica 18: 413-417.
Erns WHO, Verkleij JAC, Schat H (1992) Metal tolerance in plants. Acta Bot Neerl 41: 229-248.
Florijn PJ, Nelemans JA, VAN Beusichem ML (2000) Cadmium uptake by lettuce varieties. Neth J Agri Sci 39: 103-114.
Grant C A, Buckley WT, Bailey LD, Selles F (1998) Cadmium accumulation in crops. Can J Plant Sci 78: 1-17.
Greger ML, Kautskey T, Sandberg A (1995) A tentative model of Cd uptake in Potamogeton petinatus in relation to salinity. Environ Exp Bot 35: 215–225.
Grill E, Winnacker EL, Zenk MH (1985) Phytochelatins: The principal heavy-metal complexing peptides of higher plants. Science 230: 674-676.
Grill E, Thumann J, Winnacker E-L, Zenk MH (1988) Induction of heavy metal binding phytochelatins by inoculation of cell cultures in standard media. Plant Cell Rep 7: 375-378.
Grill E, Löffler S, Winnacker EL, Zenk MH (1989) Phytochelatins, the heavy-metal-binding peptides of plants are synthesized from glutathione by a specific ☐-glutamylcysteine dipeptidyl transpeptidase (phytochelatin synthase). Proc Natl Acad Sci USA 86: 6838-6842.
Guo DF (1994) Environmental sources of Pb and Cd and their toxicity to man and animals. Adv Environ Sci 2: 71-76.
Ha SB, Simith AP, Howden R (1999) Phytochelatin synthase genes from Arabidopsis and the yeast *Schizosaccharomyces pombe*. Plant Cell 11: 1153-1163.
Haag-Kerwer A, Schafer HJ, Heiss S, Walter C, Rausch T (1999) Cadmium exposure in *Brassica juncea* causes a decline in transpiration rate and leaf expansion without effect on photosynthesis. J Exp Bot 50: 1827-1835
Hagemeyer J, Waisel Y (1989) Uptake of Cd2+ and Fe2+ by excised roots of *Tamariz aphylla*. Physoil Plant 77: 247-253.
Hardiman RT, Jacoby B (1984) Absorption and translocation of Cd in bush beans. Physiol Plant 61: 470-474.
Hart JJ, Welch RM, Noevell WA, Sullivan LA, Kochian LV (1998) Characterization of cadmium biding, uptake, and translocation in intact seedlings of bread and durum wheat cultivars. Plant Physiol 116: 1413-1420.
Hassen M J, Zhang GP, Wu FB (2005) Zinc alleviates growth inhibition and oxidative stress caused by cadmium toxicity in rice. J Plant Nutr Soil Sci: In press
Helmisaari HS, Makkonen K, Olsson M, Viksna A, Mälkönen E (1999) Fine-root growth, mortality and heavy metal concentrations in limed and fertilized Pinus silvestris (L.) stands in the vicinity of a Cu-Ni smelter in SW Finland. Plant Soil 209: 193-200.
Hernandez LE, Lozano-Rodriguez E, Garate A, Carpena-Ruiz R (1998) Influence of cadmium on the uptake, tissue accumulation and subcellular distribution of manganese in pea seedlings. Plant Sci 132: 139-151.
Herren T, Feller U (1994) Transfer of zinc from xylem to phloem in the peduncle of wheat. J Plant Nutr 17: 1587-1598.
Heuillet E, Moreau A, Halpern S, Jeanne N, Puiseax-Dao S (1986) Binding to a thiol-molecule in vacuoles of Dunaliella bioculata contaminated with $CdCl_2$, electron probe microanalysis. Biol Cell 58: 79-86
Hong RY, Pu CH (1991) Effects of cadmium on the growth and physiological and biochemical relations of wheat seedlings. Acta Agric Biogeali-Sinica 6. 70-75.
Howden R, Andersen CR, Goldsbrough PB, Cobbett CS (1995a) A cadmium-sensitive, glutathione-deficient mutant of *Arabidopsis thaliana*. Plant Physiol 107: 1067-1073.
Howden R, Goldsbrough PB, Andersen CR, Cobbett CS (1995b) Cadmium-sensitive, cad1 mutants of *Arabidopsis thaliana* are phytochelatin deficient. Plant Physiol 107: 1059-1066.
Huang CY, Bazzaz FA (1974) The inhibition of soybean metabolism by cadmium and lead. Plant Physiol 54: 122-124.
Ivonin VM, Shumakova GE (1991) Effect of industrial pollution on the condition of roadside shelterbelts. Izvest Vysshikh Uch Zav Les Z 6: 12-17.
Jalil A, Selles F, Clarke JM (1994) Effect of cadmium on growth and the uptake of cadmium and other elements by durum wheat. J Plant Nutr 17: 1839-1858.
Kahle H (1993) Response of roots of trees to heavy metals. Environ Exp Bot 33: 99-119.
Khan DH, Duckett JG, Frankland B, Kirkham JB (1984) An X-ray microanalytical study of the distribution of cadmium in roots of *Zea mays* L. J Plant Physiol 115: 19-28.

Kneer R, Zenk MH (1992) Phytochelatins protect plant enzymes from heavy metal poisoning. Phytochemistry 31: 2663-2667.

Kondo N, Isobe M, Imai K, Goto T (1985) Synthesis of metallothionein-like peptides Cadystin A and B occurring in a fission yeast and their isomers. Agri Biol Chem 49: 71-83.

Krotz RM, Evangelou BP, Wanger GJ (1989) Relationships between cadmium, zinc, Cd-peptide, and organic acid in tobacco suspension cells. Plant Physiol 91: 780-787

Krupa Z, Baszynski T (1995) Some aspects of heavy metals toxicity towards photosynthetic apparatus-direct and indirect effects on light and dark reactions. Acta Physiol Plant 17: 177-190

Kuo S, McNeal BL (1984) Effects of pH and phosphate on cadmium sorption by hydrous ferrous oxide. Soil Sci Soc Am J 48: 1040-1044.

Larsson EH, Bornman JF, Asp H (1998) Influence of UV-B radiation and Cd^{2+} on chlorophll flurescence, growth and nutrient content in *Brassica napus*. J Exp Bot 49: 1031-1039

Li MM, Jiang H, Hou WQ (1998) Study on heavy metal biosorption of yeasts. Mycosystema 17: 367-373.

LiY, Wang HX, Wu YS (1992) Effects of cadmium and iron on the some physiological indications in leaves of tobacco. Acta Ecol Sinica 12: 147-154.

Lin CC, Kao CH (2000) Effect of NaCl stress on H_2O_2 metabolism in rice leaves. Plant Growth Regul 30: 151-155.

Liu H, Ling QL, Xiang ZY, Shang KJ (1998) Regulatory action of a calmodulin binding protein on activity of the plasma membrane H^+-ATPase in wheat. Acta Phytophysiol Sinica 24: 91-94.

Lozano-Rodríguez E, Hernandez LE, Bonay P, Carpena-Ruiz RO (1997) Distribution of cadmium in shoot and root tissues of maize and pea plants: Physiological disturbance. J Exp Bot 306: 123-128.

Meharg AA (1994) Integrated tolerance mechanisms: Constitutive and adaptive plant responses to elevated metal concentrations in the environment. Plant Cell Environ 17: 969-993.

Meng FQ, Shi YJ, Wu WL (2000) Developmental of soil environmental quality standards of heavy metal for non-polluted agricultural products in China. Agro Environ Prot 19: 356-359.

Noctor G, Arisi ACM, Jouanin L, Foyer CH (1998) Manipulation of glutathione and amino acid biosynthesis in the chloroplast. Plant Physiol 118: 471-482.

Padmaja K, Prasad DDK, Prasad ARK (1990) Inhibition of chlorophyll synthesis in *Phaseolus vulgaris* seedlings by cadmium acetate. Photosynthetica 24:399-405.

Pawlik B, Skowionski T (1994) Transport and toxicity of cadmium: Its regulation in the cyanobacterium *Synechocystis aquntilis*. Environ Exp Bot 34: 225-233.

Pearson JN, Rengel Z (1995) Uptake and distribution of 65Zn and 54Mn in wheat grown at sufficient and deficient levels of Zn and Mn. II. During grain development. J Exp Bot 46: 841-845.

Pertjt CM, van de Geijn SC (1978) In vivo measurement of cadmium transport and accumulation on the stem of intact tomato plants. I. Long distance transport and local accumulation. Planta 138: 137-143.

Popelka JC, Schubert S, Schulz R, Hansen AP (1996) Cadmium uptake and translocation during reproductive development of peanut (*Arachis hypogaea* L.). Angew Bot 70: 140-143.

Punz WF, Sieghardt H (1993) The response of roots of herbaceous plant species to heavy metals. Environ Exp Bot 33: 85-98.

Rauser WE, Ackerley CA (1987) Localization of cadmium in granules within differentiating and mature roots cells. Can J Bot 65: 643-646.

Root RA, Miller RJ, Koeppe DE (1975) Uptake of cadmium its toxicity, and effect on the iron ratio in hydroponically grown corn. J Environ Qual 4: 473- 476.

Ryan JA, Pahren HR, Lucas JB (1982) Controlling cadmium in the human food chain: A review and rationale based on health effects. Environ Res 18: 251-302.

Salt DE, Prince RC, Pickering IJ, Raskin I (1995) Mechanisms of cadmium mobility and accumulation in Indian mustard. Plant Physiol 109: 1427-1433.

Salt DE, Wagner GJ (1993) Cadmium transport across tonoplast of vesicles from oat roots. J Biol Chem 268: 12297-12302.

Sandalio LM, Dalurzo HC, Gómez M, Romero-Puertas MC, del Río LA (2001) Cadmium-induced changes in the growth and oxidative metabolism of pea plants. J Exp Bot 52: 2115-2126.

Senden MHMN, Wolterbeek H (1990) The effect of citric acid on the transport of cadmium through xylem vessels of tomato stem-leaf systems. Acta Bot Neerl 39: 297-303.

Smith GC, Brennan EG (1983) Cadmium-zinc interactions in tomato plants. Phytopathol 73: 879-882.
Stobart AR, Gritths WT, Ameen-Bukhari J, Shewood RP (1985) The effect of Cd2+ on the biosynthesis of chlorophyll in leaves of barley. Physiol Plant 63: 293-298.
Sun SC, Wang HX, Li QR (1985) Preliminary studies on physiological changes and injury mechanism in aquatic vascular plants treated with cadmium. Acta Phytophysiol Sinica 11: 113-121.
Tang M (1998) Progress in study on VA-mycorrhizal fungi in enhancing plant resistance to saline-alkali and heavy metals. Soils : 251-254.
Truong PNV, Claridge J (1996) Effects of heavy metals toxicities on vetiver growth. Vetiver Newsletter 15: 32-36.
Tu SI, Nungesser E, Brauer D (1989) Characterization of the effects of divalent cations on the coupled activities of the H^+-ATPase in tonoplast vesicles. Plant Physiol 10: 1636-1643.
Vatamaniuk OK, Mari S, Lu YP, Rea PA (1999) AtPCS1, a phytochelatin synthase from *Arobidopsis*: Isolation and in vitro reconstitution. Proc Natl Acad Sci USA 96: 7110-7115.
van Assche F, Clijsters H (1985) Inhibition of photosynthesis by heavy metals. Photosynthesis Res 7: 31-40.
Vögeli-Lange R, Wanger GJ (1990) Subcellular localization of cadmium and cadmium-binding peptides in tobacco leaves. Plant Physiol 92: 1086-1093.
Wallace A, Romney EM, Alexander GV, Souti SM, Patel PM (1977) Some interactions in plants among cadmium, other heavy metals and chelating agents. Agron J 69: 18-20.
Wang X, Wu YY (1995) Effect of modification treatments on behaviour of heavy metals in combined polluted soil. Chinese J Appl Ecol 6: 440-444.
Wang X, Wu YY (1997) Behaviour property of heavy metals in soil-rice system. Chinese J Appl Ecol 16: 10-14.
White MC, Chaney RL, Decker AM (1981) Metal complexation in xylem fluid. III. Electrophoretic evidence. Plant Physiol 67: 311-315.
Wlckllff C, Cvans HJ, Carrter KR (1980) Cadmium effects on the nitrogen fixation system of red alder. J Environ Qual 9: 180-184.
Wong MK, Chuah GK, Koh LL, Ang KP, Hew CS (1984) The uptake of cadmium by *Brassica chinensis* and its effect on plant zinc and iron distribution. Environ Exp Bot 24: 189-195.
Wu FB, Wu HX, Zhang GP, Bachir DML (2004) Difference in growth and yield in response to cadmium toxicity in cotton genotypes. J Plant Nutr Soil Sci 167: 85-90.
Wu FB, Zhang GP (2002a) Genotypic variation in kernel heavy metal concentrations in barley and as affected by soil factors. J Plant Nutr 25: 1163-1173.
Wu FB, Zhang GP (2002b) Genotypic differences in effect of Cd on growth and mineral concentrations in barley seedlings. Bull Environ Contam Toxicol 69: 219-227.
Wu FB, Zhang GP (2002c) Alleviation of cadmium-toxicity by application of zinc and ascorbic acid in barley. J Plant Nutr 25: 2745–2761.
Wu FB, Dong J, Qian QQ, Zhang GP (2005) Subcellular distribution and chemical form of Cd and Cd–Zn interaction in different barley genotypes. Chemosphere. In Press
Wu FB, Zhang GP, Yu JS (2003a) Genotypic differences in effect of Cd on photosynthesis and chlorophyll fluorescence of barley (*Hordeum vulgare* L). Bull Environ Contam Toxicol 71: 1272-1281.
Wu FB, Qian QQ, Zhang GP (2003b) Genotypic differences in effect of cadmium on growth parameters of barley during ontogenesis. Comm Soil Sci Plant Anal 34: 2021–2034.
Wu FB, Zhang GP, Dominy P (2003c) Four barley genotypes respond differently to cadmium: Lipid peroxidation and activities of antioxidant capacity. Environ Exp Bot 50: 67-78.
Wu FB, Zhang GP, Yu JS (2003d) Interaction of cadmium and four microelements for uptake and translocation in different barley genotypes. Comm Soil Sci Plant Anal 34: 2003–2020.
Xia ZL (1991) Soil Environment Capacity and its Information System. Beijing: Meteorological Press. Pp..108-114.
Xian X (1989) Effect of chemical form of cadmium, zinc and lead in polluted soil, and on their uptake by cabbage plants. Plant Soil 113: 257-264.
Xu JL, Yang JR (1995) Heavy Metals in Land Ecosystem. Beijing: China Environ Sci Press 1: 24-36.
Yang JR, He JQ, Jiang WR (1995) Effect of Cd pollution on the physiology and biochemistry of plant. Agro Environ Prot 14: 193-197.

Yang MG, Lin XY, Yang XE (1998) Impact of Cd on growth and nutrient accumulation of different plant species. Chinese J Appl Ecol 9: 89-94.

Yang XE, Yang JM (1996) Transfer of cadmium from agricultural soils to human food chain. Guangdong Trace Elements Sci 3: 1-13.

Zeng W, Hemmasi B (1992) Solution synthesis of phytochelatins isopepetides from the plant kingdom. Liebigs Ann Chem 12: 311-315.

Zenk MH (1996) Heavy metal detoxification in higher plants. Gene 179: 21-30.

Zhang JB, Huang WN (2000) Advances on physiological and ecological effects of cadmium on plants (in Chinese). Acta Ecol Sinica 20: 514-523.

Zhang XY, Chen MZ (1995) Studies on the relation between mycorrhiza of pine and mucilage layer in Rhizosphere with electron microprobe analysis. J Nanjing For Univ 19: 1-5.

Zhu YL, Pilon-Smits EAH, Tarun AS (1999a) Cadmium tolerance and accumulation in Indian mustard is enhanced by overexpressing γ-glutamylcysteine synthetase. Plant Physiol 121: 1169-1177.

Zhu YL, Pilon-Smits EAH, Jouanin L, Terry T (1999b) Overexpression of glutathione synthetase in *Brassica juncea* enhances cadmium accumulation and tolerance. Plant Physiol 119: 73-79.

Cadmium Toxicity and Tolerance in Plants
Editors: Nafees A. Khan and Samiullah
Copyright © 2006, Narosa Publishing House, New Delhi, India

Cadmium Phytoextraction from Contaminated Soils

Andon Vassilev[1], Ivan Yordanov[2] and Jaco Vangronsveld[3]*

[1] Agricultural University of Plovdiv, Dept. Plant Physiology and Biochemistry, 12 Mendeleev St., 4000 Plovdiv, Bulgaria
[2] Acad. M. Popov Institute of Plant Physiology, Acad. G. Bonchev St., Bl.21, 1113, Sofia, Bulgaria
[3] Environmental Biology, Center for Environmental Sciences, Limburgs Universitair Centrum, Universitaire Campus, B-3590 Diepenbeek, Belgium

1. INTRODUCTION

A part of the agricultural soils all over the world are slightly to moderately contaminated by Cd due to large-scale use of super phosphate fertilisers, sewage sludge application as well as atmospheric deposition of smelters dust. Due to high Cd mobility in the soil-plant system it can easily enter into food chain and can create risk for human and environmental health (Grant et al., 1998; Ryan et al., 1982). Since the Ministry of Health and Welfare of Japan has admitted in 1968 that "*Itai-Itai* disease" was caused by primarily by Cd, a significant research attention has been oriented to the risks presented by this heavy metal. Now it is well established that Cd can cause carcinogenic effects in humans (Ariza et al., 1999) and well-known toxicity effects in microorganisms, animals and plants.

Increasing international concern about the risks associated with long-term consumption of food with increased Cd concentrations has led the international food standards organisation, Codex Committee on Food Additives and Contaminants, to propose a 0.1 mg Cd kg^{-1} limit for cereals, pulses and legumes (Harris and Taylor, 2001). Consequently, several strategies have been proposed for the successful management of the Cd-contaminated agricultural soils. One approach, applicable on slightly contaminated soils, is aiming to screen and use low Cd-accumulating genotypes of crops, known to accumulate unacceptable high Cd levels in grain (Archibald et al., 2001). The second approach recommends profitable use of non-food crops, the so-called "adaptable agriculture" (Zheljazkov and Nielsen, 1996; Yankov et al., 2000). The third option is directed towards Cd phytoextraction, representing use of plants for metal removal from contaminated soils (Chaney, 1983; Robinson et al., 1998).

Generally, metal phytoextraction is an environmentally friendly approach, based on the natural or "induced" ability of plants to take up metals from soil and to concentrate them in the harvestable parts (Salt et al., 1998). There is some evidence that metal phytoextraction is a promising technology applicable to slightly or moderately contaminated soils as an alternative to the *ex situ* decontamination techniques, which are very expensive and unacceptable from ecological point of view (McGrath et al., 2001). Presently, metal phytoextraction is still in its infancy stage. A significant multidisciplinary research effort is now underway to contribute to its further development.

---

[*] Corresponding Author (a_vasilev2001@yahoo.com)

Agricultural soils may be contaminated with just Cd (by phosphate fertilisers) or a mixture of heavy metals, where in the most cases Cd is the first metal of concern. There is an expectation that Cd phytoextraction from agricultural soils could be easily implemented due to both lower Cd contamination and higher Cd mobility in the soil-plant system as compared to the other target metals, for example, to Zn (Robinson et al., 1998; Hammer and Keller, 2003). On the other hand, the significant public attention on Cd-induced health risk may contribute to the faster development of this approach.

Generally, the metal phytoextraction protocol consists of the following elements: (1) plant cultivation on the contaminated site, (2) removal of harvested metal-enriched biomass, (3) post-harvest treatments to reduce biomass volume, (4) biomass disposal as a hazardous waste, and (5) eventual recuperation of metals from the metal-enriched biomass (Blaylock and Huang, 2000). The utilization of metal-rich biomass for generation of electricity is a more recently suggested option that could help make phytoextraction more cost effective (Vassilev et al., 2004).

Two groups of plant species are considered for metal phytoextraction: (1) hyperaccumulator species, able to accumulate and tolerate extraordinary metal levels and (2) high biomass producing species (referred in the text as biomass plants) compensating lower metal accumulation by high biomass yields. An opinion exists that phytoextraction will be more economically feasible if, in addition to metal removal, plants produce biomass with an added economical value. For example, the biomass of fibres, oil or fragrance producing crops could be used to recover these valuable products (Schwitzguébel et al., 2002).

In most countries the permissible threshold value for Cd in agricultural soils is about 2 mg kg^{-1} soil. Higher soil Cd concentrations may lead to unacceptable Cd levels in the grain of some crops, for example durum wheat and sunflower grains, putting under pressure their market (Harris and Taylor, 2001). The best way to solve this problem is to decrease soil Cd concentration to the permissible limit extracting a certain amount of Cd from the soil. If we assume that soil, penetrated by roots to a depth of 15 cm has bulk density about 1.3 and Cd contamination 5 mg kg^{-1}, its Cd content will total 10 kg Cd ha^{-1}. Thus, to decrease soil Cd level from 5 to 2 mg kg^{-1}, it will be necessary to extract about 6 kg Cd ha^{-1}.

2. POTENTIAL OF NATURAL HYPERACCUMULATORS FOR Cd PHYTOEXTRACTION

Hyperaccumulation of Cd is a rare phenomenon in higher plants. So far, only *Thlaspi caerulescens* J. & C. Presl (*Brassicaceae*), known also as pennycress (Fig. 1), has been identified as Cd hyperaccumulator able to meet the criterion for Cd hyperaccumulation (100 mg kg^{-1}) shoot dry weight (Baker et al., 2000). For comparison, more than 300 taxa are hyperaccumulating Ni and about 20 – Zn, Cu and Co. Other possible Cd hyperaccumulators could be *Arabidopsis halleri* (L) O'Kane & Al-Shenbaz and *Sedum alfredii* Hance, which were found to hyperaccumulate Cd in hydroponical conditions (Küpper et al., 2000; Yang et al., 2004), but according to the strong hyperaccumulation definition, this ability has to be proved in its natural habitat, not at artificial conditions (Reeves and Baker, 2000). Hyperaccumulator plants are usually found on metalliferous soils, where the natural exposure to a surplus of various metals has driven the evolution of metal hyperaccumulation as well as plant resistance to heavy metals (Ernst, 1998).

The ability of *T. caerulescens* to hyperaccumulate Cd (and Zn) is known for a long time (Ernst, 1968, and references therein). Robinson et al. (1998) have found Cd accumulation in the leaves of *T. caerulescens* up to 1600 mg Cd kg^{-1} DW without detectable decrease of its dry biomass up to 50 mg extractable Cd kg^{-1} soil. Recently, Lombi et al. (2000) found that at hydroponics one French population of *T. caerulescens* (Ganges ecotype) was able to accumulate Cd in the shoots, over 3000 mg kg^{-1} without biomass reduction. Moreover, in field trials, this population was shown to accumulate up to 500 mg Cd kg^{-1} in the shoot at 12 mg Cd kg^{-1} soil, which is encouraging for Cd phytoextraction from agricultural soils.

Fig. 1. *Thlaspi caerulescens* J. & C. Presl

Very different values have been reported about dry mass potential of *T. caerulescens*. Robinson et al. (1998) estimated that dry biomass production of this species averaging 2.6 t ha^{-1}, whereas in field trials Kayser et al. (2000) found less than 1 t ha^{-1}. Recently Hammer and Keller (2003) have observed in field trials that an annual removal (three harvests) of 3-6 t DM ha^{-1} of *T. caerulescens* would be expected, that confirmed the earlier suggestion of Bennett et al. (1998) that the yield of nitrogen fertilised crop of *T. caerulescens* could be easily increased by a factor of 2 - 3 without significant reduction in Cd tissue concentration.

Robinson et al. (1998) accepting dry mass potential of *T. caerulescens* of about 5 t ha^{-1} calculated that soil contaminated by 10 mg Cd kg^{-1} would be cleaned in only 2 years. Using mass balances. Schwartz et al. (2003) calculated that two crops of *T. caerulescens* extracted about 9% of the total Cd added to the soil as a contaminated sludge. However, the extraction potential strongly depends on the population origin, and on the soil Cd concentration. Under optimal growth conditions Saxena *et al.* (1999) reported that the Cd extraction potential of *T. caerulescens* should be 2 kg ha^{-1} yr^{-1}, whereas Hammer and Keller (2003) calculated lower values – 130 and 540 g Cd ha^{-1} yr^{-1} on calcareous and acidic soil, respectively. Zhao et al. (2003) using target biomass of 5 t ha^{-1} and assuming the soil metal concentration occurs only in the active rooting zone (0-20 cm) calculated that 9 crops would be required to reduce soil Cd concentration from 5 to 3 mg Cd kg^{-1}.

In spite of the high Cd accumulation capacity of *T. caerulescens* as well as the possibilities to increase its dry biomass yield to a sufficient level, an opinion exists that this species is not suitable for metal phytoextraction due to its rosette characteristics, making mechanical harvesting difficult (Ernst,

1998). Another negative trait of *T. caerulescens* is its low resistance to both hot and dry environments (Kayser et al., 2000), which will restrict its wider dissemination. Recently, Hammer and Keller (2003) suggested that it might be possible to overcome the harvesting problem using the plantation techniques for edible corn salad (*Valerianella locusta*) as it has similar habits to that of *Thlaspi*.

3. POTENTIAL OF BIOMASS PLANTS FOR Cd PHYTOEXTRACTION
3.1. Woody Plants

The ideal plant for metal phytoextraction has to be highly productive in biomass and to assimilate and translocate to shoots a significant part of metals of concern. Additional favourable traits are fast growth, easy propagation and a deep extended root system. Some tree species, mainly willow (*Salix*) and poplar (*Populus*) exhibit these traits and are in use in phytoremediation programs. Short-rotation coppice of willow has shown particular promise as a renewable energy crop possessing the ability to accumulate high levels of some metals, for example Zn and Cd (Riddell-Black et al., 1997; Greger and Landberg, 1999; Pulford et al., 2002). In fact, some *Salix* genotypes are good Cd accumulators, but not hyperaccumulators as even at high contamination levels in hydroponics conditions (for example 15 µM Cd), their shoot Cd concentrations rarely exceeds 50 mg kg^{-1} (Vassilev et al., in press). On the other hand, their high stem yield and high evapotranspiration rate may result in a significant Cd removal.

Due to the relatively high variation in both shoot Cd concentrations and stem yield, very different values about Cd removal by *Salix* have been calculated in the literature: 222 g Cd ha^{-1} yr^{-1} (Felix, 1997), 61.7 g Cd ha^{-1} yr^{-1} (Pulford et al., 2002), 1060 g Cd ha^{-1} yr^{-1} (Robinson et al., 2000) and 24 – 34 g Cd ha^{-1} yr^{-1} (Hammer et al., 2003).

For the phytoextraction aim, the tolerance of the willow plants to heavy metals is a very important characteristic, as they have to be able to grow continuously on Cd-contaminated land and to deliver an additional economical value (dry matter for energy, chip wood and paper production). Studying more than 150 willow clones Greger and Landberg (1999) observed a large variation in their tolerance to Cd. A similar conclusion may be drawn from the works of Punshon and Dickinson (1999) and Vyslouzilova et al. (2003). Obviously, the optimization of *Salix* potential for Cd extraction depends on the choice of appropriate genotype combining high shoot Cd accumulation, sufficient tolerance as well as high biomass productivity. For that purpose, a hydroponics screening technique to assess metal tolerance of willow genotypes has been developed and its efficiency has been proved in field trials (Watson et al., 2003). Besides the easily measured growth parameters used in these screening studies, some physiological indicators (antioxidative enzyme activities, leaf gas-exchange, photosynthetic pigments, etc.) should complement the former ones for more accuracy (Vassilev et al., in press).

3.2. Non-Woody Plants

The use of non-woody biomass plants has been introduced in the phytoextraction concept in order to overcome the limitation related to the low biomass of some metal hyperaccumulators. The primarily interest is focused on Indian mustard (*Brassica juncea*), oilseed rape (*Brassica napus*), tobacco (*Nicotiana tabaccum*), flax (*Linum usitatissimum*), peppermint (*Mentha piperita*), cotton (*Gossipium hirsutum*), maize (*Zea mays*), sunflower (*Helianthus annuus*), etc. (Zheljazkov and Nielsen, 1996; Zheljazkov et al., 1999; Yankov and Taxsin, 1999; Yankov et al., 2000; Schmidt, 2003; Griga et al., 2002). These crops could make metal phytoextraction more economically feasible as, in addition to metal removal, they produce biomass with an added economical value (Schwitzguébel et al., 2002).

To increase the metal uptake by biomass plants, metal solubility in the soil should be enhanced following the chemically-assisted approach (Salt et al., 1998). This approach is based mostly on EDTA (ethylene-diamine-tetraacetic acid) application, but it has been recently criticised for some toxic properties of the chemical itself as well as for the possible leaching of metals to groundwater (Crčman et al., 2001; Geebelen et al., 2002; Wenzel et al., 2003). On the other hand, the EDTA application dramatically enhances mostly Pb phytoaccumulation (Blaylock et al., 1997), whereas its effect on Cd solubility and plant uptake is relatively low. However, the effect is dependant on both species and soil properties. For example, the application of 0.5 g EDTA kg^{-1} soil increased Cd solubility 10-fold, but only by 50% its leaf concentration in maize (Sapundjieva et al., 2003). At higher rates (2 g kg^{-1} soil) EDTA caused a temporary increase in Cd uptake by poplar, but also resulted in significant reduction in growth, as well as abscission of leaves (Robinson et al., 2000). Using 2.85 g EDTA kg^{-1} an industrially polluted by 5 mg Cd kg^{-1} soil, Crčman et al. (2001) did not found significant changes in leaf Cd concentration in cabbage, but the leaching rate was high.

Other amendments used for metal phytoextraction are NTA (nitrilotriacetic acid), organic acids and inorganic agents. Kayser et al. (2000) applied NTA to the rooting zone of several plants (9 injections with a total 9.2 g NTA m^{-2}) and found that Cd removal was doubled, reaching the highest value in tobacco (48 g Cd ha^{-1}). Nigam et al. (2001) reported that the application of citric and malic acids to a Cd-spiked soil enhanced Cd accumulation by corn. Among the recently proposed inorganic amendments are the elemental sulphur and ammonium sulphate. There is some evidence that these amendments may decrease transiently the rhizosphere pH and subsequently enhance metal phytoavailability (including Cd) on carbonate rich soils. As sulphur is gradually oxidised by sulphur-oxidising bacteria, its application doesn't have significant negative effects on the soil environment (Kayser et al., 2000). Kayser et al. (1999) achieved in this way a significant increase of Cd phytoavailability and a 27-fold Cd-concentration (84 mg Cd kg^{-1} DW) in *Brassica juncea*. On the contrary, Hammer et al. (2003) did not observe a significant effect of the sulphur application on cumulative Cd extraction by *Salix viminalis* and explained this by the negative effect of the extra acidity on plant growth. Puschenreiter et al. (2001) have recently reported a positive effect of ammonium sulphate application on phytoavailability and plant uptake for several metals. Results obtained from own experiments with oilseed rape and beans grown in Cd-spiked soil indicated a positive effect of ammonium sulphate, but not of elemental sulphur, on plant behavior and metal removal (Vassilev et al., unpublished data).

McGrath et al. (2001) suggested that the enhanced plant loading by metals as Cd, Cu and Zn might induce phytotoxicity problems in biomass crops, which in turn may limit Cd phytoextraction potential. As the crops envisaged for Cd removal should grow continuously on the contaminated soil, their ability to withstand Cd is a very important trait. To avoid acute phytotoxicity problems and to achieve higher metal uptake, Barocsi et al. (2003) suggested to apply EDTA in multiple doses, in several small increments, providing time for plants to initiate their adaptation mechanisms and raise their damage threshold.

4. LIMITATIONS TO Cd PHYTOEXTRACTION

The acceptable duration of a phytoextraction process is suggested to be not more than 5-10 years (Robinson et al., 1998; Blaylock and Huang, 2000). The results presented in Table 1 show that, in general, and even supposing that a linear exhaustion of Cd from the soil should occur, plant Cd removal is not high enough to meet this requirement. On the other hand, it should be mentioned that all these calculations are made assuming homogenous Cd distribution in the soil, which rarely appears in a real situation. Recently Podar et al. (2004) using a simple, but elegant experimental design have shown that at heterogeneous Cd and Zn distribution, the total metal removal by *Brassica juncea* is much higher.

Table 1. Cd phytoextraction potential of several plant species

Plant species	Leaf Cd concentration (mg kg^{-1})	Reference	Possible DM yield (t ha^{-1})	Possible Cd removal (g ha^{-1} yr^{-1})
		Hyperaccumulators		
T. caerulescens	1600	Robinson et al., 1998	2.6 – 5.2	4160-8320
T. caerulescens	120 – 280	Hammer and Keller, 2003	0.85 – 1.8	130 - 540
		Woody biomass plants		
Salix viminalis	3 – 40	Riddell-Black et al., 1997	10 – 16	80 – 220
Salix viminalis	10 – 20	Pulford et al., 2002	8	80 – 160
Salix viminalis	2 – 6	Hammer et al., 2003	25 – 35	24 – 34
Populus deltoides x P. yunnanensis + EDTA	53	Robinson et al., 2000	20	1060
		Non-woody biomass plants		
Nicotiana tabaccum	9 – 40 (120)	Kayser et al., 1999	9 – 13	90 – 115
Brassica juncea	73	Blaylock et al., 1997	4 – 24	60 – 80
Zea mays	59	Sapundjieva et al., 2003	10	400 – 500
Zea mays	1-9	Kayser et al., 2000	10	11 – 74

Presently, the model of phytoextraction crop is still not consolidated. Several authors have considered hyperaccumulation as more important trait than biomass yield (Chaney et al., 1997), while others do not accept this opinion (Ebbs et al., 1997; Kayser et al., 2000; Ernst, 1998). On the other hand, the use of the hyperaccumulators, such as *T. caerulescens* is still problematic, at least in terms of its sensitivity to heat and drought. At present, it is a preferable model plant for studying the molecular biology of Zn and Cd hyperaccumulation and tolerance (Lasat et al., 1998; Saier, 2000; Zhao et al., 2002). For metal phytoextraction purpose Barcelo et al. (2001) have recommended the search for new hyperaccumulators with better survival strategies at different climatic conditions.

The efficiency of Cd phytoextraction by biomass crops is mainly limited by low Cd concentrations in the harvestable parts. If higher shoot Cd concentrations would be achieved by means of chemically-assisted approach, Cd phytoextraction potential might be limited by phytotoxicity problems. Chronic phytotoxicity at industrially contaminated sites more often is due to physiological disorders provoked by multiple metals contamination than to single Cd ions and, at this situation, metals can react in synergistic, antagonistic or additive manner.

Generally, plant defence against excess heavy metals includes two strategies: (1) heavy metal detoxification and (2) coping with metal-induced oxidative stress. Several mechanisms for metal detoxification in plants exist, the best known are complexation with phytochelatins (PCs) and metallothioneins (MTs). Briefly, metal detoxification is achieved by metal binding to the cell wall, complexation with PCs followed by compartmentalisation in the vacuole and complexation with MTs and glutathione (GSH) in the cytosol (Vögeli-Lange and Wagner, 1996, and references therein). There is evidence that heavy metals, including Cd, can induce oxidative stress in plant cells (Hendry et al., 1992; Lagriffoul et al., 1998; Piqueras et al., 1999; Di Cagno et al., 2001). The current knowledge about plant defence against heavy metals reveals that besides of antioxidative enzymes, an important role can be played by the metabolites glutathione (GSH) and ascorbate (AsA) (Finckh and Kunert, 1985; Howden et al., 1995) and the enzymes involved in glutathione synthesis and the ascorbate-glutathion cycle.

When the plant defence network is not able to maintain the concentration of free metal ions in the cytosol below the toxic threshold level, metal-induced over-production of the reactive oxygen species results in numerous toxic effects. Among them the most known are lipid peroxidation (Sandalio et al., 2001), altered enzymes structure and activities (Van Assche and Clijsters, 1990), disruption in membrane integrity (Vangronsveld and Clijsters, 1994), etc. Further these primary cell toxic effects are multiplied on the cardinal physiological processes in plants, such as water relations, mineral nutrition, photosynthesis, etc. and in turn lead to growth retardation and phytotoxicity (for review see: Clijsters and Van Assche, 1985; Baszynski, 1986; Barcelo and Poschenrieder, 1990; Breckle, 1991; Siedleska, 1995).

5. APPROACHES FOR OPTIMISATION OF Cd PHYTOEXTRACTION
5.1. Biotechnological Approach

The genetic engineering is expected to contribute to the further development of metal phytoextraction improving plant metal tolerance and/or uptake (Salt et al., 1998; Kärenlampi et al., 2000). Some promising results concerning Cd phytoextraction tested at a laboratory scale have been already obtained.

Firstly, the attention was focussed on the overexpression of MTs as a means to increase Cd tolerance. The first report about stable integration of human MTs into tobacco and oilseed rape showed that the growth of transformed seedling was unaffected up to 100 µM Cd (Misra and Gedamu, 1989). The expression of mammalian MTs in transformed *Nicotiana tabacum* plants also resulted in improved Cd resistance (Pan et al., 1994). *Brassica oleraceae* expressing yeast MTs gene has better tolerance to Cd than that of the wild-type (Hasegawa et al., 1997).

Secondly, modification or over-expression of the enzymes involved in the synthesis of GSH and PCs has been considered as a good approach to enhance Cd tolerance and accumulation. GSH is synthesised from its constituent aminoacids in two steps by ATP-dependant enzymatic reactions catalysed by γ–glutamylcysteine synthetase (γ–ECS) and glutathione synthetase (GS). Transgenic *Brassica juncea* plants overexpressing bacterial (*Escherichia coli*) GS gene (gsh2) were found to have both higher shoot Cd concentration (25% increase) and enhanced Cd tolerance (Zhu et al., 1999a). Following an expression of another gene encoding γ–ECS (gsh1) the authors found that *Brassica juncea* transformants have higher Cd tolerance and up to 50 - 70% higher Cd concentration (Zhu et al., 1999b). Recently, Arisi et al. (2000) have reported that poplars overexpressing bacterial γ–ECS showed better Cd accumulation, but not improved Cd tolerance. Pilon-Smith et al. (2000) overexposed *E. coli* gene encoding the enzyme GR in the cytosol (cytGR) and the plastids (cpGR) of *Brassica juncea*, hypothesising that the enhanced GR activity will alleviate Cd-induced oxidative stress. They found Cd tolerance at whole plant level was not affected, but was enhanced at the chloroplast level as judged by chlorophyll a fluorescence parameters. This effect was explained by the observations of lower Cd content in the chloroplast and higher GSH in cpGR plants as compared to the untransformed plants.

Another biotechnological approach for improving performance of the metal-extracting plants is the overexpression of metal transporters. The first results showed some evidence for increased metal accumulation and plant tolerance: Cd in the roots of tobacco overexpressing CAX2 under the control of a 35S promoter (Hirschi et al., 2000), Ni in tobacco overexpressing NtCBP4 (Arazi et al., 1999) and Zn in *A. thaliana* overexpressing ZAT1 (Van der Zaal et al., 1999).

However, these promising results from the biotechnological research efforts should also be studied and proved on Cd-contaminated soils. On the other hand, the use of genetically modified plants for metal phytoextraction strongly depends of their public acceptance.

5.2. Agronomy-Based Approach: Some Evidence from Lab Scale Experiments

The economically profitable Cd phytoextraction from agricultural soils, contaminated by multiple metals, may be limited by lower yields. Thus, one approach that could improve Cd phytoextraction is to improve plant's capacities to cope with chronic metal phytotoxicity.

As heavy metals can induce essential nutrient deficiency and even decrease concentrations of several macronutrients in plants (Siedleska, 1995), it seems possible to reverse or reduce (at least partly) some of the metals-induced negative effects on the plants by optimization of mineral nutrition.

Chen and Huerta (1997) showed that sulphur (S) is a critical nutritional factor for reduction of Cd toxicity. These authors observed that the negative effects of Cd on growth and photosynthesis are stronger at barley plants supplied with 0.1 mM S than in plants receiving 1 mM S. The positive effect of higher S nutrition on barley resistance to Cd was expressed by weaker effects on plant morphology, leaf gas exchange and chlorophyll fluorescence parameters. A positive effect of S nutrition on Cd detoxification in sugar beet plants has also been established (Popovic et al., 1996). It has been found that at sub-optimal S nutrition Cd-exposed plants preferably allocate S to PCs synthesis, which may provoke S deficiency (McMahon and Anderson, 1998). Probably, the improved S nutrition allows a more adequate plant defence response to Cd, but also prevents S deficiency. Recently El-Shintinawy (1999) showed that exogenous GSH counteracted all retardation effects in soybean seedling induced by Cd. He provided data allowing to conclude the dual function of GSH: first, as being an antioxidant and oxy radicals scavenger, it protects chloroplasts from oxidative damage by trapping the hydroxyl radicals, and secondly, as a substrate for PCs synthesis that mainly sequesters and detoxifies excess Cd ions. When both functions are realised, repairing membrane damage, and sequestration of Cd are achieved by GSH, the electron transport rate as well as photosynthetic parameters are not disturbed and this results in a photosynthetic performance similar to that in the control plants.

Panković et al. (2000) have shown that optimal nitrogen (N) supply decreased the inhibitory effects of Cd on photosynthesis of sunflower plants. The authors studied the effects of Cd on sunflower plants grown at optimal, sub-optimal and supra-optimal N supply. They found the lowest inhibition of photosynthetic activity by Cd at optimal N supply, when N investment in soluble proteins and Rubisco were at their maximum. Higher N supplies did not alleviate the toxic Cd effect; therefore the authors concluded that N nutrition can be manipulated as a means of decreasing Cd phytotoxicity. Recently, we have confirmed that N and S supply partially ameliorated chronic Cd toxicity in young bean plants without effect on Cd phytoaccumulation (Vassilev et al., unpublished data).

Instead of mineral nutrition, there exists some evidence that plant ability to withstand Cd may be improved by plant growth regulating substances. Recently, Ghorbanli et al. (1999) showed that exogenous gibberellins positively affected growth rate and reduced the toxic effects of Cd on soybean plants. They suggested that this effect might result from cooperative effects on leaf area development, photosynthetic rate, modification in partitioning of photosynthates, etc. It was also shown that the pre-treatment of Cd-exposed leaf discs with exogenously added polyamine spermin reverted the lipid peroxidation almost to the control values (Groppa et al., 2001). Other promising results are achieved by treating Cd-suffering plants with a natural product, named biomin (Kamenova-Jouhimenko et al., 1997/1998). The addition of biomin improved photosynthetic performance of Cd-exposed pea plants through a stabilisation of chloroplast ultrastructure as well as the increased share of the C_4 pathway for CO_2 fixation at the expense of the C_3 pathway.

In conclusion, the information presented here shows plants in a new vision – as environmental counterbalances to heavy metal contamination. In particular, Cd phytoextraction from agricultural plants seems to be a promising approach for soil decontamination. However, its practical implementation requires significant optimization. As phytoextraction needs a quite interdisciplinary approach such improvements should be addressed to many plant and soil sciences.

References

Arazi T, Sunkar R, Kaplan B, Fromm H (1999) A tobacco plasma membrane clamouring -binding transporter confers Ni^{2+} tolerance and Pb^{2+} hypersensitivity in transgenic plants. Plant J 20: 171-182.

Archibald D, Marinates E, Buckley W, Clarke J, Taylor G (2001) A rapid, seedling–based bioassay for identifying low cadmium-accumulating individuals of durum wheat (*Triticum turgidum* L.). Euphytica 117: 175-182.

Arisi ACM, Mocquot B, Lagriffoul A, Mench M, Foyer CH, Jouanin L (2000) Responses to cadmium in leaves of transformed poplars overexpressing γ-glutamylcysteine synthetase. Physiol Plant 109: 143-149.

Ariza ME, Bijur GN, Williams MV (1999) Environmental Metal Pollutants, Reactive Oxygen Intermediaries and Genotoxicity. Molecular Approaches to Determine Mechanisms of Toxicity. Kluwer Academic Publishers, Boston, Dordrecht, London

Baker AJM, McGrath SP, Reeves RD, Smith JAC (2000) Metal hyperaccumulator plants: a review of the ecology and physiology of a biochemical resource for phytoremediation of metal-polluted soils. In: Terry N, Bañuelos G (eds.). Phytoremediation of Contaminated Soil and Water. Lewis Publishers, Boca Raton, Florida, USA, pp. 85-107.

Barcelo J, Poschenrieder C (1990) Plant water relations as affected by heavy metal stress: A review. J Plant Nutr 13: 1-37.

Barcelo J, Poschenrieder C, Lombini A, Llugany M, Bech J, Dinelli E (2001) Mediterranean plant species for phytoremediation. Proceedings of the workshop of COST action 837l; Madrid, 5-7.04. 2001. (http://lbewww.epfl.ch/COST837)

Barocsi A, Csintalan Z, Kocsanyl L, Dushenkov S, Kuperberg JM, Kucharski R and Richter P (2003) Optimizing phytoremediation of heavy metal-contaminated soil by exploiting plant's stress adaptation. Int J Phytoremed 5: 13-23.

Baszynski T (1986) Interference of Cd^{2+} in functioning of the photosynthetic apparatus of higher plants. Acta Soc Bot Pol 55: 291-304.

Bennett FA, Tyler EK, Brooks RR, Gregg PEH, Stewart RB (1998) Fertilisation of hyperaccumulators to enhance their potential for phytoremediation and phytomining. In: Brooks R (ed.). Plants that Hyperaccumulate Heavy Metals. CAB International, Wallingford, pp. 54-59.

Blaylock M, Huang J (2000) Phytoextraction of metals. In: Raskin I, Ensley B (eds.). Phytoremediation of Toxic Metals: Using Plants to Clean up the Environment. John Wiley & Sons, Inc. New York, pp. 53-69.

Blaylock MJ, Salt DE, Dushenkov S, Zakharova O, Gussman C, Kapulnik Y, Ensley BD, Raskin I (1997) Enhanced accumulation of Pb in Indian mustard by soil-applied chelating agents. Environ Sci Tech 31: 860-865.

Breckle SW (1991) Growth under stress: heavy metals. In: Waisel Y, Eshel A, Kafkafi X (eds.). Plant Roots: The Hidden Half. Marcel Dekker, New York, pp. 351-373.

Chaney R (1983) Plant uptake of inorganic waste constituents. In: Parr P, Marsh P, Kla J (eds.). Land Treatment of Hazardous Wastes. Noyes Data, Park Ridge, NJ, pp. 50-76.

Chaney R, Malik M, Li Y, Brown S, Brewer E, Angle J, Baker AJM (1997) Phytoremediation of soil metals. Curr Opin Biotech 8: 279-284.

Chen Y, Huerta A (1997) Effects of sulfur nutrition on photosynthesis in Cd-treated barley seedlings. J Plant Nutr 20: 845-855.

Clijsters H, Van Assche F (1985) Inhibition of photosynthesis by heavy metals. Photosynthesis Res 7: 31-40.

Crčman H, Velikonja-Bolta Š, Vodnik D, Kos B, Leštan D (2001) EDTA enhanced heavy metal phytoextraction: metal accumulation, leaching and toxicity. Plant Soil 235: 105-114.

Di Cagno R, Guidi L, De Gara L, Soldatini GF (2001) Combined cadmium and ozone treatments affect photosynthesis and ascorbate-dependent defences in sunflower. New Phytol 151: 627-636.

Ebbs SD, Lasat M, Brady D, Cornish J, Gordon R, Kochian LV (1997) Phytoextraction of cadmium and zinc from a contaminated soil. J Environ Qual 26: 1424-1430.

El-Shintinawy F (1999) Glutathione counteracts the inhibitory effect induced by cadmium on photosynthetic process in soybean. Photosynthetica 36: 171-179.

Ernst WHO (1968) Der einfluss der Phosphatversorgung sowie die Wirkung von ionogem and chelatisiertem Zink auf die Zink-and Phosphataufnahme einiger Schwermetallpflanzen. Physiol Plant 21: 323-333.

Ernst WHO (1998) The Origin and Ecology of Contaminated, Stabilized and Non-Pristine Soils. In: Vangronsveld J, Cunningham S (eds.). Metal-Contaminated Soils: Situ Inactivation and Phytorestoration. Springer-Verlag and RG Landes Company, pp. 17-25.

Felix H (1997) Vor-Ort-Reinigung schwermetallbelasteter Böden mit Hilfe von metallakkumulierenden Pflanzen (Hyperakkumulatoren). Terra Tech 2: 47-49.

Finckh BF, Kunert KJ (1985) Vitamins C and E: Antioxidative system against herbicide-induced lipid peroxidation in higher plants. J Agron Food Chem 33: 574-577.

Geebelen W, Vangronsveld J, Adriano DC, Van Poucke LC, Clijsters H (2002) Effects of Pb-EDTA and EDTA on oxidative stress reactions and mineral uptake in *Phaseolus vulgaris*. Physiol Plant 115: 371-383.

Ghorbanli M, Kaveh SH, Sepehr MF (1999) Effects of cadmium and gibberelin on growth and photosynthesis of *Glycine max*. Photosynthetica 37: 627-631.

Grant C, Buckley W, Bailey L, Selles F (1998) Cadmium accumulation in crops. Can J Plant Sci 78: 1-17.

Greger M, Landberg T (1999) Use of willow in phytoextraction. Int J Phytoremed 1: 115-123.

Griga M, Bjelkova M, Tejklova E (2002) Potential of flax (*Linum usitatissimum*) for heavy metal extraction and industrial processing of contaminated biomass – a review. Proceedings of the 4th workshop of COST action 837; Bordeaux, April 25th-26th, 2002.

Groppa M, Tomaro M, Behavides M (2001) Polyamines as protectors against cadmium or copper-induced oxidative damage in sunflower leaf discs. Plant Sci 161: 481-488.

Hammer D, Keller C (2003) Phytoextraction of Cd and Zn with *Thlaspi caerulescens* in field trials. Soil Use Manag 19: 144-149.

Hammer D, Kayser A, Keller C (2003) Phytoextraction of Cd and Zn with *Salix viminalis* in field trials. Soil Use Manag 19: 187-192.

Harris NS, Taylor GJ (2001) Remobilization of cadmium in maturing shoots of near isogenic lines of durum wheat that differ in grain cadmium accumulation. J Exp Bot 52: 1473-1481.

Hasegawa I, Terada E, Sunairi M, Wakita H, Shinmachi F, Noguchi A, Nakajima M, Yakazi J (1997) Genetic improvement of heavy metal tolerance in plants by transfer of the yeast metallothionein gene (*CUP1*). Plant Soil 196: 277-281.

Hendry GAF, Baker AJM, Ewart CF (1992) Cadmium tolerance and toxicity, oxygen radical processes and molecular damage in cadmium tolerant and cadmium-sensitive clones of *Holcus lanatus* L. Acta Bot. Neerl 41: 271-281.

Hirschi KD, Korenkov VD, Wilganovski NL, Wagner GJ (2000) Expression of Arabidopsis CAX2 in tobacco. Altered metal accumulation and increased manganese tolerance. Plant Physiol 124: 125-134.

Howden R, Goldsbrough PB, Andersen CR, Cobbettt CS (1995) Cadmium sensitive, *cad1* mutants of *Arabidopsis thaliana* are phytochelatin deficient. Plant Physiol 107: 1059–1066.

Kamenova-Jouhimenko SM, Markovska YK, Georgieva VT (1997/1998) Effects of biomin and algae suspensions on the activities of carboxylating and decarboxylating enzymes in cadmium-treated pea plants. Biol Plant 40: 405-410.

Kärenlampi S, Schat H, Vangronsveld J, Verkleij JAC, van der Lelie D, Mergeay M, Tervahauta AI (2000) Genetic engineering in the improvement of plants for phytoremediation of metal polluted soils. Environ Poll 107: 225-231.

Kayser A, Wenger K, Keller A, Attinger W, Felix H, Gupta SK, Schulin R (2000) Enhancement of phytoextraction of Zn, Cd, and Cu from calcareous soil: the use of NTA and sulfur amendments. Environ Sci Technol 34: 1778-1783.

Kayser A, Schulin R and Felix H. 1999. Mobilisation of Zn and Cd in three Swiss soils by use of elemental sulphur. Proceedings of the 5t^{h} International Conference on the Biogeochemistry of Trace Elements,

Vienna, Austria, 788-789.

Küpper H, Lombi E, Zhao FJ, McGrath SP (2000) Cellular compartmentation of cadmium and zinc in relation to other elements in the hyperaccumulator *Arabidopsis halleri*. Planta 212: 75-84.

Lagriffoul A, Mocquot B, Mench M, Vangronsveld J (1998) Cadmium toxicity effects on growth, mineral and chlorophyll contents, and activities of stress related enzymes in young maize plants (*Zea mays* L.). Plant Soil 200: 241-250.

Lasat M, Baker AJM, Kochian L (1998) Altered Zn compartmentation in the root symplast and stimulated Zn absorption into the leaf as mechanisms involved in Zn hyperaccumulation in *Thlaspi caerulescens*. Plant Physiol 118: 875-883.

Lombi E, Zhao F, Dunham S, McGrath SP (2000) Cadmium accumulation in populations of *Thlaspi caerulescens* and *Thlaspi goesingense*. New Phytol 145: 11-20.

McGrath SP, Zhao FJ, Lombi E (2001) Plant and rhizosphere processes involved in phytoremediation of metal-contaminated soils. Plant Soil 232: 207-214.

McMahon PJ, Anderson JW (1998) Preferential allocation of sulphur into γ–glutamylcysteinyl peptides in wheat plants grown at low sulphur nutrition in the presence of cadmium. Physiol Plant 104: 440-448.

Misra S, Gedamu L (1989) Heavy metal tolerant transgenic *Brassica napus* L. and *Nicotiana tabacum* L. plants. Theo Appl Genet 78: 161-168.

Nigam R, Srivastava S, Prakash S, Srivastava MM (2001) Cadmium mobilization and plant availability – the impact of organic acids commonly exuded from roots. Plant Soil 230: 107-113.

Pan A, Yang M, Tie F, Li L, Chen Z, Ru B (1994) Expression of mouse metallothionein-I gene confers cadmium resistance in transgenic tobacco plants. Plant Mol Biol 24: 341-352.

Panković D, Plesničar M, Arsenijević-Maksimović I, Petrović N, Sakač Z, Kastori R (2000) Effects of nitrogen nutrition on photosynthesis in Cd-treated sunflower plants. Ann Bot 86: 841-847.

Pilon-Smith EAH, Zhu YL, Sears T, Terry N (2000) Overexpression of glutathione reductase in *Brassica juncea*: Effects on cadmium accumulation and tolerance. Physiol Plant 110: 455-460.

Piqueras A, Olmos E, Martines-Solano JR, Hellin E (1999) Cd-induced oxidative burst in tobacco BY2 cells: time course, subcellular localisation and antioxidative response. Free Radical Res 31: 33-38.

Podar D, Ramsey MH, Hutchings MJ (2004) Effect of cadmium, zinc, and substrate heterogeneity on yield, shoot metal concentration and metal uptake by *Brassica juncea*: implications for human health risk assessment and phytoremediation. New Phytol 163: 313-324.

Popovic M, Kevresan S, Kandrac J, Nicolic J, Petrovic N, Kastori R (1996) The role of sulphur in detoxification of cadmium in young sugar beet plants. Biol Plant 38: 281-287.

Punshon T, Dickinson N (1999) Heavy metal resistance and accumulation characteristics in willow. Int J Phytomed 1: 361-385.

Puschenreiter M, Stoger G, Lombi E, Horak O, Wenzel W (2001) Phytoextraction of heavy metal contaminated soils with *Thlaspi goesingense* and *Amaranthus hybridus*: Rhizosphere manipulation using EDTA and ammonium sulfate. J Plant Nutr Soil Sci 164: 615-621.

Pulford ID, Riddell-Black D, Stewart C (2002) Heavy metal uptake by willow clones from sewage sludge-treated soil: the potential for phytoremediation. Int J Phytomed 4: 59-72.

Reeves R, Baker AJM (2000) Metal accumulating plants. In: Raskin I, Ensley B (eds.). Phytoremediation of Toxic Metals: Using Plants to Clean up the Environment. John Wiley & Sons, Inc., pp. 193-229.

Riddell-Black D, Pulford ID, Stewart C (1997) Clonal variation in heavy metal uptake by willow. Aspects of Applied Biology 49, Biomass and energy crops. The association of Applied Biologists, c/o Horticultural Research International, Wellesbourne, Warwick, CV35 9EF, pp. 327-334.

Robinson BH, Meblanc L, Petit D, Brooks RR, Kirkman JH, Gregg PEH (1998) The potential of *Thlaspi caerulescens* for phytoremediation of contaminated soils. Plant Soil 203: 47-56.

Robinson BH, Mills T, Petit D, Fung L, Green S, Clothier B (2000) Natural and induced cadmium-accumulation in poplar and willow: Implications for phytoremediation. Plant Soil 227: 301-306.

Ryan J, Pahren H, Lucas J (1982) Controlling cadmium in the human food chain. A review and rationale based on health effects. Environ Res 27: 251-302.

Saier MH (2000) A functional-phylogenetic classification system for transmembrane solute transporters. Microbiol Mol Biol Rev 64: 354-411.

Salt DE, Smith RD, Raskin I (1998) Phytoremediation. Ann Rev Plant Physiol Mol Biol 49: 643-668.

Sandalio LM, Dalurzo HC, Gomez M, Romero-Puertas MC, del Rio LA (2001) Cadmium-induced changes in the growth and oxidative metabolism of pea plants. J Exp Bot 52: 2115-2126.

Sapundjieva K, Kartalska Y, Vassilev A, Krastev S, Kuzmanova Y (2003) Effects of EDTA on metal solubility in the soil, metal uptake and performance of maize plants, and soil microorganisms. Bulgarian J Agric Sci 9: 1-5.

Saxena P, KrishnaRaj S, Dan T, Perras M, Vettakkorumakankav N (1999) Phytoremediation of heavy metals contaminated and polluted soils. In: Prasad MNV, Hagemaer J (eds.). Heavy metal Stress in Plants: From Molecules to Ecosystems. Springer-Verlag Berlin Heidelberg, pp. 305-329.

Schmidt U (2003) Enhancing phytoextraction: The effects of chemical soil manipulation on mobility, plant accumulation, and leaching of heavy metals. J Environ Qual 32: 1939-1954.

Schwartz C, Echevarria G, Morel J-L (2003) Phytoextraction of cadmium with *Thlaspi caerulescens*. Plant Soil 249: 27-35.

Schwitzguébel J-P, van der Lelie D, Baker AJM, Glass D, Vangronsveld J (2002) Phytoremediation: European and American Trends. Successes, obstacles and needs. J Soils Sediments 1: 1-9.

Siedleska A (1995) Some aspects of interactions between heavy metals and plant mineral nutrients. Acta Soc Bot Poloniae 64: 265-272.

Van Assche F, Clijsters H (1990) Effects of metals on enzyme activity in plants. Plant Cell Environ 13: 195-206.

Van der Zaal BJ, Neuteboom LW, Pina JE, Chardonnens AN, Schat H, Verkeij JAC, Hooykaas PJJ (1999) Overexpression of a novel *Arabidopsis* gene related to putative zinc-transporter genes from animals can lead to enhanced zinc resistance and accumulation. Plant Physiol 199: 1047-1055.

Vangronsveld J, Clijsters H (1994) Toxic effects of metals. In: Farago ME (ed.). Plants and the Chemical Elements. Biochemistry, Uptake, Tolerance and Toxicity. VCH Publishers, Weinheim, Germany, pp. 150-177.

Vassilev A, Schwitzguebel J-P, Thewys T, van der Lelie D, Vangronsveld J (2004) The use of plants for remediation of metal contaminated soils. The Scientific World J 4: 9-34.

Vassilev A, Perez-Sanz A, Semane B, Carleer R, Vangronsveld J (2004) Cadmium accumulation and tolerance of two *Salix* genotypes, hydroponically grown in presence of cadmium. J Plant Nutr: (in press)

Vögeli-Lange R, Wagner GJ (1996) Relationship between cadmium, glutathione and cadmium-binding peptides (phytochelatins) in leaves of intact tobacco seedlings. Plant Sci 114: 11-18.

Vyslouzilova M, Tlustos P, Szakova J (2003) Cadmium and zinc phytoextraction potential of seven clones of *Salix* spp. planted on heavy metal contaminated soils. Plant Soil Environ 49: 542-547.

Watson C, Pulford ID, Riddell-Black D (2003) Screening of willow species for resistance to heavy metals: comparison of performance in a hydroponics system and field trials. Int J Phytoremed 5: 351-365.

Wenzel W, Unterbrunner R, Sommer P, Sacco P (2003) Chelate-assisted phytoextraction using canola (*Brassica napus* L.) in outdoors pot and lysimeter experiments. Plant Soil 249: 83-96.

Yang XE, Long XX, Ye HB, He ZL, Calvert DV, Stoffella PJ (2004) Cadmium tolerance and hyperaccumulation in a new Zn-hyperaccumulating plant species (*Sedium alfredii* Hance). Plant Soil 259: 181-189.

Yankov B, Taxsin N (1999) Accumulation and distribution of Pb, Cu, Zn and Cd in sunflower (*Helianthus annuus* L.) grown in an industrially polluted region. Helia 24: 131-136.

Yankov B, Delibaltova V, Bojinov M (2000) Content of Cu, Zn, Cd and Pb in the vegetative organs of cotton cultivars grown in industrially polluted regions. Plant Sci (Bg) 37: 525-531.

Zhao FJ, Lombi E, McGrath PS (2003) Assessing the potential for zinc and cadmium phytoremediation with the hyperaccumulator *Thlaspi caerulescens*. Plant Soil 249: 37-43.

Zhao FJ, Hamon RE, Lombi E, McLaughlin M (2002) Characteristics of cadmium uptake in two contrasting ecotypes of the hyperaccumulator *Thlaspi caerulescens*. J Exp Bot 53: 535-543.

Zheljazkov V, Nielsen N (1996) Effect of heavy metals on peppermint and cornmint. Plant Soil 178: 59-66.

Zheljazkov V, Zheljazkova E, Craker LE, Yankov B, Georgieva T, Kolev T, Kovatcheva N, Stanev S, Margina A (1999) Heavy metal uptake by mint. Acta Hort 500: 111-117.

Zhu Y, Pilon-Smits EAN, Jouanin L, Terry N (1999a) Overexpression of glutathione synthetase in *Brassica juncea* enhances cadmium tolerance and accumulation. Plant Physiol 119: 73-79.

Zhu Y, Pilon-Smits EAN, Tarum A, Weber SU, Jouanin L, Terry N (1999b) Cadmium tolerance and accumulation in Indian mustard is enhanced by overexpressiing γ–glutamylcysteine synthetase. Plant Physiol 121: 1169-1177.

Cadmium Detoxification in Roots of *Pisum sativum* Seedlings: The Role of Phytochelatins in Metal Stress Coping

Ana Isabel Gusmão Lima and *Figueira Eetelvina Maria De Almeida Paula*

Centre for cell Biology, Biology Department, University of Aveiro, Universidade de Aveiro, Portugal

1. INTRODUCTION

Heavy metals such as cadmium (Cd) are a group of metals with density higher than 5.0 g cm^{-3}, which have become ubiquitous environmental pollutants, particularly in areas with high anthropogenic pressure. Cadmium is a widespread heavy metal, which enters the environment mainly through industrial processes and the application of phosphate fertilizers (Alloway and Steinnes, 1999; Zhu et al., 1999; Mann et al., 2002). Because of its high plant-soil mobility, Cd can be highly accumulated in plant tissues; therefore its transfer to the food chain imposes a serious threat and has become a major public concern (Wagner, 1993; Mann et al., 2002).

One of the most common plant responses to metal stress is the synthesis of low-weighted isopeptides, commonly known as phytochelatins (PCs) (Zenk, 1996; Rauser, 1999; Cobbett and Goldsbrough, 2002). PCs are a family of non-protein thiol peptides with the general structure (γ-Glu-Cys)$_n$-Gly and are synthesised in a wide variety of plant species, algae, yeast and nematodes (Zenk, 1996; Rauser, 1999; Cobbett and Goldsbrough, 2002). Many reports (Grill et al., 1989; De Knecht et al., 1994; Guarsson et al., 1996; Yan et al., 2000; Xiang et al., 2001) have demonstrated that PCs are enzymatically synthesised by the constitutive enzyme PC synthase, which catalyzes the transpeptidation of the γEC moieties of glutathione (GSH, γGlu-Cys-Gly) to another molecule GSH or to elongating PCs (Grill et al., 1987, 1989; Yoshimura et al., 1990; Hayashi et al., 1991; De Knecht et al., 1995; Klapheck et al., 1995). In some plant families, where the terminal amino acid Gly is absent or replaced by either β-Ala, Ser, Glu or Gln, structural variants may be found instead or in addition to PCs (Grill et al., 1986; Klapheck, 1988; Zenk, 1996; Rauser, 1999).

When in the presence of toxic metal concentrations, particularly Cd, PCs form complexes with the metal ions, hence preventing them from interfering with the cellular metabolism (Vögelli–Lange and Wagner, 1990; Ortiz et al., 1995, Rauser, 2000). PC-Cd complexes may present different weights and different metal and sulphide accumulation abilities (Howden et al., 1995; Maitani et al., 1996; Rauser, 2000; Hu et al., 2001; Souza and Rauser, 2003). Considering these facts, Ortiz et al. (1995) suggested a model for Cd sequestration, based on a two step process, in which Cd is first sequestrated by PCs in the cytosol, forming a low molecular weight complex (LMW), which is then transferred into the vacuole, through HTM1 (Ortiz et al., 1995; Vatamaniuk et al., 2000) or a similar MgATP-dependent transporter (Salt and Rauser, 1995), where more Cd and sulphide ions are added (Speiser et al., 1992), hence forming a higher weighted complex (HMW). This final complex is more stable in the vacuole's acidic environment and has a higher ability to bound metal ions when compared to LMW complexes (Ortiz et al., 1995; Zenk, 1996).

---

* Corresponding Author (anagusmao@portugalmail.com)

Although they are considered to be the principal mechanism responsible for Cd detoxification in the majority of plant species (Grill et al., 1987; Kneer and Zenk, 1992; Rauser, 1995, 1999; Zenk, 1996), as supported by numerous inhibitory (Grill et al., 1987; Reese and Wagner, 1987; Gussarsson et al., 1996), biochemical (Kneer and Zenk, 1992) and mutant analysis studies (Mutoh and Hayashi, 1988; Howden et al., 1995; Ha et al., 1999), there are several questions concerning their efficiency in metal detoxification in plants. For instance, PCs do not appear to be involved in metal tolerance in the hyperaccumulator *Thlaspi carulnescens* (Ebbs et al., 2002) or in hypertolerant populations of several plant species (Schat and Kalf, 2002). Some reports (Leopold et al., 1999; di Toppi and Gabrielli, 1999; Piechalak et al., 2002), suggest that PCs only play a transient role in metal detoxification and are not always associated with higher tolerance.

Although PCs have been widely studied in several plant families, their real importance in the metal detoxification process, i.e. their quantitative role and efficiency in metal chelation and detoxification, is poorly elucidated and has been object of some controversy. There are several specific factors that are of significant importance when evaluating the real role of PCs in metal stress coping, in which the first is the metal concentration and the degree of exposure. Most investigators apply extremely high Cd concentrations in their exposures, hence reporting overstressed situations, rather than environmentally representative ones. Most works concerning Cd and PC studies have not taken this fact into consideration, which may lead to controversial conclusions about PCs and their real role in tolerance to different degrees of stress, as well as its impact on plant development.

Another important factor to take into evaluation is the process of metal chelation itself, and its efficiency in intracellular Cd detoxification. Some authors have suggested that sulphur incorporation and the formation of stable metal complexes may be more important than the capacity of PC synthesis *per se* (Reese et al., 1988; Reese and Winge, 1988; De Knecht et al., 1992, 1994). Hence, the knowledge of the metal complexion process and the metal-binding features of these peptides are a key factor to understand their role in Cd detoxification. Although some aspects of the Ortiz et al. (1995) model have already been demonstrated in higher plants (Salt and Rauser, 1995; Howden et al., 1995; Maitani et al., 1996; Rauser, 2000; Hu et al., 2001; Souza and Rauser, 2003), a full characterization of the metal-binding complexes is widely neglected. In addition, the total amount of metal that in fact is sequestrated by PCs is usually not taken into consideration. In addition, some questions remain to be elucidated, particularly those concerning the kinetics of thiol and sulphide content in these two complexes and their fate during a more prolonged exposure. These are very important questions to take into consideration when evaluating the real role of PCs in metal stress coping and may also provide information on how the chelation process occurs, its differences among plant species and its relation to tolerance ability. These considerations are of major significance when evaluating the importance of PCs in the metal detoxification process and should be addressed for each individual plant species.

The present report concerns the evaluation of the role of PCs in the metal detoxification process in roots of *Pisum sativum* L. seedlings. Two different approaches are presented, one in which the importance of the degree of exposure and the metal coping ability is evaluated and another one in which the isolation of the Cd-PC complexes, the characterisation of the LMW and HMW complexes and the understanding of the actual process of Cd detoxification, focusing on the biochemical maturation of Cd-PC complexes and its relation to the metal chelation ability are focused.

2. PLANT RESPONSES TO DIFFERENT Cd CONCENTRATIONS: RELATIONSHIP BETWEEN TOXICITY LEVELS, THIOL POOL ALTERATIONS AND TOLERANCE

Plant adaptation to cytotoxic amounts of Cd is generally thought to be associated with the ability to produce PCs with different chain lengths and to form metal-binding complexes (Grill et al., 1985;

Jackson et al., 1987). It has been hypothesised that Cd concentrations and time of exposure are *grosso modo* correlated with the amount of PCs produced and with the increase in repeated units of γ-GluCys (Grill et al., 1987; Mehra et al., 1995; di Toppi and Gabrielli, 1999). In Lima et al. (2004) the role of PCs during an extended 9-day period of exposure to a wide range of Cd concentrations was evaluated. The main goals were to mimic different degrees of exposures, ranging from those observed in moderately contaminated (1 and 3 µM) to highly contaminated soils (30 µM) (Wagner, 1993), but also including concentrations usually employed in PCs and Cd accumulation studies (30, 60 and 120 µM), especially in *Pisum sativum* (Klapheck et al., 1995; di Toppi and Gabrielli, 1999; Sandalio et al., 2001). The relationship between Cys, GSH and hGSH pool alterations and stress tolerance ability was also addressed. In addition, we established a qualitative relationship between the levels of stress, polythiol synthesis and root growth ability, in order to evaluate the impact of the individual PC or hPC-mediated tolerance in root development, particularly in this highly sensitive stage of growth.

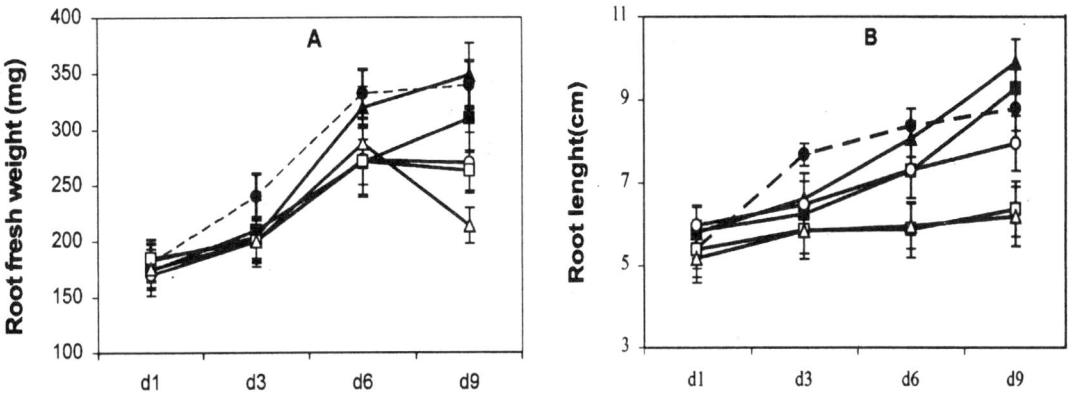

Fig. 1. Influence of Cd on root fresh weight (A) and length (B) of *Pisum sativum* plants during a 9-day exposure to 0 (●), 1 (■), 3 (▲), 30 (○), 60 (□) and 120 (□) µM Cd. Data are the mean values of at least 10 replicate experiments. Vertical bars represent standard errors. *In* Lima et al. (2004).

2.1. Plant Tolerance and Cadmium Accumulation Under Different Metal Exposures

It has been widely described that increasing Cd concentrations induce significant growth inhibition in pea plants when exposed to heavy metals (Lozano-Rodriguez et al., 1997; Sandalio et al., 2001; Piechalak et al., 2002). Data presented in Figs. 1A, 1B and 2 show root growth responses to different Cd exposures (1, 3, 30, 60 and 120 µM Cd). Data demonstrate that higher metal exposures yield a linear increase in root tissue Cd accumulation, hence a higher degree of toxicity. However, it is easily observed that the inhibitory effects are not similar in all concentrations. Despite the presence of intracellular Cd concentrations, plants exposed to 1 and 3 µM are not significantly inhibited, whereas in higher exposures (30, 60 and 120 µM), growth inhibition is consistently higher with the increase in metal concentration. Several works demonstrated similar results in other plant species exposed to more "realistic" concentrations of 1 and 3 µM Cd (Meuwly and Rauser, 1992; Souza and Rauser, 2003). When considering that these Cd concentrations are themselves representative of high Cd environmental exposures and that despite the high metal accumulation rates, plants did not present any phytotoxic symptoms, these data are of specific importance because firstly, they demonstrate the danger of metal contamination in plant foods and its subsequent transfer to the food-chain and secondly, they show that the evaluation of the tolerance mechanisms should be addressed in these tolerable concentrations, where they are still effective as a resistance mechanism, rather than in those were the plant can not cope with stress.

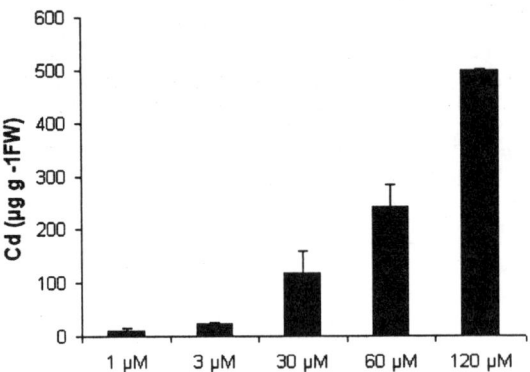

Fig. 2. Intracellular cadmium contents of roots of *Pisum sativum* plants subjected to 9 days of exposure to 1, 3, 30, 60 and 120 µM cadmium concentrations. Data are the mean values of at least 3 replicate experiments with standard errors. *In* Lima et al. (2004).

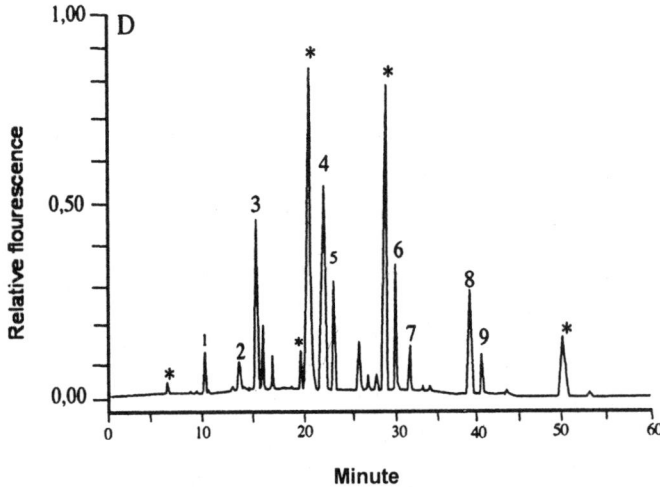

Fig. 3. Example of a chromatographic profile of HPLC analysis of thiol compounds produced in roots of *Pisum sativum* plants exposed 9 days to 1 µM Cd, using mBBr derivatisation. Isolated peaks are as follows: (**1**) Cys; (**2**) GSH; (**3**) hGSH; (**4**) PC_2; (**5**) hPC_2; (**6**) PC_3; (**7**) hPC_3; (**8**) PC_4 and (**9**) hPC_4. Peaks marked with * correspond to mBBr hydrolysis peaks. *In* Lima et al. (2004).

2.2. Thiol Pool Alterations and Different Degrees of Stress
2.2.1. *Constitutive Thiols*

In pea plants, the presence of 1 µM Cd is sufficient to stimulate the synthesis of the two isoforms, PCs and hPCs, with two to four repeated oligomeric units (Fig. 3). However, different degrees of exposure induce important alterations in both constitutive and non-constitutive thiols. When in the presence of Cd, the constitutive monothiol levels of exposed roots change. Fig 4 (A, B and C) illustrates the GSH, hGSH and Cys levels in *P. sativum* roots exposed to 1, 3, 30, 60 and 120 µM Cd concentrations. Besides their role in polythiol synthesis and plant metabolism, these monothiols can also sequester metal (Perrin and Watt, 1971; Oven et al., 2002a, b). It has been reported that GSH plays a key role in the detoxification of Cd and is involved in vital functions, including plant morphogenesis and cell division (Perrin and Watt, 1971; May et al., 1998). Exposure of plant seedlings and cell cultures to Cd results in a rapid and extensive decline in the GSH and Cys pools (Grill et al., 1985; Delhaize and Adams, 1989; Rüegsegger et al., 1990; Tukendorf and Rauser, 1990) due to a high cellular requirement for SH compounds (Steffens, 1990; Nocito et al., 2002), although there is an increase in γEC or GSH synthase genes expression after Cd exposure (Yong et al., 1999; Zhu et al., 1999).

Although hGSH is the precursor of hPCs, this tripeptide is not highly consumed during the exposure periods. In *P. sativum* plants, the presence of excess Cd reduces GSH concentrations, yet raises hGSH levels (Fig. 4). These responses have also been reported for the same species by Klapheck et al. (1995), Piechalack et al. (2002) and in *Phaseolus* by Klapheck (1988) and Tukendorf et al. (1997). The reason for this fact might lie on the process of PC synthesis itself. Klapheck et al. (1994, 1995) have demonstrated that PC and hPC biosynthesis can be relied on the same enzyme, PC synthase, which possesses a higher specificity to GSH. This fact can explain the observed differences in the consumption patterns of GSH and hGSH found in this and other works (Klapheck et al., 1995) and more importantly, can justify the lower synthesis of hPCs. In fact, data presented suggest that hPCs synthesis was stimulated when hGSH supplanted GSH levels, as substantiated by the increase of hGSH/GSH ratios observed in all concentrations after 6 and 9 days of exposure.

2.2.2. *Synthesised Thiols*

Fig. 5 presents the total polythiol accumulations, PC and hPCs with different chain lengths, induced by the different Cd concentrations. Increasing Cd amounts elicit higher accumulations of these PCs forms, however, hPCs are always in lower concentrations. When comparing the polythiol synthesis under the different Cd exposures, important differences are observed in *P. sativum* roots. In lower, concentrations (1, 3 and 30 µM), tolerance to increasing intracellular Cd levels is accompanied by a higher polythiol accumulation. Similar profiles have already been described in *P. sativum* plants exposed to Pb (Piechalack et al., 2002) and Cd (Klapheck et al., 1995) and in other plant species (Hayashi and Nakagaua, 1988; Delhaize and Adams, 1989; Grill et al., 1987; Tukendorf e Rauser, 1990) where higher degrees of metal exposure, achieved by higher Cd concentrations or longer exposures, were associated to an increase in thiol production and an accumulation of the longer chains.

Earlier reports have suggested that the shorter PCs are the precursors of the longer ones (Grill et al., 1987; Robbinson et al., 1988; Tukendorf and Rauser, 1990), and tolerance ability may be correlated to a faster synthesis of longer-chained PCs. Tukendorf and Rauser (1990) and Grill et al. (1987) showed a shift in PC_2 dominance towards PC_3 after 3 days of exposure in maize plants, which are more tolerant to Cd than pea plants (Lozano-Rodriguez et al., 1997; Piechalack et al., 2002). In *P. sativum*, the dimeric forms prevail in every exposure, as demonstrated in Fig. 5. Since a higher synthesis of longer chains, which provide more stable complexes, allows a more efficient metal complexation (Hayashi and Nakagaua, 1988; Delhaize and Adams, 1989), it seems plausible to suggest that the lower synthesis of the polythiols with n=3 and 4 may be related to the lower tolerance of pea

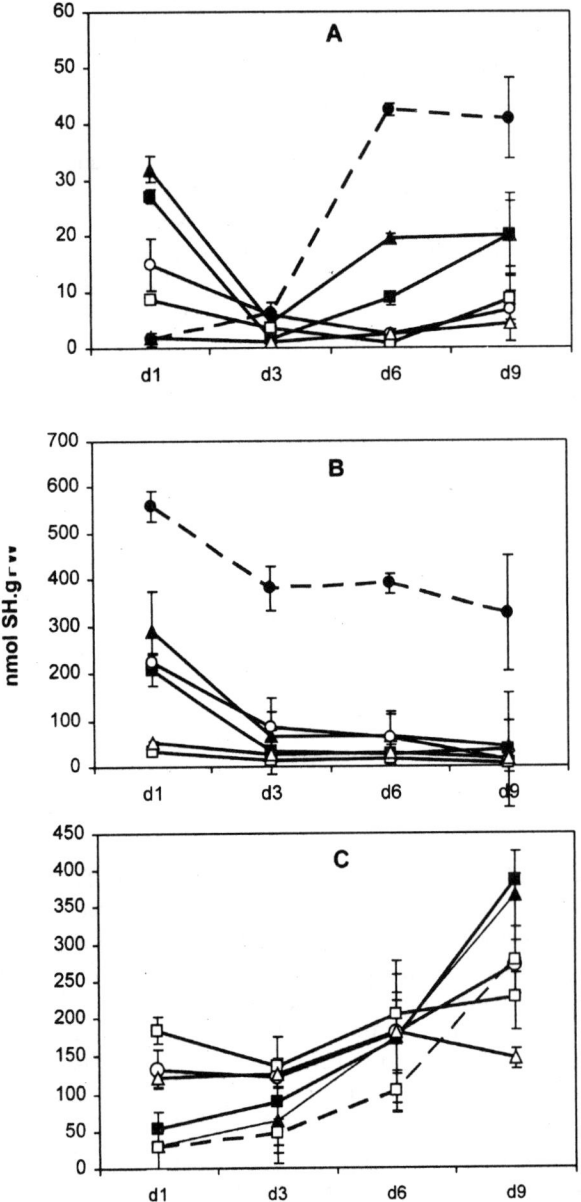

Fig. 4. Influence of cadmium on constitutive monothiols Cys (A), GSH (B) and hGSH (C) content in acid soluble extracts of roots of *Pisum sativum* plants exposed to a 9-day period o 0 (●), 1 (■), 3 (▲), 30 (○), 60 (□) and 120 (□) µCd. Data of mean values ±SE of at least 3 replicate experiments are presented. *In* Lima et al. (2004).

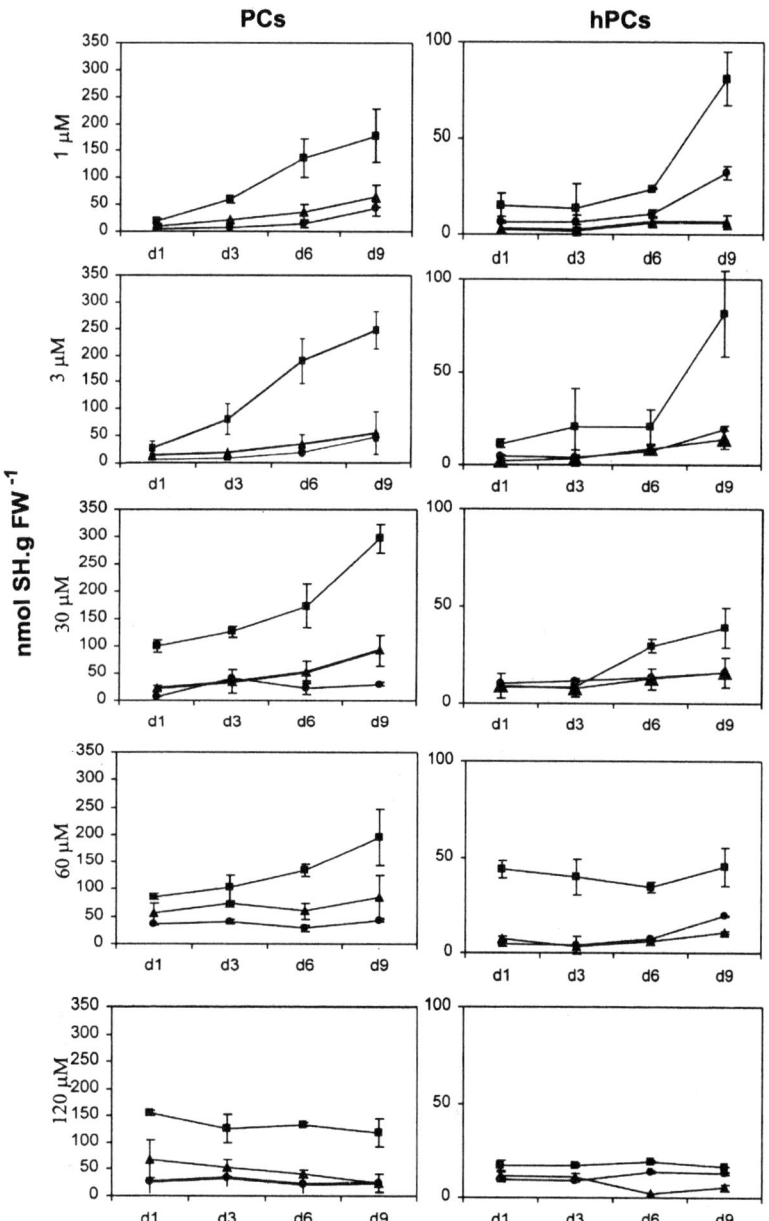

Fig. 5. Influence of cadmium on polythiols PCs and hPCs with n=2 (■), n=3 (▲) and n=4 (•) synthesis and accumulation in roots of *Pisum sativum* plants submitted to a 9-day exposure to 1, 3, 30, 60 and 120 μM Cd. Data are the mean values of at least 3 replicate experiments. Vertical bars represent standard errors. *In* Lima et al. (2004).

plants when compared with other species such as *Zea mays* and *Phaseolus vulgaris* (Lozano-Rodriguez et al., 1997; Piechalack et al., 2002), despite of the high thiol synthesis capacity observed in *P. sativum* (Rüeggserger et al., 1990; Piechalack et al., 2002).

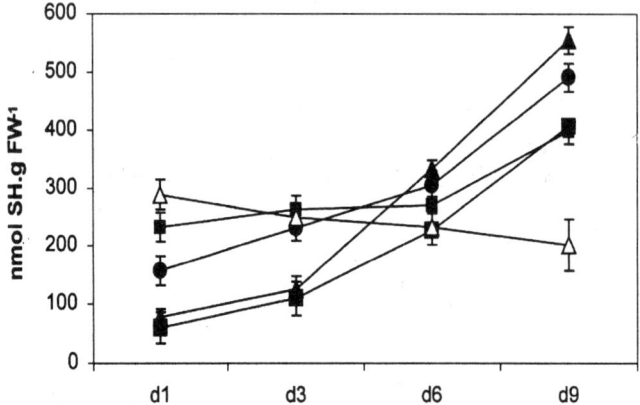

Fig. 6. Influence of Cd on total PC and hPC concentrations in roots of *Pisum sativum* plants exposed to 1 (■), 3 (▲), 30 (○), 60 (□) and 120 (□) µM Cd during a 9-day exposure period. Data are the mean values of at least 3 replicate experiments with standard errors. *In* Lima et al. (2004).

One important consideration to take from the results presented for *P. sativum* is that in higher stress levels (60 and 120 µM), a reduction in thiol production is observed (Fig. 6). This is an important fact, since it emphasises the importance of the exposure periods when evaluating the role of PCs in metal stress coping. Other authors have also observed a decrease in total PC or PC/metal ratios when plants were grown in the continuous presence of high metal concentrations (Keltjens and Van Beusichem, 1998; Sneller et al., 1999). It has been suggested that the PC disappearance in *Vicia faba* plants after a Pb exposure of 4 days indicates that these polythiols only possess a transient role in coping with metal stress (Piechalack et al., 2002),. Leopold et al. (1999) also observed that in *Lycopersicon esculentum* and *Silene vulgaris* cultures exposed to 0.1 mM Cd^{2+} and 0.1 mM Cu^{2+}, the metal binding complexes disappeared after the 7th and 14th days of treatment. According to these authors, PC formation may only have a partial role in metal resistance and in higher degrees of tolerance other mechanisms may be activated. However, in *P. sativum* plants these does not seem to be the case, since root growth was more affected when PC and hPC levels decreased.

Other works have reported significant reductions in enzyme activities and in the photosynthetic capacity, as well as an increase in oxidative stress when *P. sativum* plants were exposed to 20, 50 or 100 µM Cd (Rüegsegger et al., 1990; Lozano-Rodriguez et al., 1997; Sandalio et al., 2001; Dixit et al., 2001). Hence, it seems plausible that the toxic effects observed in the higher exposures are a result of a metabolic limitation of some of the elements necessary for GSH and hGSH synthesis (Tukendorf et al., 1997), thus for PC formation and for root growth (Siedlecka et al., 1997). This is corroborated by the decreasing levels of Cys, which is necessary for thiol formation and by the significant reduction of GSH, particularly in the higher Cd exposures. Several reports have confirmed that in plants, GSH plays a crucial role as an antioxidant and is involved in the transport and storage of sulphur, the synthesis of proteins and DNA and plant growth, as well as tolerance to biotic and abiotic stresses (Hausladen and Halsher, 1993; Rennenberg, 1997; May et al., 1998; Matamouros et al., 1999).

Tukendorf and Rauser (1990) also reported that the PC reduction after higher periods of exposure was related to the high GSH consumption. Therefore, the GSH deficit would not only limit PC synthesis but also the plant development. It is also plausible to assume that hGSH only begins to be consumed when there was a significant decrease in GSH levels, which possibly represents a backup mechanism for polythiol formation. Meuwly and Rauser (1992) have already suggested that after the GSH pool is depleted, Cd exposed roots might use alternative routes for the continued formation of PCs and other thiols. In fact, Fig. 4 shows that under the higher concentrations (30, 60 and 120 µM), in the 9th day of treatment hGSH levels were highly decreased when compared to other exposures, which was concomitant to an increase in hPC levels (Fig. 5). These data are indicative of the plant's effort to maintain the resistance mechanism active, even in detriment of its growth.

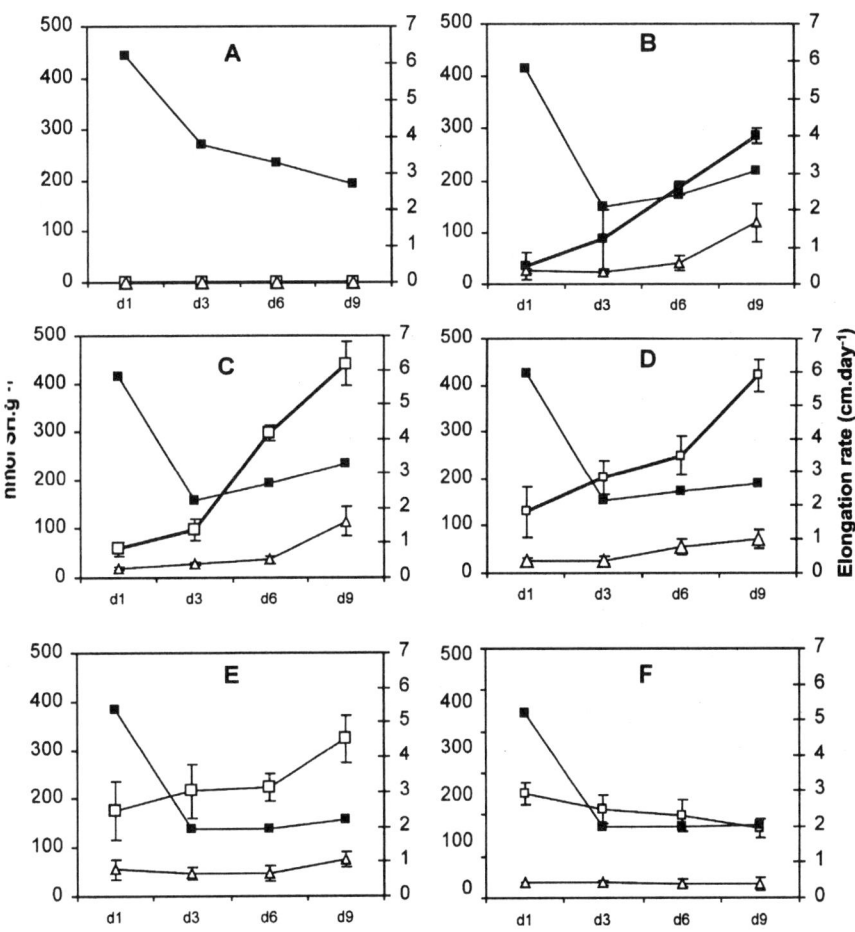

Fig. 7. Relationship between root elongation rate (■) and individual PC (□) and hPC (Δ) accumulation in roots of *Pisum sativum* plants submitted to a 9-day exposure period to 0 (A), 1 (B), 3 (C), 30 (D), 60 (E) and 120 (F) µM Cd. Mean values of at least 3 replicate experiments are presented. Vertical bars represent standard errors. *In* Lima et al. (2004).

2.2.3. Relation Between PC Production and Growth Ability

A comparison between root growth and polythiol accumulation (Fig. 7A-F) shows that there is a relation between root growth and polythiol production, particularly with PC synthesis. In controls (Fig. 7 A) root elongation rates steadily declined as plants were developing. In the presence of increasing Cd concentrations (Fig.7 B-F) these rates were more reduced compared to controls during the first 3 days, but recoveries were observed in all exposures except in the presence of 120 μM. Furthermore, higher recovery rates were associated with higher polythiol accumulations. Meuwly and Rauser (1992) stated that thiol incorporation is made at the cost of sulphur-containing proteins that are necessary for development. According to these authors, the stable amount of thiols in control root results from a balance between the input and the output of GSH, through the incorporation of cystein. Exposure to Cd disturbs this balance because of the high consumption of these thiol groups. Therefore, less reduced sulphur is available for root growth which may be reflected by an inability to recover growth rates. For the lower Cd concentrations, the timing for total root elongation rates recovery matched the increase in PC production. According to Rauser and Meuwly (1995) this timing corresponds to a stabilisation of the molar ratio of PC/Cd, suggesting that enough Cd is complexed, allowing the root to elongate in a new equilibrium. This is substantiated by the stabilisation of the different polythiols, during increasing times of exposure. In plants where a progressive decrease in all thiol levels was observed, roots were not able to grow, providing evidence for the importance of PCs and hPCs in Cd tolerance ability in *Pisum sativum*. In more tolerant species, where a reduction in polythiol synthesis was also reported (Keltjens and Van Beusichem, 1998; Sneller et al., 1999; Piechalack et al., 2002), tolerance to high Cd stress levels is likely to be related to the ability to trigger mechanisms other than PC synthesis, as described by di Toppi and Gabrielli (1999), in a "fan-shape" response. The high metabolic cost of PC synthesis might also explain why phytochelatins do not play a major role in heavy metal hyperaccumulators (Ebbs et al., 2002, Schat et al., 2002, Küpper et al., 2004).

3. THE IMPORTANCE OF THE METAL CHELATION PROCESS: CHARACTERIZATION OF THE Cd-PC COMPLEXES

Although it seems clear that in more toxic exposures, plants are not able to tolerate intracellular Cd concentrations, results presented previously obviate that in the lower concentrations, particularly under 3 μM, plants were able to maintain growth, due to a high synthesis of PCs and hPCS. However, how the cells were able to maintain equilibrium was no elucidated. Because the efficiency of PC-Cd complex formation has been pointed out as more important than PC synthesis itself, it becomes crucial to characterise the process of metal complexation under the tolerable concentrations, in order to understand how it is processed and how it maintains cellular integrity. In Lima and Figueira (2005) we isolated the γEC peptide complexes in *Pisum sativum* to characterize the process of Cd detoxification during an exposure of 1, 3, 5, 7, 10 and 15 days to 3 μM Cd, focusing on the process of maturation of Cd-PC complexes and its relation to the metal chelation ability. The choice of 3 μM Cd was based on the prior study, where this concentration promoted a higher polythiol production, without inducing any noticeable toxic symptoms, thus suggesting equilibrium between metal exposure and the process of metal detoxification. A more prolonged time of exposure of 15 days was used to obtain more information on the fate of the two complexes, on the biochemical alterations that they undergo and how they are connected to tolerance and root development. The kinetics of metal and sulphide accumulation and the alterations in thiol content in both PC-Cd complexes were determined and a relationship between the metal tolerance process, complex formation and root development was established. Furthermore, the quantitative importance of PC-Cd complexes in metal detoxification in this species was demonstrated.

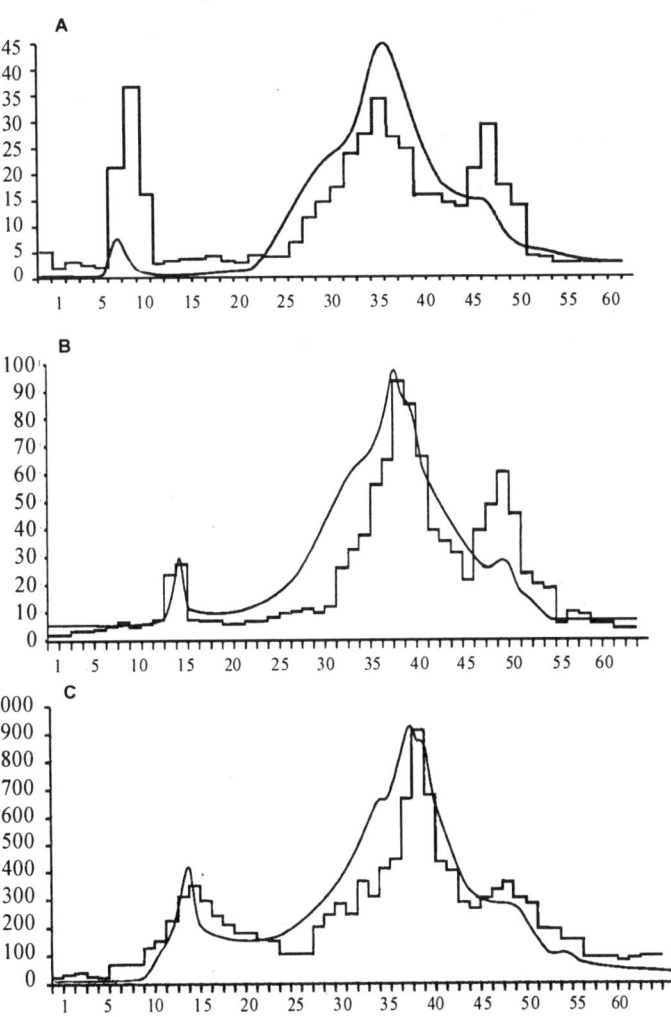

Fig. 8. Examples of gel filtration chromatography of the buffer-soluble Cd extracted from roots of *Pisum sativum* after exposure to 3µM Cd during 3 (A), 7 (B) and 15 d (C). The six-buffer extracts were lyophilised and injected in a Sephacryl S-100 chromatography column (32.6 i.d. x 11.2cm). The gel bed was equilibrated with elution buffer 10 mM HEPES (pH 8.0) and 300 mM KCl and elution was achieved at a flow rate of 0.8 ml min^{-1}, at room temperature. The continuous trace indicates the absorvance at 254 nm. The histograms associated with each chromatogram show the Cd concentrations in the fractions, distributed by peaks I, II and III. *In* Lima and Figueira (2005).

3.1. Cadmium Partitioning and Cd-Binding Complexes

One way to analyse and to assess the quantitative function of PC complexes is to fully extract the metal binding complexes with buffer, resolve the metal extracts into PC-based complexes through gel filtration and subsequently calculate the total amount of intracellular metal that is in a complexed form (Rauser and Meuwly, 1995; Rauser, 2000, 2003). This type of methodology was used to provide the necessary tools for metal-PC complex characterization.

In *P. sativum* plants, intracellular buffer-soluble-Cd co-elutes with three distinct protein peaks. Fig.8 (A, B and C) demonstrates three examples of the chromatographic profiles of buffer-soluble Cd extracted as described in Rauser (2000) and corresponding to the 3rd, 7th and 15th days of exposure. The Cd distribution among the collected fractions is represented by the histograms. Only a very small proportion of free Cd was eluted after the protein peaks, which demonstrates that practically all Cd was present in a complexed form. Peak I has been referred to as non-specific adsorption of Cd to higher weighted proteins (Rauser and Meuwly, 1995; Rauser, 2000) and looses its importance throughout the time of exposure. On the other hand, the biochemical and metal binding features of peaks II and III identify them as the HMW and LMW complexes, respectively (Ortiz et al., 1995; Rauser, 2000; 2003) and as presented by Fig. 8, their importance becomes increasingly higher with time.

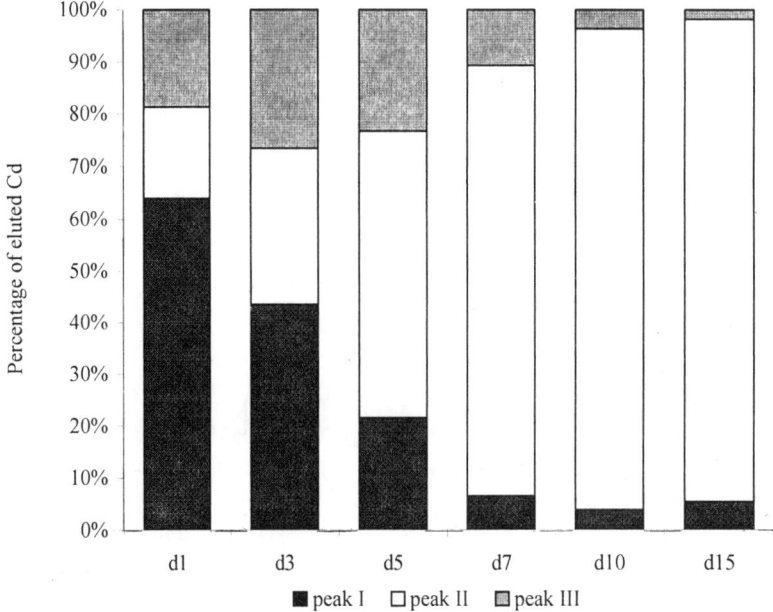

Fig. 9. Cd partioning through Cd-binding complexes isolated through gel filtration chromatography of buffer-soluble Cd extracted from roots of *Pisum sativum* during a 15-d exposure to 3 μM Cd. Data are the mean values of 3 replicate experiments. *In* Lima and Figueira (2005).

3.2. The Dynamic Nature of Cd Distribution Between the LMW and HMW Complexes

Several studies show that PC-Cd complexes are a part of a dynamic process. For example, Murasugi *et al.* (1981) showed that in *Schizosaccharomyces pombe,* there was a higher proportion of the LMW to the HMW form during the first 5 hours. However, this was altered after 9.5 hours to a prominent HMW peak, with only a partial shoulder as the LMW complex. In plant cultures, the dynamic change from LMW complexes to a dominant HMW complex has only been shown for maize, and was evident after 3 days of exposure to 3 µM Cd (Rauser, 2003). Fig. 9 demonstrates that in *P. sativum*, the importance of these two peaks is in fact time-dependent, but is different from those presented for maize plants (Rauser, 2000). By the first day, the LMW presented a higher metal accumulation, however, by the third day, both complexes accounted for similar amounts of Cd, which changed for a higher dominance of the HMW peak after the 5th day and forward. Hence, these results not only support the model proposed by Ortiz et al. (1995), where the LMW complex is synthesised prior to the HMW, but also demonstrated that the dynamic relation between the two complexes varies among species.

3.3. Biochemical Characterization of the Cd-Binding Complexes: The Process of Maturation of LMW and HMW Complexes
3.3.1. *Thiol Composition*

The γEC peptides found in *P. sativum* Cd complexes comprise the monothiols Cys, GSH, hGSH and the non-protein thiols PCs and hPCs bearing two to four oligomeric repeats (Fig. 10). The prevalence of the n_3 and n_4 isoforms in the HMW complex corroborates the increasing affinity of Cd for longer peptides (Hayashi et al., 1988) and the reduction in the total amount of SHs in the LMW complex is consistent with the Cd reduction in this complex.

These data provide important information about the residence time of PC-Cd complexes. In a more prolonged exposure, Cd complexed as the LMW form decreases, whereas the amount of metal complexed in the form of HMW increases. However, at the 15th day, the total thiol amount in the HMW form is reduced; even though Cd amounts were still increasing. Leopold et al. (1999) demonstrated that in *Silene vulgaris*, the HMW complex disappeared between the 7th and 14th d of exposure to 100 µM Cu and Cd, and that the majority of metals appeared in the LMW area. These authors concluded that the HMW complex was transitory. Zenk (1996) and Sneller *et al.* (1999) have suggested that after the Cd sequestration in the vacuole there is breakdown of vacuolar PCs in the acidic environment and that free Cd can be complexed by vacuolar organic acids, particularly when roots are more developed and vacuolated, hence presenting higher levels of citrate, malate and oxalate (di Toppi and Gabrielli, 1999), thus explaining the reduction in PC amounts, but with the maintenance of a high metal accumulation in the vacuole. According to di Toppi and Gabrielli (1999), PCs may be dissociated in the HMW complex, hence leaving Cd ions to be complexed by the vacuolar organic acids or amino-acids. Apo-PCs can then return to the cytosol and continue their "shuttle role". In *P. sativum* roots, although the amount of SH compounds decreased in the last day of exposure, the total Cd complexed in the HMW complex was still increasing. These results suggest that there may be a recycling of the SH compounds stored in the HMW complex, but the complex itself remains stablewhen considering the amount of metal sequestrated, which suggests that there is no dissociation of the HMW complex. One possible interpretation for the decline in thiol content between days 10 and 15 is also the formation of Cd-S crystallites with time. Scarano and Morelli (2003) demonstrated the presence of Cd-S crystallites in tomato plants and in marine microalgae exposed to Cd revealing the

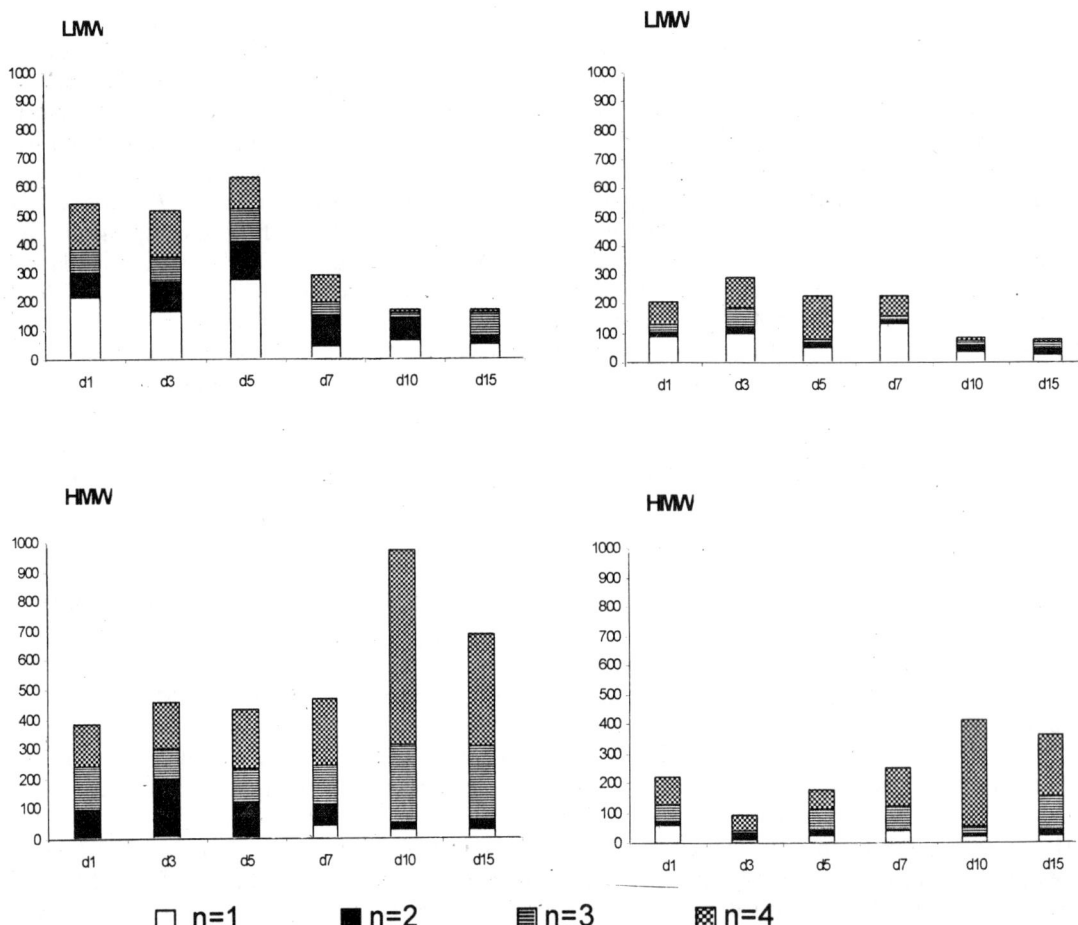

Fig. 10. Thiol profiles in PC-Cd complexes in HMW and LMW complexes collected through gel filtration chromatography of the buffer-soluble Cd extracted from roots of *Pisum sativum* during a 15-d exposure to 3 µM Cd and analysed by HPLC with fluorescence labelling. Data are presented as the mean of 3 replicate experiments. *In* Lima and Figueira (2005).

formation of Cd-S crystallites with a PC coating. In *Candida glabrata*, Cd-S crystallites are formed in the cytosol and accumulated in the vacuole (Mehra et al., 1994). It is possible that PCs may be the initial nucleating molecules for growth of Cd-S crystallites, and that with time, further enlargement of a Cd-S crystallite core could then proceed within the phytochelatin "casing".

Table 1. SH, Cd and acid labile sulphide molar ratios in PC Cd-binding complexes from peaks II and III isolated through gel filtration chromatography of buffer-soluble Cd extracted from roots of *Pisum sativum* during a 15-d exposure to 3 µM Cd. Data are the mean values of at least 3 replicate experiments. *In* Lima and Figueira (2005).

	SH/Cd		S/Cd		SH+S/Cd		S/SH	
	LMW	HMW	LMW	HMW	LMW	HMW	LMW	HMW
d1	3.95	3.38	0.03	3.31	3.98	6.68	0.17	0.16
d3	2.52	1.42	0.46	3.38	2.98	4.80	0.36	0.18
d5	2.78	0.89	0.78	0.63	3.56	1.53	0.37	0.28
d7	2.48	0.43	0.76	0.63	3.23	1.06	0.31	0.31
d10	2.51	0.50	1.10	0.38	3.61	0.88	0.17	0.54
d15	1.97	0.27	1.03	0.06	3.00	0.34	0.24	0.62

3.3.2. Acid Labile Sulphide Incorporation and Metal Sequestration Ability

In some plant species and yeasts sulphide ions have a very important role in the PC detoxification efficiency (Murasugi et al., 1983; Mehra et al., 1994). The incorporation of labile sulphur in the HMW complex does not only increased the amount of Cd complexed per PC molecule but also increases its stability (Cobbett and Goldsbrough, 2002). In order to demonstrate the role of sulphide accumulation in Cd complexation in both complexes, Lima and Figueira (2005) determined the amount of S^{2-} per mol of SH present as well as the molar rations of SH/Cd, S/Cd and SH+S/Cd, in the HMW and LMW complexes, during the 15 d of exposure (Table 1). Results presented are consistent with the fact that the process of bio-mineralization enhances the Cd-binding ability of higher weighted complexes (Murasugi et al., 1983; Mehra et al., 1994). The increase in estimated molar S/SH ratios indicate that there was S incorporation in both complexes, particularly in the vacuolar form, and the S/Cd ratios show that sulphide ions played a significant role in metal chelation, although SHs were the predominant Cd ligands. Data also demonstrate that although enhanced Cd sequestration ability was first achieved by an increase in thiol chain length, after the $5^{th}/7^{th}$ days of exposure, sulphide incorporation was more effective in decreasing the amount of SH needed for Cd sequestration. This is consistent with the hypothesis suggested by Kneer and Zenk (1996) that state that S^{2-} incorporation allows a reduction in PC concentration, but maintaining Cd complexation, so cells can reuse the needed amino acids for primary metabolism, without releasing toxic Cd ions (Kneer and Zenk, 1996). Unexpectedly, from that day on, there was a higher amount of Cd per SH+S molecules than would be theoretically expected, particularly comparing to other reports (Rauser & Meuwly, 1995; Kneer and Zenk, 1996; Rauser, 2000). In the final d of exposure, Cd concentration was still increasing in the eluted HMW complex, but this trend was not accompanied by an increase in the S or SH concentrations. These findings suggest that after a certain period of exposure, other molecules may be interfering with metal chelation, or that Cd ions are not only bound to the SH moieties in the PC molecules. Since these alterations were specific of the HMW complex, the hypothesis of the formation of S crystallites in PC-coated Cd complexes is consistent with these results. It seems plausible to suggest that with time, cells are capable of reducing the amount of SH needed in the vacuolar complex, and to reuse the needed thiols for other purposes. In previous reports in the same species, Lima et al. (2004) showed that the high SH consumption for PC synthesis becomes itself a limiting factor in higher degrees of Cd stress. Hence, we might be in the presence of an efficient way for maintaining the thiol-based mechanism active, while reducing its detrimental effects to the cell. This process of maturation of the HMW Cd-complexes goes towards a higher degree of biochemical complexity and metal chelation ability and seems to be time-dependent.

4. THE QUANTITATIVE ROLE OF THE LMW AND HMW COMPLEXES IN TOTAL INTRACELLULAR METAL DETOXIFICATION AND ITS RELATION TO ROOT AGING

In Fig. 11 the amount of metal sequestrated by the LMW and HMW complexes with the amount of metal absorbed by the plant is compared throughout the 15-days exposure.

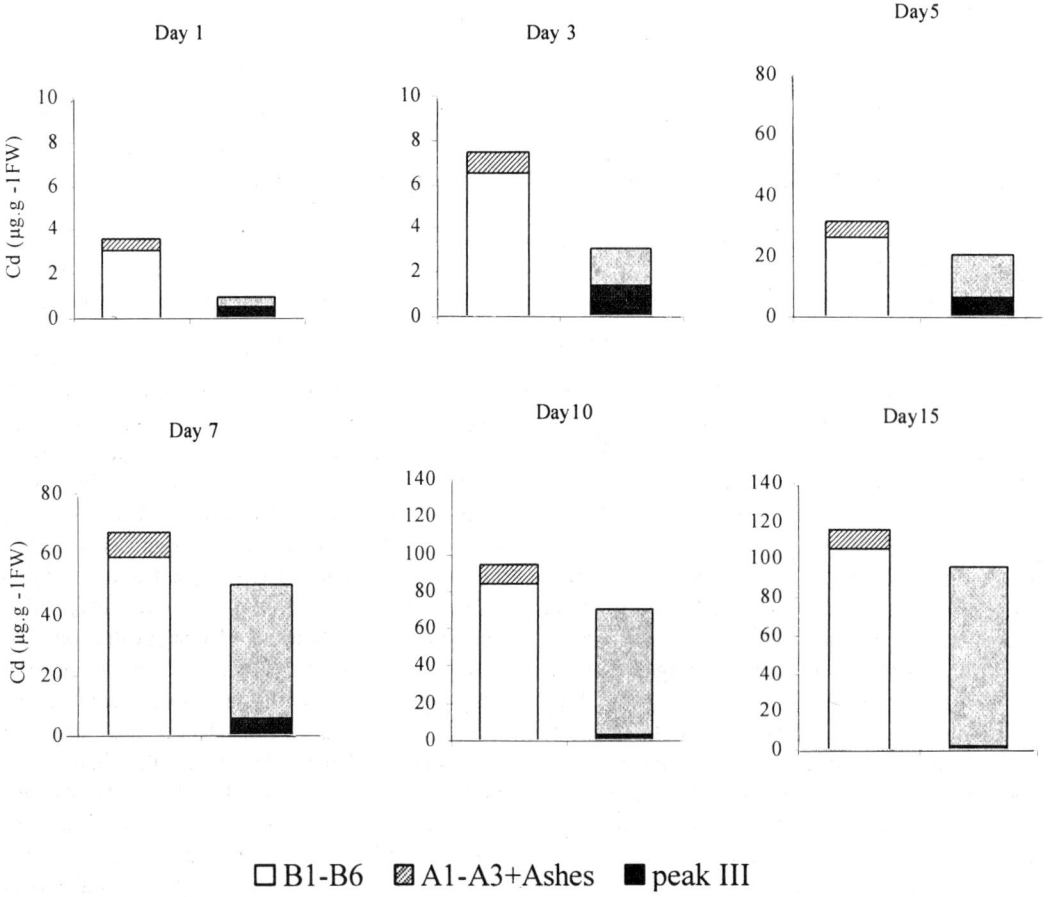

Fig.11. Comparison between the Cd sequestered by the two LMW and HMW PC-Cd complexes and the total amount of metal absorbed by roots. Total amounts of Cd absorbed by roots are represented by buffer extracts 1 to 6 plus the acid extracts and ash. Complexed Cd is given by the sum of Cd bounded to high and low molecular weight Cd-binding complexes. Data are the mean values of 3 replicate experiments. *In* Lima and Figueira (2005).

It is observed that the total Cd present in a complexed form increased gradually over time. During the first day of exposure, the sum of the two complexes would account for 29% (0.9 µg Cd) of the total intracellular Cd. The role of these complexes was progressively more important after the 7th day. After 15 days of exposure, 82.9% of the intracellular Cd was present as LMW or HMW complexes. The amount of metal chelated by the LMW complex was prevalent during the first 3 days, but lost its importance progressively over time, stabilizing during the 5th and 7th days, after which HMW complexes dominated.

Results are, therefore, specific examples of the dynamic process that characterises metal complexation in this species and its relation to root development. When analysing the shift of dominance between the two complexes during the time of exposure, it became apparent that root growth recoveries were simultaneous to the increase of HMW complexes. Recent works suggest that the LMW complex is not able, *per se*, to efficiently detoxify Cd and that the HMW is essential for the sequestration process (Cobbett and Goldsbrough, 2002). Our results indicated a later dominance of HMW when compared to maize plants exposed to the same Cd concentration for 6 days (Rauser, 2000). While in *P. sativum* the HMW complex only became dominant after 5 days, by day 4, the same complex already exceed by far the LMW in *Z. mays*. It previously demonstrated that the lower synthesis of the polythiols with n=3 and 4 may be related to the lower tolerance capacity of pea plants when compared with other species (Lima et al., 2004). The analysis of PC-Cd complexes brings more light into this question, suggesting that the reduced level of longer olygomers in *P. sativum* is related to a lower synthesis of the HMW complex during the initial exposure. In addition, they are consistent with the hypothesis that the process of Cd-complex formation as well as sulphide incorporation is more important than PC synthesis itself. The high SH/Cd molar ratios found in the first days of exposure corroborate that the metal sequestration capacity was not fully achieved in the LMW complex, particularly in the initial stages of exposure.

Concerning the dynamic status of the two complexes during the time of exposure, the accumulation patterns of the cytosolic LMW and the vacuolar HMW complexes are related to the degree of cellular vacuolization (Rauser, 2000). While in the first stages of root growth, the presence of young cells would limit a higher accumulation of the vacuolar form (Rauser, 2000), during the 5th and 7th days, the increasing dominance of the HMW form and stabilization of the LMW complexes is a reflex of the heterogeneity of cells present in the samples collected. Fig. 11 shows that until the 7th day of exposure, roots still present high elongation rates. This means that although in the older part of the collected roots, the more vacuolated cells provided high quantities of HMW complexes, the percentage of younger cells was still enough to yield high levels of LMW complexes. After the 10th day, the HMW form was clearly dominant, whereas the LMW was much reduced, which is consistent with the presence of more mature and highly vacuolated cells, as corroborated by the reduction in root elongation rates, showing the importance of root development stage in the ability to form metal binding complexes.

5. CONCLUSIONS: THE IMPORTANCE OF PCs IN METAL DETOXIFICATION IN *PISUM SATIVUM* PLANTS

The present report allows us to better understand the complexicity of the metal detoxification process and the importance of γEC peptides in intracellular metal complexation in roots of *P. sativum*. Several considerations may be withdrawn from our results. Firstly, in exposures to "more realistic" concentrations (1 and 3 µM), plants are able to cope with stress without displaying any noticeable phytotoxic effects, and enhance the risk of food-chain transfer. In more toxic exposures, higher levels of stress induced a reduction in thiol production. This emphasises that the concentration and time of exposure are important factors when evaluating the true role of PCs in metal stress coping. In addition, it also provides evidence that the thiol-based tolerance is only efficient in lower degrees of stress. The reason for this fact is possibly related to the high consumption of monothiols GSH and hGSH, which would limit PC synthesis, compromising plant development. The high metabolic cost of PC synthesis implies that, in higher degrees of metal stress, only plants able to trigger other mechanisms, which can efficiently replace PC synthesis, will be able to tolerate metal exposure.

Nevertheless, in lower and more realistic concentrations, tolerance is in fact relied on polythiol production, providing evidence of the importance of PCs and hPCs in *Pisum sativum's* tolerance to environmental Cd exposures. In the presence of a 3 µM Cd concentration, PCs are the major intracellular Cd chelators, evidencing their importance in metal detoxification in this species. Both LMW and HMW PC-Cd complexes are synthesized at different stages of plant development and their metal complexation abilities are dependent on their sulphide and thiol status. The shift to a HMW complex's dominance is related to a higher degree of tolerance and its timing differed from other plant species, which might explain the lower Cd-tolerance of pea plants, when compared for example to maize. With longer times of exposure, the HMW complex undergoes a process of maturation, with important biochemical alterations, in which sulphide incorporation, in parallel to an increase in the PC oligomeric polymerization, enhanced the Cd sequestration ability. This process of maturation apparently goes beyond the simple SH/S-Cd bound and possibly involves a time-dependent formation of Cd-S crystallite, hence suggesting a capacity for reducing the amount of SH needed in the vacuolar complex. We also demonstrated the importance of root development in metal stress coping.

Taken all results into consideration, it is possible to conclude that PCs and hPCs have a decisive role in Cd detoxification in *Pisum sativum*, at any of the studied concentrations. Their efficiency is nevertheless limited, at some extent, by the GSH and Cys status as well as a rapid and effective formation of the HMW complex. Under more prolonged or higher exposures, their high metabolic cost becomes a setback that plants have to overcome and limits their efficiency as a metal resistance mechanism. In high exposures, the inability to cope with the high GSH consumption, needed for a high PC synthesis, renders the plants unable to survive. Although it has been suggested that with time PCs are a transient mechanism, our studies with *P. sativum* show that under more tolerable exposures, alternative mechanisms may be developed in order to reduce the thiol consumption and minimise its damages to the plant, hence maintaining the PC-based mechanism active. The maturation process of the HMW complex seems to demonstrate this trend. Increasing sulphide incorporation in the Cd-PC complexes was proven to reduce the number of SH groups needed for Cd chelation and with time there are important alterations in the complexes, which allow the decrease of needed SH groups. Results presented in pea plants suggest the presence of other molecules assisting metal chelation after a more prolonged exposure, such as S-Cd crystallites, which may corroborate the hypothesis of a thiol recycling mechanism rather than a PC replacing mechanism. We can conclude that PCs play a crucial role in Cd stress adaptation and that at least during the first 15-days of exposure, they are responsible for the detention of the majority of intracellular Cd ions.

References

Alloway BJ, Steinnes E (1999) Anthropogenic additions of cadmium to soils. In: Machlaughin MJ, Singh BR (eds.) Cadmium in soils and plants. Kluwer Academic Publishers, pp. 97-118.

Cobbett CS, Goldsbrough P (2002) Phytochelatins and methalothioneins: roles in heavy metal detoxification and homeostasis. Ann Rev Plant Biol 53: 100301-135154.

Delhaize E, Adams TL (1989) Poly (γ-glutamilcysteinyl) glycine synthesis in *Dattura inoxia* and binding of cadmium. Plant Physiol 89: 700-706.

De Knecht JA, Koeovets PLM, Verkleij JAC, Ernst WHO (1992) Evidence against a role of phytochelatins in naturally selected increased cadmium tolerance in *Silene vulgaris* (Moench) Garcle. New Phytol 122: 681-688.

De Knecht JA, von Dillene M, Koevoets RM, Schat JH, Verkleij JAC, Ernst W (1994) Phytochelatins in cadmium-tolerant *Silene vulgaris*. Plant Physiol 104: 255-261.

De Knecht JA, Van Baren N, Ten Bookum, WM, Wong Fong Sang HM, Koeovets, PLM, Schat, Verkleij, JAC (1995) Synthesis and degradation of phytochelatins in cadmium sensitive and cadmium tolerant *Silene vulgaris*. Plant Physiol 104: 255-261.

di Toppi SL, Gabrielli R (1999) Response to cadmium in higher plants. Environ Exp Bot 41: 105-130.

Dixit V, Pandey V, Shyam R (2001) Differential antioxidative responses to cadmium in roots and leaves of pea *Pisum sativum* L. (cv Azad). J Exp Bot 52: 1101-1109.

Ebbs S, Lau I, Ahner B, Kochian L (2002) Phytochelatin synthesis is not responsible for Cd tolerance in the Zn/Cd hyperaccumulator *Thlaspi caerulescens* (J. & C. Presl). Planta 214: 635-640.

Grill E, Löffer S, Winnacker E-L, Zenk MH (1989) Phytochelatins, the heavy-metal-binding peptides of plants are synthesized from glutathione from a specific γ-glutamilcysteine dipeptil transpeptidase (phytochelatin synthase). Proc Natl Acad Sci USA 86: 6838-6842.

Grill E, Winnacker EL, Zenk MH (1985) Phytochelatins: The principal heavy-metal complexing peptides of higher plants. Science 230: 674-676.

Grill E, Winnacker EL, Zenk MH (1986) Synthesis of seven different homologous phytochelatins in metal exposed *Schizosaccharomices pombe* cells. FEBS Lett 197: 115-119.

Grill E, Winnaker EL, Zenk MH (1987) Phythochelatins, a class of heavy metal complexing peptides from plants, are functionally analogous to metalothioneins. Proc Natl Acad Sci USA 84: 439-443.

Gussarsson M, Asp H, Adalsteinsson S, Jensén P (1996) Enhancement of cadmium effects on growth and nutrient composition of birch (*Betula pendulata*) by buthionine sulphoximine (BSO). J Exp Bot 47: 211-215.

Ha SB, Smith AP, Howden R, Dietrich WM, Bugg S, O'Connell MJ, Goldsbrough, PB, Cobbett CS (1999) Phytochelatin synthase genes from *Arabidiopsis* and the yeast *Schizosaccharomyces pombe*. Plant Cell 11: 1153-1163.

Hausladen A, Halsher RG (1993) Glutathione. In: Alscher RG, Hess, JL (eds.) Antioxidants in Higher Plants. CRC Press, Boca Raton, pp. 1-30.

Hayashi Y, Nakagaua CW (1988) The change in cystein components in Cd-binding peptides from the fission east during their induction by cadmium. Biochem Cell Biol 66: 268-274.

Hayashi Y, Nakagaua CW, Mutoh N, Isoobe M, Goto T (1991) Two pathways in the biosynthesis of cadistins (γEC)$_n$G in the cell free system of the fission east. Biochem Cell Biol 69: 115-121.

Howden R, Goldsbrough PB, Andersen CR, Cobbett CS (1995) Cadmium-sensitive, *cad1* mutants of *Arabidopsis thaliana* are phytochelatin deficient. Plant Physiol 107: 1059-1066.

Hu S, Ken WKL, Wu M (2001) Cadmium sequestration in *Chlamidomonas reinhardtii*. Plant Sci 161: 987-996.

Jackson PJ, Unkefer CJ, Doolen JA, Watt K, Robbinson NJ (1987) Poly (γglutamilcisteynil) glycine: Its role in cadmium resistance in plant cells. Proc Natl Acad Sci USA 84: 6619-6623.

Keltjens WG, Van Beusichem ML (1998) Phytochelatins as a biomarker for heavy metal toxicity in maize, single effects of copper and cadmium. J Plant Nutr 21: 635-648.

Klapheck S (1988) Homoglutathione: Isolation, quantification and occurrence in legumes. Physiol Plant 74: 727-732.

Klapheck S, Schlun S, Bergman L (1995) Synthesis of pytochelatins and homo-phytochelatins in *Pisum sativum* L. Plant Physiol 107: 515-521.

Klapheck S, Fliegner W, Zimmer I (1994) Hydroxy-methil-phytochelatins [γ-glutamilcisteine.$_n$-serine] are metal-induced peptides of the poaceae. Plant Physiol 104: 1325-1332.

Kneer R, Zenk MH (1992) Phytochelatins protect plant enzymes from heavy metal poisoning. Biochemistry 31: 2663-2667.

Kneer R, Zenk MH (1996) The formation of Cd-phytochelatin complexes in plant cell cultures Phytochemistry 44: 69-74.

Küpper H, Mijovilovich A, Meyer-Klaucke W, Kroneck PMH (2004) Tissue- and age-dependent differences in the complexation of cadmium and zinc in the cadmium/zinc hyperaccumulator *Thlaspi caerulescens* (Ganges Ecotype) revealed by X-ray absorption spectroscopy. Plant Physiol 134: 748-757.

Leopold I, Gunther D, Schmit J, Neumann D (1999) Phytochelatins and heavy metal tolerance. Phytochemistry 50: 1323-1328.

Lima AIG, Pereira, SAI, Figueira EMAP, Caldeira GCN, Caldeira HDQ (2004) Cadmium detoxification in roots of *Pisum sativum* seedlings: Relationship between toxicity levels, thiol pool alterations and growth. Environ Exp Bot *In press*.

Lima AIG, Figueira EMAP (2005) Time-maturation and biochemical characterisation of high and low molecular weight cadmium-phytochelatin complexes in roots of *Pisum sativum*. Submitted for publication.

Lozano-Rodriguez E, Hérnandez LE, Bonay P, Carpena-Rui RO (1997) Distribution of cadmium in root tissues of maize and pea plants: Physiological disturbances. J Exp Bot 306: 123-128.

Maitani T, Kubota H, Sato K, Yamada T (1996) The composition of metals bound to class III metalothionein (phytochelatins and their desglycyl peptide) induced by various metals in root cultures of *Rubia tinctorum*. Plant Physiol 110: 1145-1150.

Mann SS, Rate AW, Gilkes RJ (2002) Cadmium accumulation in agricultural soils in Western Australia. Water Air Soil Poll 141: 281-297.

Matamouros AM, Moran JF, Iturbe-Ormaetxe I, Rubio MC, Becana M (1999) Glutathione and homoglutathione synthesis in legume root nodules. Plant Physiol 21: 879-888.

May MJ, Vernoux T, Leaver C, Montagu M, Inzé D (1998) Glutathione homeostasis in plants: Implications for environmental sensing and plant development. J Exp Bot 49: 649-667.

Mehra RJ, Kodati R, Abdulah R (1995) Chain-length-dependent Pb(II) coordination in phytchelatins. Biochem Biophy Res Comm 215: 730-736.

Mehra RJ, Mulchandani P, Hunter TC (1994) Role of CdS quantum crystallites in cadmium resistance in *Candida glabrata*. Biochem Biophys Res Comm 200: 1193-1200.

Meuwly P, Rauser WE (1992) Alteration of thiol pools in roots and shoots of maize seedlings exposed to cadmium. Adaptation and developing cost. Plant Physiol 99: 8-15

Murasugi A, Wada C, Hayashy Y (1983) Occurrence of acid-labile sulphide in cadmium binding peptides from fission east. J Biol Chem 93: 661-667.

Murasugi A, Wada C, Hayashi Y (1981) Cadmium-binding peptide induced in fission yeast, *Schizosaccharomyces pombe* J Biochem 90: 1561-1564.

Mehra RK, Kodati R, Abdulah R (1995) Chain-length-dependent (PbII) coordination in phytchelatins. Biochem Biophy Res Comm 215: 730-736.

Meuwly P, Rauser WE (1992) Alteration of thiol pools in roots and shoots of maize seedlings exposed to cadmium: Adaptation and developing cost. Plant Physiol 99: 8-15.

Mutoh N, Hayashi Y (1988) Isolation of mutants of *Schizosaccharomices pombe* unable to synthesize cadystins, small cadmium-binding peptides. Biochem Biophy Res Comm 151: 32-39.

Nocito FF, Pirovano L, Maurizio C, Sacchi GA (2002) Cadmium-induced sulphate uptake in maize roots. Plant Physiol 129: 1872-1879.

Ortiz, DF, Ruscitti KF, McCue KF, Ow DW (1995) Transport of metal-binding peptides by HMT1, a fission yeast ABC-type vacuolar membrane protein. J Biol Chem 27: 4721-4728.

Oven M, Grill E, Golan-Goldhirsh A, Kutchan TM, Zenk MH (2002a) Increase in free cysteine and citric acid in plant cells exposed to cobalt ions. Phytochemistry 60: 467-474.

Oven M, Page JE, Zenk MH, Kutchan TH (2002b) Molecular characterisation of the homo-phytochelatin synthase of soybean *Glycine max*: Relation to phytochelatin synthase. J Biol Chem 277: 4747-4754.

Perrin DD, Watt AE (1971) Complex formation of zinc and cadmium with glutathione. Biochem Biophy Acta 230: 96-104.

Piechalak A, Tomaszewska B, Baralkiewicz D, Malecka A (2002) Accumulation and detoxification of lead ions in legumes. Phytochemistry 60: 153-152.

Rauser WE (1999) Structure and function of metal chelators produced by plants: The case of amino acids, organic acids, phytin and methalothioneins. Cell Biochem Biophys 31: 1-31.

Rauser WE (2000) Roots of maize seedlings retain most of their cadmium through two complexes. J Plant Physiol 156: 545-551.

Rauser WE (2003) Phytochelatin-based complexes bind various amounts of cadmium in maize seedlings depending on the time of exposure, the concentration of cadmium and the tissue. New Phytol 158: 269-278.

Rauser WE, Meuwly P (1995) Retention of cadmium in roots of maize seedlings. Role of complexation of phytochelatins and related thiol peptides. Plant Physiol 109: 195-202.

Reese RN, White CA, Winge DR (1992) Cadmium-sulfide crystallites in Cd-(γ-EC)nG peptide complexes from tomato. Plant Physiol 107: 225-229.

Reese RN, Wagner GJ (1987) Effects of buthionine sulfoximine on Cd-binding peptide levels in suspension-cultured tobacco cells treated with Cd, Zn or Cu. Plant Physiol 84: 574-577.

Reese RN, White CA, Winge DR (1988) Cadmium-sulfide crystallites in Cd-(γ-EC)nG peptide complexes from tomato. Plant Physiol 98: 225-229.

Reese NR, Winge DR (1988) Sulfide stabilisation of the cadmium-γ-glutamyl peptide complex *Schizisaccharomices pombe*. J Biol Chem 263: 12832-12835.

Rennenberg H (1997) Molecular approaches to glutathione biosynthesis In: Cram WJ, DeKok LJ, Stulem I, Brunold C, Rennenberg H (eds.) Sulphur Metabolism in Higher Plants. Backhuys Publishers, Leiden, The Netherlands, pp. 59-70.

Rüeggserger D, Schumtz D, Brunold C (1990) Regulation of Glutathione synthesis by cadmium in *Pisum sativum* L. Plant Physiol 93: 1579-1584.

Robbinson NJ, Ratliff J, Anderson E, Delhaize JM, Berger JM, Jackson PJ (1988) Biosynthesis of poly (γ-glumilcysteinyl) glycine in cadmium-tolerant *Datura innoxia* Mill cells. Plant Sci 56: 197-204.

Salt DE, Rauser WE (1995) MgATP-dependent transport of phytochelatins across the tonoplast of oat roots Plant Physiol 107: 1293-1301.

Sandalio LM, Daurzo HC, Gómez M, Romro-Puertas MC, delRio LA (2001) Cadmium induced changes in the growth and oxidative metabolism of pea plants. J Exp Bot 364: 2115-2126.

Scarano G, Morelli E (2003) Properties of phytochelatin-coated CdS nanocrystallites formed in a marine phytoplanktonic alga (*Phaeodactylum tricornutum*, Bohlin) in response to Cd. Plant Sci 165: 803-810.

Schat H, Kalf MMA (2002) The role of phytochelatins in constitutive and adaptative heavy metal tolerances in hyperacumulator and non-hyperacumulator metallophytes. J Exp Bot 53: 2381-2392.

Siedlecka A, Krupa Z, Samuelsson G, Öquist G, Gardetröm P (1997) Primary carbon metabolism in *Phaseolus vulgaris* plant under Cd/Fe interaction. Plant Physiol Biochem 35: 951-957.

Sneller FEC, Noordover ECM, Bookum WMT, Schat H, Bedaux JJM, Verkleij JAC (1999) Quantitative relashionship between phytochelatins accumulation and growth inhibition during prolonged exposure to cadmium in *Silene vulgaris*. Ecotoxicology 8: 167-175.

Speiser DM, Abrahamson SL, Banuelos G, Ow DW (1992) *Brassica juncea* produces a phytochelatin-cadmium-sulfide complex. Plant Physiol 99: 817-821.

Souza JF, Rauser WE (2003) Maize and radish sequester excess cadmium and zinc in different ways. Plant Sci 165: 1009-1022.

Steffens JC (1990) The heavy metal-binding peptides of plants. Ann Rev Plant Physiol Plant Mol Biol 41: 553-575.

Tukendorf A, Rauser WE (1990) Changes in glutathione and phytochelatins in roots of maize seedlings exposed to cadmium. Plant Sci 70: 155-166.

Tukendorf A, Skórzynska-Polit E, Baszýnski T (1997) Homophytochelatin accumulation in Cd-treated runner bean plants is correlated to their growth stage. Plant Sci 129: 21-28.

Vatamaniuk OK, Mari S, Lu Y-P, Rea PA (2000) Mechanisms of heavy metal ion activation of phytochelatin (PC) synthase: blocked thiols are sufficient for PC synthase-catalysed transpeptidation of glutathione and related thiol peptides. J Biol Chem 275: 31451-31459.

Wagner GJ (1993) Accumulation of cadmium in crop plants and it consequences to human health. Adv Agron 51: 173-212.

Vögelli-Lange R, Wagner GJ (1990) Subcellular localisation of cadmium and cadmium-binding peptides in tobacco leaves. Plant Physiol 92: 1086-1093.

Yan S, Tsay C, Chen Y (2000) Isolation and characterisation of phytochelatin synthase in rice seedlings. Proc Natl Acad Sci USA 244: 202-207.

Yong, LZ, Pilon-Smits EAH, Tarun AS, Weber SU, Juanin L, Terry N (1999) Cadmium tolerance in Indian mustard is enhanced by overexpressing γ-glutamilcysteine synthase. Plant Physiol 119: 1169-1177.

Yoshimura E, Kabuyama Y, Yamazaki S, Toda S (1990) Activity of poly γ-(glutamilcisteinil)glycine synthesis in crude extract of fission east, *Schizosaccaromices pombe*. Agric Biol Chem 54: 3025-3026.

Xiang C, Werner BL, Christensen EM, Oliver DJ (2001) The biological functions of glutathione revisited in *Arabidiopsis* transgenic plants with altered glutathione levels. Plant Physiol 126: 564-574.

Zenk MH (1996) Heavy metal detoxification in higher plants-a review. Gene 179: 21-30.

Zhu YL, Pilon-Smits AH, Jouarin L, Terry N (1999) Overexpresion of glutathione synthase in Indian mustard enhances cadmium accumulation and tolerance. Plant Physiol 119: 73-80.

Cadmium Toxicity and Tolerance in Plants
Editors: Nafees A. Khan and Samiullah
Copyright © 2006, Narosa Publishing House, New Delhi, India

Purification of Glutamate Dehydrogenase Isoenzymes from Control and Cadmium Treated Tomato Leaf

Chiraz Chaffei[1]*, Celine Masclaux-Daubresse[2], Houda Gouia[1] and Mohamed Habib Ghorbel[1]

1- Unité de Recherche : Nutrition et Métabolisme Azotés et Protéines de Stress (UR/09), Département de Biologie, Faculté des Sciences de Tunis.
2- Institut National de Recherche Agronomique : Unité de Nutrition Azotée des Plantes, INRA Versailles, Route de Saint-Cyr, 78026 Versailles CEDEX-France.

1. INTRODUCTION

Cadmium (Cd) is a toxic heavy metal that represents a serious environmental pollutant for both animals and plants. As with other metals, Cd is known to interact with proteins, influencing protein activity. Although Cd has been shown to be genotoxic metal, the molecular mechanism of Cd toxicity to plants is not well understood. In plants, Cd is known to disturb photosynthesis (Chugh and Swahney, 1999; Baszynski et al., 1980; Gouia et al., 2003), carbohydrate metabolism (Greger and Bertell, 1992) and several enzyme activities especially of nitrogen metabolism. These enzymes are differentially affected by Cd stress (Chugh et al., 1992; Gouia et al., 2002; Chaffei et al., 20003a, b). Nitrate reductase (NR, EC 1.6.6.1), nitrite reductase (NiR, EC 1.6.6.4), the GS2 and the Fd-GOGAT (of GS/GOGAT cycle) activities are significantly decreased (Ouariti et al., 1997; Gouia et al., 2000; Chaffei et al., 2003a). In contrast, glutamate dehydrogenase (NADH-GDH, EC 1.4.1.2) shows a substantial rise in activity under Cd stress (Boussama et al., 1999; Gouia et al., 2000; Chaffei et al., 2003a).

Glutamate dehydrogenase (EC 1.4.1.2) was long considered as a key enzyme for the assimilation of ammonia into amino acids but this role has been questioned since GOGAT (EC 1.4.7.1) was discovered (Miflin and Lea, 1976; Suzuki and Gadal, 1982). The GS/GOGAT pathway is thought to be the favoured process for ammonia assimilation at normal intracellular concentrations (Miflin and Lea, 1976; Kurganov, 2000). The main objection to a major role of GDH in nitrogen metabolism is its apparent low affinity for ammonia and its low availability in plant cells (Miflin and Lea, 1977). It was suggested that GDH may be a catabolic enzyme, active in the deamination of glutamate with partial functioning in the synthesis of glutamate (Wallsgrove et al., 1983; Wallsgrove et al., 1987). The synthesis of glutamate from oxoglutarate and ammonium mediated by the enzyme GDH has recently been re-evaluated; Yamaya et al. (1984) showed that the levels of ammonia in mitochondria are in the range of 5 to 10 mM and that mitochondria can tolerate these relatively high concentrations. The main objection to a major role for GDH in nitrogen metabolism is its low affinity for ammonia. The relevance of the large amounts of GDH in plants and the conditions under which the enzyme may be involved in the nitrogen assimilation process has yet to be defined. The possible role of GDH in the

---

* Corresponding Author (chiraz_ch2001@yahoo.fr)

ammonia detoxification process is supported by the finding that the enzyme is induced by high levels of ammonia (Miflin and Habash, 2002; Dubois et al., 2003; Tercé-Laforgue et al., 2004).

The aminating activity of GDH is quite sensitive to air and soil pollutants and as such it has been advocated to be the best enzymic indicator of pollution stress (Syntichaki et al., 1996; Mohan and Hosetti, 1997).

Recently, Munoz-Blanco and Cardenas (1989) working with *C. reinhardtii* under different trophic and stress conditions, Skopelitis and Roubelakis-Angelakis (2002) working with plants under differentially exogenous nitrogen environment, Gouia et al. (2003) on bean seedlings under cadmium stress have shown that GDH aminating activity is adaptive, being involved in the maintenance of intracellular levels of L-glutamate, when it cannot be maintained by the GS/GOGAT cycle. Although the exact mechanism of increasing GDH by pollutants is not known, some postulates have been advanced. The pollutants may also increase enzyme activity through altered membrane permeability. It has also been shown that SO_2 changes the isozyme pattern and electrophoretic mobility of GDH (Pahlich and Joy, 1971).

L-Glutamate dehydrogenase, with a molecular weight of 208 000 to 270 000, is composed of four to six subunits, contain a free – SH group at the active centre, and associated with metal ions. Some isoenzymes of GDH are inducible and vary according to the nutritional and environmental status of the tissues. The level and activity of enzyme is regulated by age, light/dark regime, organic and inorganic nitrogen, carbon and energy status, growth regulators and some other factors. The enzyme seems to be important in assimilation of ammonia under stress conditions such as dark, high temperature, salinity, water stress, environmental pollution (heavy metals), senescence and other abnormalities (Srivastava and Singh, 1987; Lasa et al., 2002).

Loulakakis and Roubelakis-Angelakis (1991, 1996), Cammaerts and Jacobs (1985) showed that GDH from grapevine and *Arabidopsis thaliana* tissues has a haxameric structure consisting of two subunit-polypeptides with similar antigenic properties but with different molecular weights and charges (Loulakakais and Roubelakis-Angelakis, 1991). This is interpreted as reflecting the differential association of two subunit types in a hexameric protein. In maize, it has been suggested that both subunits are encoded by different genes located in separate loci. However, it has recently been reported that extensive library screening yielded only single maize GDH cDNA species, failing to isolate the cDNA encoding for the second subunit (Sakakibara et al., 1995).

The occurrence of GDH and the existence of multiple molecular forms of the enzyme have been reported in a number of plants (Hartmann et al., 1973; Yue, 1969). A GDH pattern of seven isoenzymes is mostly represented among the higher plants (Chou and Splittstoesser, 1972). Leaf GDH responds to various environmental cues including water, temperature, herbicides and heavy metals (Osuji and Madu, 1997; Masclaux-Daubresse et al., 2002; Chaffei at al, 2003).

Activity staining of polyacrylamide gel electrophoresed plant extracts reveals the presence of upto seven GDH isoforms in many plants tissues (Scheid et al., 1980; Nauen and Hartmann, 1980; Cammaerts and Jacobs, 1985; Loulakakis and Roubelakis-Angelakis, 1990; Magalhaes and Huber, 1991). This is interpreted as reflecting the differential association of two subunit types in a hexameric protein. The two subunits have been described as having very similar mass, charge and antigenic properties (Pahlich et al., 1980; Kindt et al., 1980; Loulakakis and Roubelakis-Angelakis, 1990, 1991; Loulakakis et al., 1994), but there is no direct evidence of their genomic organisation.

Understanding the modifications in GDH activity and isoenzymes pattern could help to clarify the physiological role of the enzyme. For this reason, in this study we have developed a preliminary report about GDH results on the effect of Cd on GDH isoenzymes, content of GDH protein and mRNA of GDH in tomato leaf, on the purification to homogeneity of the major NADH-GDH isoenzymes from the tomato leaf tissue and on the electrophoretical characteristics of the enzyme, to elucidate further the physiological significance of GDH activity under Cd stress conditions.

An increase in GDH activity during leaf senescence appears to be a general feature (Lauriere et al., 1981a, b; Cammaerts and Jacobs, 1985; Thomas, 1978; Kar and Feierabend, 1984). The increase is reported to be due to de novo synthesis of the enzyme in *Hordeum* during the first hour of senescence (Guello and Sabater, 1982). It has been suggested that ammonia produced by proteolysis during senescence (Masclaux-Daubresse et al., 2000; Limami et al., 2002) could be responsible for de novo synthesis of the enzyme (Lauriere and Daussant, 1983; Postius and Jacobi, 1976; Thomas, 1978). It has been demonstrated recently that an increase in GDH activity follows a similar control mechanism which governs the course of senescence, and kinetin, which retards senescence, also retards the increase in GDH activity (Kar and Feierabend, 1984).

This work is in progress for determining the immunological properties of GDH from tomato leaves grown at 20 µM of Cd and for determining the rate of synthesis and degradation of the enzyme and the level of its translatable mRNA under aforementioned conditions. The effect of some pollutants on the enzymatic activity, however, warrants further investigation to ascertain the possible role of this enzyme during pollution stress conditions, in order to determine whether this heavy metal, Cd is representative of the overall molecular changes in GDH protein.

A study on tomato (*Lycopersicon esculentum* Mill. Cv. 63/5F1) is presented in this chapter to show physiological functions of glutamate dehydrogenase in ammonia assimilation under Cd stress.

2. PLANT MATERIAL AND GROWTH CONDITIONS

Seeds were sterilized in 10% (v/v) hydrogen peroxide for 20 min, and washed abundantly in distilled water afterwards. After imbibition, the seeds were germinated on moistened filter paper at 25°C in the dark. After 7 days, the uniform seedlings were transferred to plastic beakers of 6 L capacity (8 plants per beaker) filled with continuously aerated, basal nutrient solutions of an initial pH 5.8-6, containing 3 mM KNO_3, 0.5 mM Ca $(NO_3)_2$, 2.4 mM KH_2PO_4, 0.5 mM $MgSO_4$, 100 µM Fe-K_2-EDTA, 30 µM H_3BO_3, 5 µM $MnSO_4$, 1 µM $CuSO_4$, 1 µM $ZnSO_4$, and 1 µM $(NH_4)_6MO_7O_{24}$. Plants were grown in a growth chamber (26°C/70 % relative humidity during the day, 20°C/90% relative humidity during the night). The photoperiod was 16 h daily with a light irradiance of 150 µmol m^{-2} s^{-1} at the canopy level. At the age of 10 days after transplant, Cd was added to the medium as $CdCl_2$ at 0 to 50 µM. After one week of Cd treatment, plants were separated into shoots and roots. Roots were rapidly washed three times in 1 L of distilled water, then samples were stored in liquid nitrogen for subsequent analysis or dried at 70°C for at least three days in order to determine both dry material and ionic contents.

2.1. Enzymatic Extraction and Assay

Enzymes were extracted from frozen leaf and root material stored at –80°C. Leaf tissues were pulverized in liquid nitrogen with the aid of mortar and pestle. The powder was suspended in 5 vols (v/w) of ice cold grinding medium consisting of 200 mM Tris-HCl, pH 8.0, 10 mM L-cysteine-HCl, 14 mM β-mercaptoethanol, 0.5 mM PMSF and 2% (w/v) PVP and was homogenized in an Omnimixer homogenizer four times for 12 s each. After filtration through four layers of cheesecloth, the filtrate was centrifuged at 10 000 g for 30 min.

All extractions were performed at 4°C. After centrifugation, the supernatant (Crude extracts) was used for enzyme assays. The glutamate dehydrogenase NADH-GDH and NAD^+-GDH activities were measured in the aminating and the deaminating direction by following the absorption change at

230 nm (Loulakakis and Roubelakis-Angelakis, 1990). The GDH activity was expressed as µmol NAD(H) or NAD$^+$ min^{-1} g FW^{-1}.

2.2. Analyses of Cd, Total Protein and Ammonium Determinations

Cd was essayed by digestion of dried plant material in a HNO$_3$/HClO$_4$ mixture (3/1, v/v) and characterised by atomic absorption spectrophotometry (Perkin-Elmer Analyst 300). Protein content in the extracts and fractions obtained after a partial purification were determined according to Bradford (1976). The values given are means of duplicate and triplicate determinations. Free ammonium in plant tissues was measured after extraction of 0.5 g at 4°C in 2 mL of 0.3 mM H$_2$SO$_4$ and 0.5% (w/v) Polyclar AT. The homogenate was then clarified by centrifugation for 15 min at 30 000 g. Ammonium was quantified by the Berthelot reaction modified method according to Weatherburn (1967).

2.3. Gel Electrophoresis, Western Blot Analysis and Gel Staining Procedure

Proteins (10 µg) were separated by SDS-PAGE. Samples were treated at 100°C for 5 min with 3% SDS and 5% β-mercaptoethanol before electrophoresis. The percentage of polyacrylamide in the running gels was 10% for GDH, GS. Denatured proteins were electrophoretically transferred to nitrocellulose membranes. Polypeptide detection was performed using polyclonal antiserum raised against grape leaf GDH (Loulakakis and Roubelakis-Angelakis, 1990). The molecular weight of the enzyme subunit was estimated by calibrating with standard mol. wt. proteins from Pharmacia: Phosphorylase b (94 000), albumin (67 000), ovalbumuin (43 000), carbonic anhydrase (30 000), trypsin inhibitor (20 100) and α-lactalbumin (14 400). Relative GDH protein amount was determined by densitometric scanning of Western blot membranes and quantifier (NIH image 1.63, public domain).

2.4. Partial Purification and Concentration of Extracts

Protein extracts were prepared from 4 g of tomato leaves and loaded onto a DEAE-cellulose column (700 mm long, 2 mm i.d.: Pharmacia) equilibrated with a buffer containing 25 mM Tris-HCl (pH 8.0), 14 mM β-mercaptoethanol and 1 mM MgCl$_2$. A linear gradient of 0 to 400 mM NaCl was used to elute the proteins. The flow rate was adjusted to 20 mL h^{-1} and 3 cm^3 fractions were collected. The NADH-GDH and NAD$^+$-GDH activities and protein content in aliquots of the collected fractions were determined according to Loulakakis and Roubelekis-Angelakis (1990). Samples from each enzyme activity peak were pooled, dialysed, concentrated and used in SDS-PAGE and Native-PAGE for comparisons between isoenzymes.

2.5. Native Polyacrylamide Gel Electrophoresis

Native PAGE was performed in slab gels containing 7 % acrylamide by the method of Davis and run at 25 volts for 20 h in a refrigerator. At the completion of electrophoresis, bands containing GDH activity (deamination) were visualized with a tetrazolium assay (Hartman, Nagel, and Ilert, 1973). After incubation at 25°C for 20 min, the gel was destained with distilled water at 4°C and photographed.

2.6. Extraction of Total RNA and Northern Blot Analysis

Total RNA was extracted from plant material stored at –80°C. Northern blot analysis was performed as described previously (Masclaux-Daubresse et al., 2000). The following 32P-labelled probes were used for mRNA detection: gdh cDNA from *N. tabacum* (Masclaux-Daubresse et al., 2002). Since tobacco and tomato genes are highly homologous, hybridisation with *gdh* probes was performed under high stringency conditions at 65°C. Filters were washed with 2x SET (0.06 M Tris-HCl pH 8, 0.3 M NaCl, 4 mM EDTA) at room temperature for 5 min and at 65°C for 10 min. Additional washing were performed successively using 1xSET and 0.5xSET at 65°C for 15 min before drying and exposure to X-ray film. The relative amounts of mRNA were determined by densitometric scanning of northern blot autoradiograms and quantified using NIH image 1.63, (public domain) quantifier.

3. OBSERVATIONS AND POSSIBLE EXPLANATION
3.1. General Plant Growth

Treatment of tomato seedlings for 7 days with different Cd concentrations resulted in growth inhibition. We observed that the biomass of leaves decreased when increasing doses of Cd were added to the culture solution (Fig. 1A, 1B).

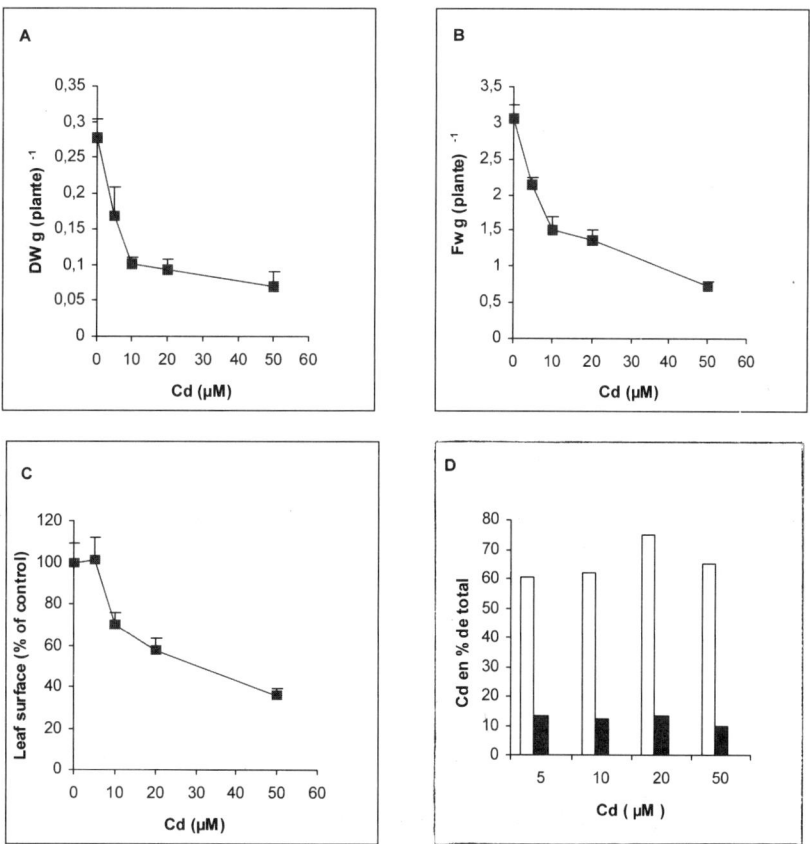

Fig. 1 A-D. A Effect of Cd in culture medium on dry weight of leaves , B effect of Cd in culture medium on fresh weight of leaves (■), C effect on leaf surface, D changes in the cadmium accumulation in leaves (■) and roots (□) of tomato plants. Values are means ± SE of five individual plants. Standard errors are not shown when they are smaller than the symbol.

The decline in fresh and dry weights was dramatic, and was observed even at the lowest Cd concentration used. An external Cd concentration of 50 µM triggered 75% of reduction in leaves growth. A progressive decrease in leaf surface was detected for plants treated with 10 µM or higher Cd concentration. Leaf surface was reduced to 60 % under the 50 µM Cd treatment (Fig. 1C). Treatment of plants during seven days with Cd was accompanied by its progressive uptake and accumulation in the plant. As shown in Fig. 1D, this accumulation was function of external metal concentration. However, for every concentration used, Cd accumulation was more important in roots than in leaves. This was confirmed by the results shown in Fig. 1D, which indicate that almost 80% of the total Cd remained confined in root systems, which represents no more than 20% of the total plant biomass. However, in spite of the high accumulation of Cd in root system, this latter remains less sensitive to this pollutant than leaves system.

3.2. Changes in Aminating, Deaminating and Specific Activities of GDH Enzyme

The aminating, deaminating and the specific activities of the GDH of tomato leaves grown on nitrate culture media with or without Cd are shown in Fig. 2A-D. In our previous work we have shown that the culture of tomato in media containing different Cd concentrations resulted in significant variation in these two activities of GDH. The activities of NADH-GDH, NAD^+-GDH and their ratios in crude extracts from tomato leaves showed significantly increase of NADH-GDH (Fig. 2A) and a decrease of NAD^+-GDH (Fig. 2B) activities when compared to control plants.

The NADH-GDH activity was increased by 3-fold under Cd treatment in leaves. These differences were reflected on the ratio of NADH-GDH to NAD^+-GDH activities. Moreover, the NAD^+/NADH ratio was decreased by Cd treatment (Fig. 2D). In the 50 µM treated plants, the ratio was equal to 25% of the control for leaves, indicating that the change in the GDH aminating and deaminating activities is favour of glutamate synthesis. The NADH-GDH (Fig. 2C) and NAD^+-GDH (Fig. 2D) specific activities in highest Cd concentration for leaf tissues were 4.5 and 1 fold, respectively than the control. The change in GDH specific activities in the treated plants suggest that ammonium assimilation is affected and both ammonia and amino acids are released by protein degradation and hydrolysis.

Fig. 3A shows ammonium accumulation in Cd treated plants. All Cd treatments showed a parallel increase in GDH activity and ammonium content. The stimulation of NADH-GDH activity in leaves tissues is closely correlated to the internal ammonium concentration measured in the corresponding tissues (r^2 = 0.97 for leaves) (Fig. 2E). Since previous studies have already demonstrated that Cd affected nitrogen uptake as well as nitrate reduction (Gouia et al., 2000), we first tested the proteolysis hypothesis by measure of the total soluble protein content of leaves (Fig. 2F). The results showed that protein total soluble decreased in leaves. This decrease reached 70% of the control value in plants treated with 50 µM Cd.

3.3. Changes in Protein and Transcripts Content of GDH Enzyme

In order to investigate the influence of Cd on nitrogen assimilation recycling we measured the activity of GDH enzyme (Masclaux-Daubresse et al., 2001) involved in ammonium detoxification under stress conditions. The polyclonal antiserum raised against grape leaf GDH (Loulakakis and Roubelakis-Angelakis, 1990) recognizes one band of GDH protein. The molecular mass of the tomato GDH is 43 KDa. Although the cross-reactivity of the GDH antiserum is lower in control and low Cd concentrations. The amount of GDH protein was not similar in the all organs of tomato, GDH content was lower in leaves than in roots. With increasing doses of Cd, the amount of GDH protein increased in the leaves by approximately 14-fold (Fig. 3A).

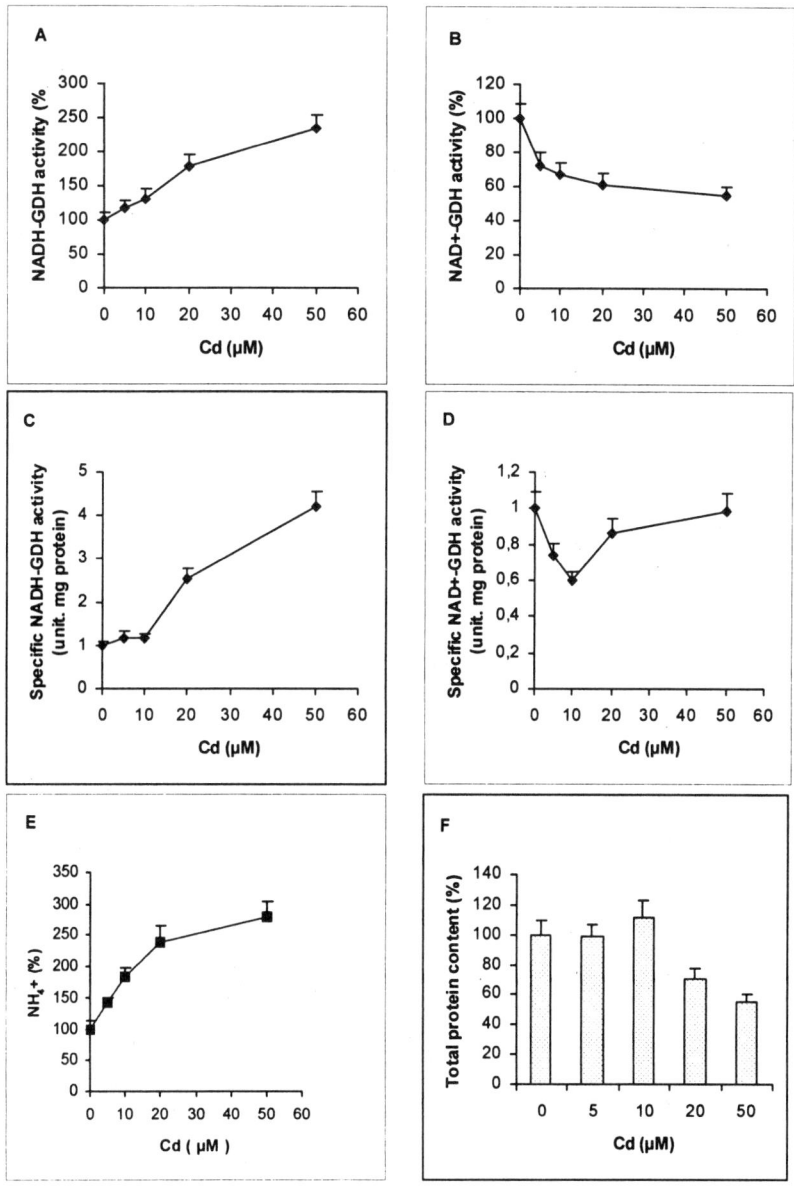

Fig. 2 A-F. Changes in activities for GDH enzyme. Changes in ammonium content and total protein content. The effect of Cd on various activities in leaves (■) was measured. A NADH-GDH activity, B NAD^+-H-GDH activity, C NADH-GDH specific activity, D NAD^+-GDH specific activity, E ammonium content, F total protein content. Values are means ± SE of five individual plants. Standard errors are not shown when they are smaller than the symbol.

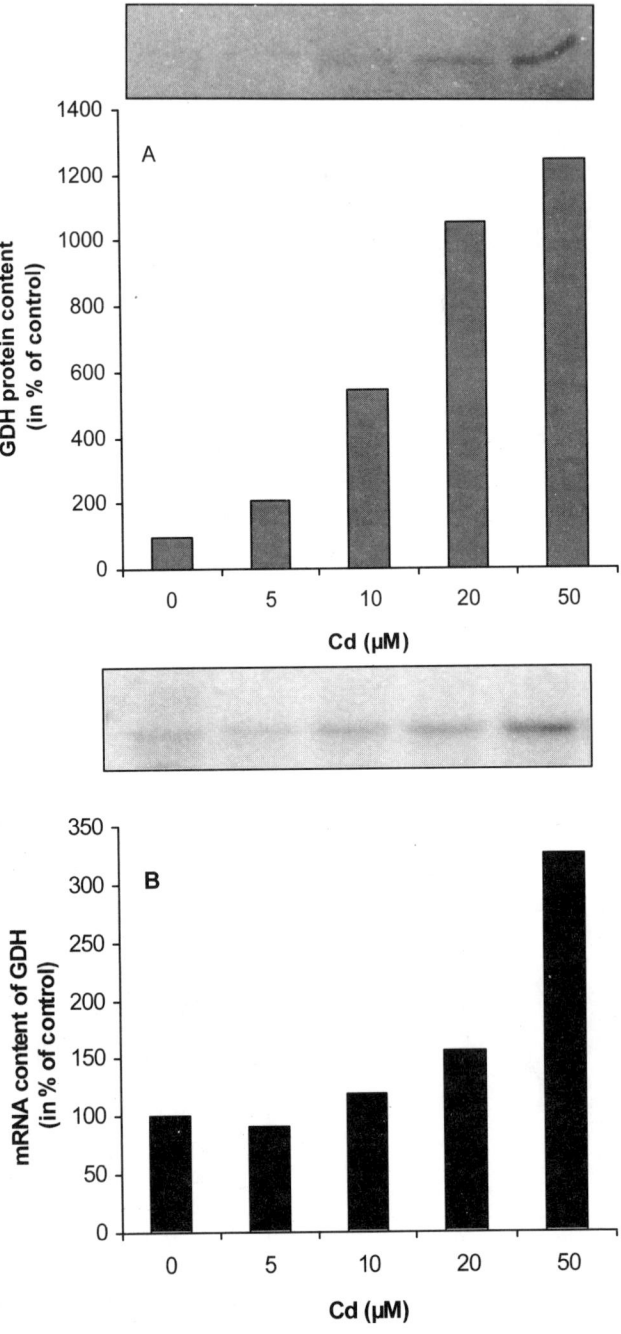

Fig. 3 A-B. A Changes with Cd concentration in the protein content for GDH enzyme. Protein gel blots for the GDH in leaves of tomato plants. An equal amount of soluble proteins (10 µg) was added per lane. B Changes in the steady-state levels of transcripts for GDH enzyme. Blots show the steady-state levels of the GDH, mRNA in leaves and roots of tomato plants. Relative values as a percentage of maximum.

To determine whether the change observed in enzyme activities and protein content are released to transcriptional control by Cd treatment, we measured the steady-state levels of the cognate transcripts (Fig. 3B). 18s rRNA was used as a constitutive control. The GDH mRNA steady-state was increased by 4-fold in the leaves of tomato, when plants were treated with increasing doses of Cd. In roots, the increase in GDH mRNA accumulation triggered by Cd treatment appeared smaller because of the pre-existing high level of transcripts in the roots of the control plants (results no shown). Cadmium treatment led also to an increase in both protein and transcripts GDH levels in leaves of tomato.

3.4. Glutamate Dehydrogenase Patterns in Crude Extracts from Leaves

The structure of GDH from animal and fungal systems has been extensively studied and the literature has been adequately reviewed. The present discussion, therefore, will be limited to more recent developments in area. The plant enzyme is thought to be a metalloprotein (Loulakakis et al., 1994). The enzyme is composed of identical subunits (α or β), or no identical subunits (α and β) (Loulakakis and Roubelakis-Angelakis, 1991). As six electrophoretic bands appear after cross-linking of the enzyme with diimidates from *Pisum* seeds, the enzyme is suggested to be a hexamer (Kindt et al., 1980).

However, using sodium dodecyl sulphate gel (Fig. 4A), the observations have demonstrated that GDH from *Lycopersicon esculentum* presented one bands in control and treated plants. Moreover, GDH band was found more greater in the treated leaves. Then the polyacrylamide gel electrophoresis and the tetrazolium techniques were employed in an attempt to investigate the isoenzymic patterns of GDH, in tomato crude extract leaves. The isoenzymic patterns of GDH are presented in Fig. 4B. The new bands with Native-Page were clearly detectable, especially in the treated leaves. The enzyme activity was also increased by the Cd treatment as shown on the zymogram. Moreover, seven GDH isoenzymes were found on both control and treated leaves.

3.5. Purification of Glutamate Dehydrogenase from Leaves of Control Plants

In order to further resolve the possible physiological role of the GDH isoenzymes, we performed a comparative study of the purified GDH activities from control and stressed leaves of each one of the 7 iso-GDHs. For this reason methods for staining aminating, deaminating activities and aminating deaminating ratio (A/D) activities are shown in Fig. 5.

The isoenzymes of GDH, as visualized by using the tetrazolium system. GDH was partially purified from control leaves tissues of *Lycopersicon esculentum*. Mill.cv.63/5F1. The successive purification steps are specified in Table I. The elution profile corresponding to DEAE-cellulose system eluted with a linear NaCl gradient is shown in Fig. 5; there is all seven glutamate dehydrogenase peak. The first peak of activity, denoted GDH1, elutes at about 0.08 M NaCl, and the last peak, GDH 7 elutes about 0.14 M NaCl. This agrees with the elution of *Vitis vinifira* GDH 7 from a DEAE-cellulose column at about 0.18 M (Loulakakis and Roubelakis-Angelakis, 1990). Isoenzyme 7 (GDH7) with the highest electophoretic mobility represented just 10 to 15% of total activity. Some analysis with PAGE and staining for GDH activity revealed an absolute correspondence between peaks and the seven isoenzymes.The isoenzyme with the lowest electrophoretic mobility (first peak of activity), which represented 30 to 40% of total activity was more expressed. The enzyme exhibited both NADH-and NAD^+-dependent activities. The highest NAD^+-GDH activity was presented by the first isoenzyme (GDH1), but a lowest A/D ratio for this isoenzyme was obtained.

Table 1. Purification of glutamate dehydrogenase from *Lycopersicon esculentum* leaves

Purification Step	Total activity (units)	Total protein (mg)	Specific activity (units mg^{-1})	Purification (-fold)
1. Crude extract	425	1302	0.415	1.0
2. Ammonium sulphate precipitation	392	467	0.95	2.5
3. DEAE-cellulose	350	19.22	18.65	80.8
4. Preparative acrylamide gel electrophoresis	332	2.009	171.2	875.0

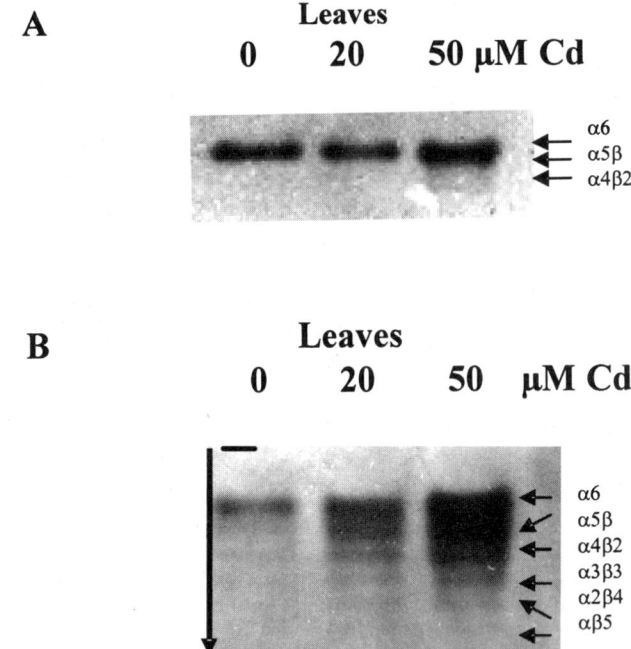

Fig. 4 A-B Plyacrylamide and native polyacrylamide before different stages of purification of the NADH-GDH from *Lycopersicon esculentum*. A Western blot analysis of GDH protein in the leaves of control and treated plants. B Native polyacrylamide gel (7.0%) of samples leaves from control and treated plants of tomato.

This result suggested a major role of GDH1 is not the assimilation of the ammonia in control leaves. In native electrophoresis, isoenzymes 2 to 7, separated by DEAE-column chromatography and stained for activity appeared as six bands (GDH2, GDH3, GDH4, GDH5, GDH6, GDH7). The tomato GDH from control tissue of leaves, showed seven bands isoenzymic pattern, the isoenzyme with the less negative charge was the more abundant.

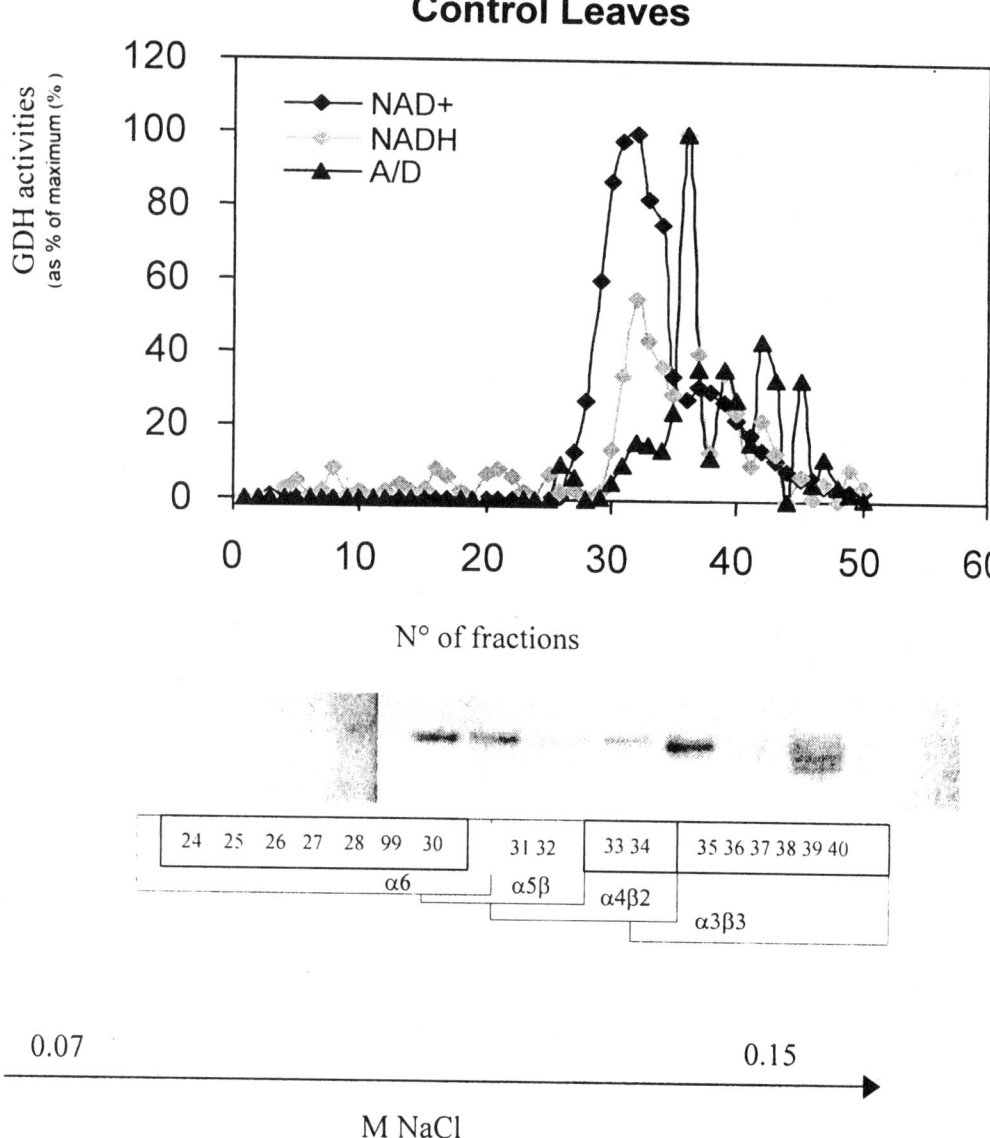

Fig. 5. Elution pattern of control tomato leaves glutamate dehydrogenase isoenzymes from a DEAE-cellulose column. The standard enzyme extract was applied to the column and elution of protein was performed by a linear gradient (0 to 0.4M) of NaCl. Fractions were collected and assayed for NADH-GDH activity, NAD^+-GDH activity and NADH-GDH/NAD^+-GDH ratio Glutamate dehydrogenase isoenzyme patterns after Native polyacrylamide gel electrophoresis of purified extract of different fractions.

3.6. Purification of Glutamate Dehydrogenase from Leaves Treated with Cadmium

To analyze whether the high content of GDH in tomato leaves accompanied by a high activity in presence of Cd, the seven isoformes were separated by chromatographie (DEAE-cellulose). They were eluted at 0.096 mM NaCl the first and at 0.14 mM NaCl the last isoenzyme. In the following the GDH isoformes were labelled with consecutive number (1–7) according to their order of elution. Sometimes these isoforms were found at distinct peaks or as shoulders in the elution gradient. The results in Fig. 6 show that GDH1, homogenate isoenzyme ($\alpha 6$) was the predominant form. Three isoformes with relatively high electrophorectic activities (GDH1, GDH2, GDH3). Furthermore, tomato GDH isoenzyme 1 was more actively in aminating and deaminating activities (Fig. 6A). A substantial values of Amination/Deamination ratios (Table II) indicate that the enzyme 1 (GDH1) characterised by a lower value of A/D ratio.

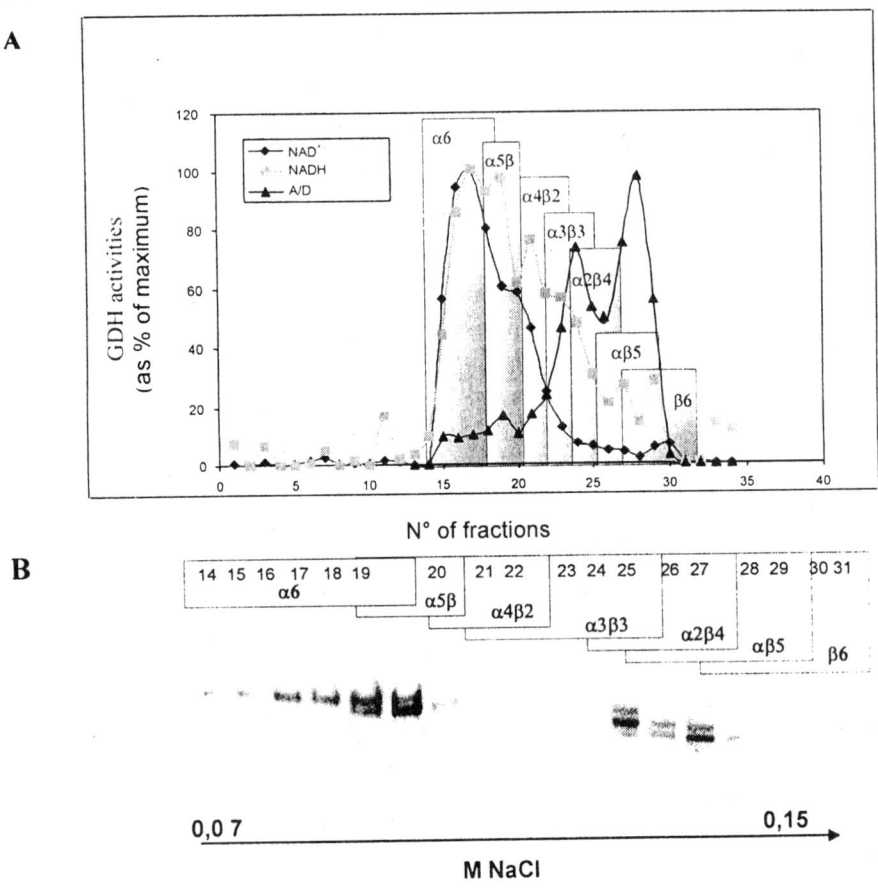

Fig. 6. Elution profiles of glutamate dehydrogenase isoforms from leaves of tomato treated with 20 µM of cadmium after separation by chromatographie in DEAE-cellulose column. Fractions were collected and assayed for NADH-GDH activity, NAD^+-GDH activity and NADH-GDH/NAD^+-GDH ratio. Glutamate dehydrogenase isoenzyme patterns after Native polyacrylamide gel electrophoresis of purified extract of different fractions.

It is logical that amination and deamination activities both were especially increased in presence of Cd. Thereafter, the resulting 20 fractions were subjected to native 7% PAGE, followed by activity staining of the electrophoresed gel (Fig. 6B).

The resolution of treated tomato GDH leaves isoenzymes to their different charge isomers was so efficient that isoenzymes 1, which was more abundant than isoenzymes 2 and isoenzymes 3 which exhibited a high aminating activities, proved by a low A/D ratio for these two isoenzymes, respectively 4.5 and 5 (Table II). In the same time a maximum value of aminating/deaminating ratio was obtained by the sixth isoenzyme (GDH6). Quantitative results from A/D values indicates that the variation of this ratio was considerable. This ratio increased considerably from 3 on the first isoenzyme to 20 at the last. In addition, the temporal change in the amount of each GDH bands was closely correlated with the course of its activities (amination, deamination, specific activity). The change in the pattern of GDH isoenzymes between control and treated leaves with 20 µM of Cd, could be attributed the more importance for isoenzyme 1 in ammonia assimilation under stress conditions.

Table 2: Amination/Désamination (A/D) ratio of the seven tomato leaves treated with 20 µM of cadmium. Values are means of three experiments and error was within 5%.

isoforme	$\alpha 6$	$\alpha 5\beta$	$\alpha 4\beta 2$	$\alpha 3\beta 3$	$\alpha 2\beta 4$	$\alpha\beta 5$	$\beta 6$
A/D	3	4.5	5-10	17	11	20	10 Environ

Generally, resistance to heavy metals in higher plants is mainly based on tolerance (Verklej et al., 1991), which implies uptake of heavy metal and the ability to "tolerate" excess of internal metal concentration. In agreement with other studies on Cd uptake (Leita et al., 1991; Chugh et al., 1992; Obata et and Umebayashi, 1993; Salt and Rauser, 1995; Schützendübel and Polle, 2002) the roots of tomato plants accumulated higher amounts of Cd than leaves. Roots thus function as a barrier than restricts the transport of Cd to leaves (Chaffei et al., 2003; Cieslinski et al., 1996). Furthermore, the accumulation of Cd in leaves did not increase with the external Cd concentration, suggesting a strong exclusion of Cd from leaves (Gouia et al., 2000; Chaffei et al., 2003). However, the capacity of plants to accumulate these ions depends on the species, the differences observed being usually associated with Cd tolerance (Obata and Umebayashi, 1993; Zhu et al., 1999b; Williams et al., 2000). According to these authors, sensitive plant species, such as tomato, that cannot accommodate high intracellular Cd, excluded it in order to avoid perturbation of the internal metabolic processes.

In response to permanent heavy metal exposure, some species have developed tolerance properties. In *Lycopersicon esculentum*, we showed that cadmium is rapidly toxic for the plant, leading to a strong decrease in plant growth observed in dry, fresh weight and leaf surface. The morphology of tomato plants was dramatically affected by Cd. We observed that the emergence of new organs was altered and that the newly emerged leaves were chlorotic (Fig. 7). We observed that roots from Cd-stressed plants were shorter and less hairy than those from controls. However, treated roots were thicker and stronger, which might have compensated in part or the loss of biomass. Indeed, it appeared that root biomass was affected to a lesser extent compared with leaves. From all these observations, it can be concluded, that fresh weight, dry weight and leaf surface are good biomarkers for estimating the senescence-like phenomena related to the stress trauma triggered by a heavy metal (Smart, 1990; Ouzounidou et al., 1997; Masclaux-Daubresse et al., 2000). This model was used to investigate the effect of Cd toxicity on nitrogen management in plants. We observed that Cd-treated plants accumulated ammonium and free amino acids in both leaves and roots. They progressively lost nitrate.

The decrease in nitrate in leaves might have been a consequence of efficient reduction activity. However, a concomitant decrease in total soluble protein and the activity of enzymes involved in primary nitrogen assimilation (NR, NiR and GS/GOGAT, results not shown) was observed (Gouia et al., 2000). Inhibition of NR, NiR and GS activity has also been described in maize, pea, bean and barley seedlings (Nussabum et al., 1988; Chugh et al., 1992; Boussama et al., 1999; Gouia et al., 2000). The decrease in nitrate might then be due to a deleterious effect of Cd on mineral absorption, may be through the alteration of transpiration. In leaves, Rubisco is the most abundant soluble protein, representing up to 50% of the total leaf soluble protein pool (Masclaux-Daubresse et al., 2001; Chiba et al., 2003). There fore, the decrease in the abundance of the Rubisco large subunit in tomato leaves during Cd stress might contribute largely to the net loss of total leaf protein observed. We have shown in previous work that nitrogen remobilization is a senescence associated process that takes place in source organs such as senescing leaves (Masclaux-Daubresse et al., 2001). Nitrogen remobilization is of major importance in plant nitrogen economy since it provides nitrogen to young, developing organs. Nitrogen remobilization is usually associated with the induction of the expression of enzymes such as specific proteases, the cytosolic GS1 (Habash et al., 2001; Becker et al., 2000) isoform and the GDH isoformes (Skopelitis and Boubelakis-Angelakis, 2002; Cammaerts and Jacobs, 1985). During leaf senescence, nitrogen remobilization is essential to support new organ emergence and growth (Masclaux-Daubresse et al., 2001). Interestingly, we observed in this and our previous work, that the nitrogen remobilization process is favoured in Cd-treated plants in which the growth and the emergence of new organs is considerably compromised, compared with untreated plants. This then suggests that GS1, GDH and proteolysis are involved for another purpose than the building of new organs.

Fig. 7. Effect of cadmium on the growth and morphology of tomato leaves. Tomato plants were grown with 50 µM $CdCl_2$ as described in materials and methods. Photograph showing necrotic spots and the loss of the green color in the all foliar surface.

Previous results studying the qualitative and quantitative changes in the content of individual amino acids can be informative. We observed that total amino acid content was increased by Cd in both leaves and roots. In leaves, we observed that the proportion of glutamate remained unchanged. This might be a consequence of buffering effects through modulations in GaBa or Pro, which are increasing. Indeed, the amino acids are direct products of Glu metabolism. Pro and GaBa are known to be two stress markers (Narayan and Nair 1990; Bown and Shelp, 1997), and their increase in stressed leaves might be considered as such. Among the main amino acids represented, glutamine (Gln) was highly increased by Cd stress. This amide is known to be highly reactive and to serve as the major nitrogen transport form in plants. In contrast, the proportions of Asn, His, Arg and minor amino acid bulk were strongly increased, thus suggesting that some products of proteolysis might be exported directly through the phloem to the roots whereas GaBa ans Asp export might be affected by Cd (Fig. 8). The sharp increase in asparagine (Asn) was observed in roots led us to question asparagine synthetase (AS) expression in both roots and leaves of tomato plants. It is interesting to note that the 3 fold increase in AS mRNA in roots of treated plants was the highest induction observed among the transcripts studied. AS is an ATPase and the active implication of Cd-stressed roots in Asn synthesis implied then availability of dedicated energy (Ireland and Lea, 1999; Devaux et al., 2003; Chevalier et al., 1996). The schematic illustration presented in Fig. 8 showed that Cd treatment induced a differential but coordinated response of the carbon and nitrogen management pathways between leaves and roots. Total and individual amino acid content as well as GS1, GDH and AS expression suggest that nitrogen is recycled and translocated from leaves to roots. We suppose that such a strategy sacrifices the aerial organs of the plants to preserve roots as a nutritional safeguard organ to ensure future recovery.

Isozymic number of the enzyme varies with plant species as well as other nutritional and environmental conditions. Seven isozymic forms have been reported in *Pisum* (Scheid et al., 1980; Turano et al., 1996; Lauriere and Daussant, 1983), *Phseolus* (Yue, 1969), *Vicia* (Fawole, 1977), *Ricinus* (Lee et al., 1976), *Arabidopsis* (Cammaerts and Jacobs, 1983) and *Vitis* (Loulakakis and Roubelakis-Angelakis, 1991). The number of isoenzymes, however, is increased by the addition of the ammonium (Loulakakis and Angelakis-Roubelakis, 1992; Knamori et al., 1972; Lauriere and Daussant, 1983) and free amino acids (Ratajczak et al., 1981). The number and banding pattern of isozymes are also influenced by growth stages (Hartman et al., 1973), plant organs (Nauen and Hartmann, 1980; Mckenzie and Lees, 1981), nitrogen nutrition (Kanamori et al., 1972; Barash et al., 1975) and stress conditions (Cammaerts and Jacobs, 1983; Loulakakis and Roubelakis-Angelakis, 1991). Individual plant organs may develop a typical GDH isozyme pattern for each organ during the growth stages which may be correlated with the known metabolic activities of those organs as well as with external nutritional and environmental conditions available to the plant. This reveals a physiological role of GDH isozymes in the regulation of nitrogen metabolism. A hypothesis has been proposed for the anabolic and catabolic role of GDH isoenzymes in plants have demonstrated from tomato leaves in this work.

From physiological studies with NAD(H)-GDH from leaves of a Solanaceous plant species, *Lycopersicon esculentum* Mill.cv.63/5F1, we observed a sharp increase in NADH-GDH and NADH-GDH specific activities, concomitantly with the decrease in NAD+-GDH and NAD+-GDH specific activities are expressed in leaves of tomato when submitted to Cd stress, which paralleled changes in GDH protein content levels. Protein gel blot analysis using an antiserum raised against grapevine GDH (Loulakakis and Roubelakis-Angelakis, 1990) demonstrated that GDH activity corresponded to a *de novo* synthesis of one protein subunit (molecular mass of 44 KDa). The increased GDH subunit content during leaf Cd stressed was well correlated with the GDH mRNA steady-state level (Fig. 3). Changes in GDH activity during stress conditions have been observed in a few plants species including

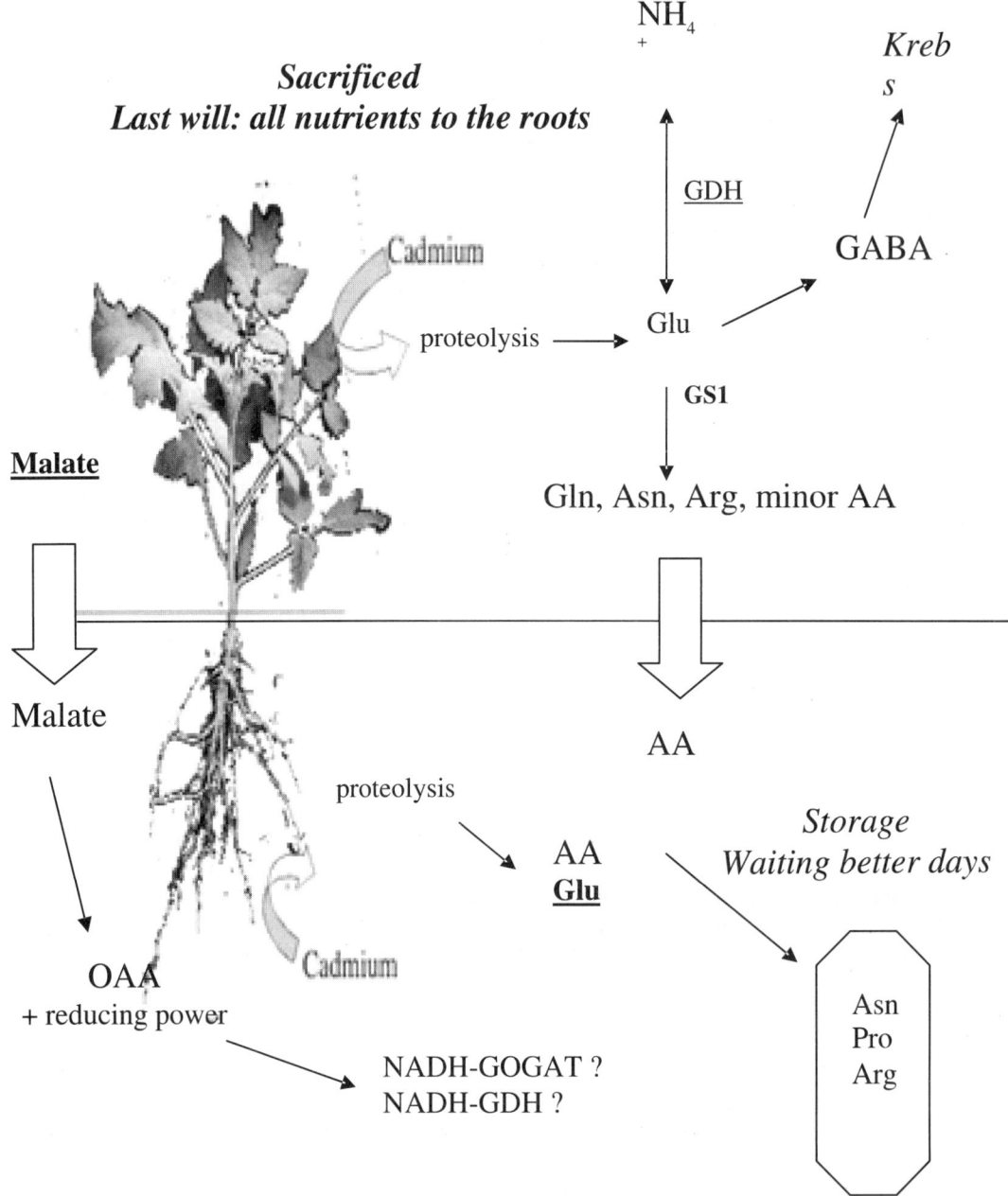

Fig. 8. Simplified scheme showing the main nitrogen metabolic pathways in tomato plant growing under cadmium stress conditions. Nitrogen and carbon flow and the possible involvement of glutamate Dehydrogenase in glutamate production which represent a major metabolite for different amino acid synthesis. The important storage of, As (asparagine), Pro (proline) and Arg (arginine) was proved in root systems. Unknown pathways are indicated by question marks.

Linum, *Vitis vinifera*, *Pisum sativum*, *Phaseolus vulgaris*, *Lycopersicon esculentum* (Loulakakis and Roubelakis-Angelakis, 1990; Chugh et al., 1992, Ouariti et al., 1997, Boussama et al., 1999).

In the present study, we clearly show that GDH protein and mRNA are detectable in very lower level in control leaves but are present in large amounts in stressed leaves. The spectacular changes in the GDH aminating and NADH-GDH specific activities in favour of the anabolic reaction might be the result of enhanced transcription of the gene followed by *de novo* synthesis of the protein. It became possible to propose a tentative model depicting the interaction and the mode of action of the various metabolic, biochemical and molecular characteristics parameters of GDH protein measured during presence of Cd in the culture medium. Investigations performed on all above results confirm the correlations observed between the activity level of the GDH and the protein content and the transcripts level under cadmium stress.

To our knowledge, this paper is the first describing exhaustively the explanation of physiological functions of glutamate dehydrogenase in ammonia assimilation under Cd stress conditions in plants requires knowledge concerning their isoenzyme patterns, activity levels in control and Cd stressed leaves as well as data about their molecular properties. The changes in GDH activity and isoenzymes pattern in tomato leaves were investigated before and after addition of Cd in the culture medium. From the data presented in this report it is clear that different seven isoforms of GDH are reported in tomato leaves at a level and activity not similar in control and Cd stressed plants. Glutamate dehydrogenase was partially purified from tomato (*Lycopersicon esculentum* Mill.cv.63/5F1) leaves tissues and its activity and isoenzymic pattern were studied. Seven anodal migrating isoenzymes were revealed after PAGE points that the enzyme could have an hexameric structure. Random association of two types of subunits with different charges in an hexameric complex responsible to production of these seven isoenzymes. Results revealed that the isoenzymes with the lowest electrophoresis mobility, which account 40 to 50% of the total activity was the more expressed both in control and stressed leaves. The common important isoenzymes between control and treated leaves is the isoerzyme 1, the more anodal isoenzymes (GDH1), which is the homo-hexamers with the second extreme isoenzymes (isoenzymes 7). In extracts purified from leaves treated with Cd shown an increase in the amination and deamination activities especially of GDH1. However, the isoenzymes 6 (GDH), which characterized by a maximum level of A/D ratio and the more cathode specific compartment. In addition excess ammonia levels were proved in extracts of Cd stressed leaves by about 5-fold than control leaves. Thus it could be suspected that leaves tissues participate in reductive amination of 2-oxoglutarate in the process of ammonia assimilation in a plant by increase under stress conditions caused by addition of Cd in the culture medium. The staining intensities of the bands which are related to the relative abundance of the isoenzymes shows that the proportions of GDH was more increased. These results indicate that the NAD^+-GDH isoenzymes have a varying anabolic and catabolic function. If indeed to which the two subunits randomly associated (Cammaerts and Jacobs, 1983, 1985), then isoenzymes 1 and 7 are homo-hexamers, and isoenzymes 2 to 6 are hybrids. If we accept a different physiological role for each of the two subunits, then the change in the NADH-GDH / NAD^+-GDH ratio could be attributed to a different participation of the catabolic subunit in the isoenzymes. The activation of GDH by divalent metallic cations (Cd^{2+}) has been attributed indirectly to high accumulation of ammonia in Cd treated leaves. The differences observed in the degree of activation of GDH isoenzymes by Cd^{2+} ions was observed for the first homologous isoenzymes (GDH1).

These biochemical and molecular results also indicate a progressively increasing degree of activation of GDH at 20 µM of Cd^{2+} by increasing an accumulation of *gdh* mRNA and furthermore support the hypothesis that the increase in the NADH-GDH activity attributed to an increase in the ammonium concentration in the extracts of Cd stressed leaves is caused by *de novo* synthesis of GDH protein. Moreover, Loulakakais and Roubelakis-Angelakis, 1992), showed that the increase of ammonium concentration in culture medium correlated with the α-subunits synthesis.

In conclusion, the present work confirms our previous report that the increase of the NADH-GDH and NADH-GDH specific activities under Cd stress were caused by *de novo* synthesis of GDH protein and especially the more anodic. Furthermore, it is now shown that ammonium induces synthesis of α-subunits of GDH (Loulakakis and Angelakis-Roubelakis, 1992). Thus, it is suggestive of an important role of GDH in ammonia assimilation under stress conditions in tomato leaves.

4. ACKNOWLEDGEMENTS

Authors thanks Dr. K.A. Roubelakis-Angelakis (University of Crete, Heraklion, Grece) for providing anti-bodies raised against the GDH protein. Thanks to Elisa Carrayol and François Gosse for technical assistance. Thanks to the Ministère des affaires Etrangères de Tunisie for financial support that allowed the visit of Mme. C. Chaffee to Versailles.

References

Barash I, Sadon T, Mor H (1975) Evidence for ammonium dependent de novo synthesis of glutamate dehydrogenase in detached oat leaves. Plant Physiol 56: 856-858.

Baszynski T, Wajda L, Krol M, Wolinska D, Krupa Z, Tukendorf A (1980) Photosynthetic activities of cadmium-treated tomato plants. Physiol Plant 48: 365-370.

Becker TW, Carrayol E, Hirel B (2000) Glutamine aynthetase and glutamate dehydrogenase isoforms in maize leaves: Localization, relative proportion and their role in ammonium assimilation or nitrogen transport. Planta 211: 800-806.

Boussama N, Ouariti O, Suzuki A, Ghorbel MH (1999) Cd-stress on nitrogen assimilation. J Plant Physiol 155: 310-317.

Bown AW, Shelp BJ (1997) The metabolism and functions of γ-aminobutyric acid. Plant Physiol 115: 1-5.

Bradford MM (1976) A rapid and sensitive method for quantification of microgram quantities of protein utilizing the principal of protein-dye binding. Anal Biochem 72: 248-254.

Cammaerts D, Jacobs M (1983) A study of the polymorphism and the genetic control of the glutamate dehydrogenase isozymes in *Arabidopsis thaliana*. Plant Sci Lett 31: 65-73.

Cammaerts D, Jacobs M (1985) A study of the role of glutamate dehydrogenase in the nitrogen metabolism of *Arabidopsis thaliana*. Planta 163: 517-526.

Chaffei C, Gouia H, Ghorbel MH (2003) Nitrogen metabolism of tomato under cadmium stress conditions. J Plant Nutr 26: 1617-1634.

Chevalier C, Bourgeois E, Just D, Raymond P (1996) Metabolic regulation of asparagine synthetase gene expression in maize (*Zea mays* L.) root tips. The Plant J 9: 1-11.

Chiba A, Ishida H, Nishizawa NK, Makino A, Mae T (2003) Exclusion of ribulose-1,5-biphosphate carboxylase/oxygenase from chloroplasts by specific bodies in naturally senescing leaves of wheat. Plant Cell Physiol 44: 914-921.

Chou KH, Splittsoesser WE (1972) Glutamate dehydrogenase from pumpkin cotyledons: Characterization and isoenzymes. Plant Physiol 49: 550-554.

Chugh LK, Gupta VK, Sawahney SK (1992) Effect of cadmium on enzymes of nitrogen metabolism in pea seedlings. Phytochemistry 31: 395-400.

Chugh LK, Swahney SK (1999) Photosynthetic activities of *Pisum sativum* seedlings grown in presence of cadmium. Plant Physiol Biochem 37: 297-303.

Cieslinski G, Van Ress KCJ, Huang PM, Kozak LM, Rozad HPM, Knott DR (1996) Cadmium uptake and bioaccumulation in selected cultivars of durum wheat and flax as affacted by soil type. Plant Soil 182: 115-124.

Dubois F, Tercé-Laforgue T, Gonzalez-Moro MB, Estavillo MB, Sangwan R, Gallais A, Hirel B (2003) Glutamate dehydrogenase in plants, is there a new story for an old enzyme ? Plant Physiol Biochem 41: 565-576.

Devaux C, Baldet P, Joubès J, Dieuaide-Noubhani M, Just D, Chevalier C, Raymond P (2003) Physiological, biochemical and molecular analysis of sugar-starvation responses in tomato roots. J Exp Bot 54: 1143-1151.

Fawole ME (1977) Glutamate dehydrogenase from *Vicia faba*. Can J Bot 55: 1850-1860.

Gouia H, Ghorbel MH, Christian M (2000) Effects of cadmium on activity of nitrate reductase and on other enzymes of the nitrate assimilation pathway in bean. Plant Physiol 38: 629-638.

Gouia H, Suzuki A, Brulfert J, Ghorbal MH (2003) Effects of cadmium on the co-ordination of nitrogen and carbon metabolism in bean seedlings. J Plant Physiol 160: 367-376.

Greger M, Bertell G (1992) Effects of Ca^{2+} and Cd^{2+} on the carbohydrate metabolism in sugar beet (*Beta vulgaris*). J Exp Bot 43: 167-173.

Guello J, Sabater B (1982) Control of some enzymes of nitrogen metabolism during senescence of detached barley (*Hordeum vulgare*. L) leaves. Plant Cell Physiol 23: 561-565.

Habash DZ, Massilah AJ, Rong HL, Wallsgrove RM, Leigh AR (2001) The role of cytosolic glutamine synthetase in wheat. Ann App Biol 138: 83-89.

Hartmann T, Nagel M, Ilert HJ (1973) Organ specific multiple forms of glutamic dehydrogenase in *Medicago sativa*. Planta 111: 119-128.

Ireland RJ, Lea PJ (1999) The enzymes of glutamine, glutamate, asparagine and aspartate metabolism in plant amino acids. In: Singh BK (ed.) Biochemistry and Biotechnology, Marcel Dekker, New York, pp. 49-109.

Kanmori T, Konishi S, Takahashi E (1972) Inducible formation of glutamate dehydrogenase in rice plant roots by the addition of ammonia to the media. Physiol Plant 26: 1-6.

Kar M, Feierabend J (1984) Changes in the activities involved in amino acid metabolism during the senescence of detached wheat leaves. Physiol Plant 62: 36-44.

Kindt R, Pahlich E, Rasched I (1980) Glutamate dehydrogenase from peas: Isolation, quaternary structure and influence of cations on activity. Europe J Biochem 112: 533-540.

Kurganov B (2000) Analysis of negative cooperativity for glutamate dehydrogenase. Biophy Chem 87: 185-199.

Lasa B, Frechilla S, Aparicio-Tejo PM, Lamsfus C (2002) Role of glutamate dehydrogenase and phosphoenolpyruvate carboxylase activity in ammonium nutrition tolerance in roots. Plant Physiol Brioche 40: 969-976.

Lauriere C, Daussant J (1983) Identification of the ammonium dependent isozyme of glutamate dehydrogenase as the form induced by senescence or darkness stress in the first leaf of wheat. Physiol Plant 58: 89-92.

Lauriere C, Weisman N, Daussant J (1981a) Glutamate dehydrogenase in the first leaf of wheat. I. Antigenic polymorphism. Physiol Plant 52: 146-150.

Lauriere C, Weisman N, Daussant J (1981b) Glutamate dehydrogenase in the first leaf of wheat. II. De novo synthesis upon darkness stress and senescence. Physiol Plant 52: 151-155.

Lee KC, Comminghan BA, Paulsen GM, Liang GH, Moore RB (1976) Effects of cadmium on respiration rate and activities of several enzymes in soybean seedlings. Plant Physiol 36: 4-6.

Leita L, Contin M, Maggioni A (1991) Distribution of cadmium and induced Cd-binding proteins in roots, stem and leaves of *Phaseolus vulgaris*. Plant Sci 77: 139-147.

Limami AM, Rouillon C, Glevarec G, Gellais A, Hirel B (2002) Genetic and physiological analysis of germination efficiency in maize in relation to nitrogen metabolism reveals the importance of the glutamine synthètase. Plant Physiol 130: 1860-1870.

Loulakakis KA, Roubelakis-Angelakis K (1990) Intracellular localization and properties of NADH-glutamate dehydrogenase from *Vitis vinifera* L.: Purification and characterization of the major leaf isoenzyme. J Exp Bot 41:1223-1230.

Loulakakis KA, Roubelakis-Angelakis K (1991) Plant NAD(H)-glutamate dehydrogenase consists of two subunit polypeptides and their participation in the seven isoenzymes occurs in an ordered ratio. Plant Physiol 106: 217-222.

Loulakakis KA, Roubelakis-Angelakis K (1992) Ammonium-induced increase in NADH-glutamate dehydrogenase activity by de-novo synthesis of the α-subunit. Planta 187: 322-327.

Loulakakis KA, Roubelakis-Angelakis KA, Kanellis AK (1994) Regulation of glutamate dehydrogenase and glutamine synthetase in avocado fruit during development and ripening. Plant Physiol 106: 217-222.

Loulakakis KA, Roubelakis-Angelakis KA (1996) The seven NAD(H) glutamate dehydrogenase isoenzymes exhibit similar anabolic and catabolic activities. Physiol Plant 96: 29-35.

Magalhaes JR, Huber DM (1991) Response of ammonium assimilation enzymes to nitrogen from treatments in different plant species. J Plant Nutr 14: 175-185.

Masclaux-Daubresse C, Quilleré I, Gallais A, Hirel B (2001) The challenge of remobilization in plant nitrogen economy: A survey of physio-agronomic and molecular approaches. Ann Appl Biol 138: 69-81.

Masclaux-Daubresse C, Valadier MH, Brugière N, Morot-Gaudry JF, Hirel B (2000) Characterization of sink/source transition in tobacco (*Nicotiana tabacum* L.) shoots in relation to nitrogen management and leaf senescence. Planta 211: 510-518.

Masclaux-Daubresse C, Valadier MH, Carrayol E, Reisdorf-Cren M, Hirel, B (2002) Diurnal changes in the expression of glutamate dehydrogenase and nitrate reductase are involved in the C/N balance of tobacco source leaves. Plant Cell Environ 25: 1451-1462.

Mckenzie EA, Lees L (1981) Glutamate dehydrogenase activity in developing soybean seed: Isolation and characterization of three forms of the enzyme. Arch Biochem Biophys 212: 290-297.

Miflin BJ, Habash DZ (2002) The role of glutamine synthetase and glutamate dehydrogenase in nitrogen assimilation and possibilities for improvement in the nitrogen utilization of crops. J Exp Bot 73: 1116-1117.

Miflin BJ, Lea PJ (1976) The pathway of nitrogen assimilation in plants. Phytochemistry 15:873-885.

Miflin BJ, Lea PJ (1977) Amino acid metabolism. Ann Rev Plant Physiol 28: 299-329.

Miflin BJ, Lea PJ (1980) Ammonia assimilation. The Biochemistry of plants, 5, Stumpf PK, Conn EE (eds.). Academic Press, New York, pp. 196-202.

Mohan BS, Hosetti BB (1997) Potential phytotoxicity of leaf and cadmium to *Lemna minor* grown in sewage stabilization ponds. Environ Poll 98: 233-238.

Munoz-Blanco J, Cardenas J (1989) Changes in glutamate dehydrogenase activity of *Chlamydomonas reinhardti* under different trophic and stress conditions. Plant Cell Environ 12: 173-182.

Narayan VS, Nair PN (1990) Metabolism enzymology and possible roles of 4-aminobutyrate in higher plants. Phytochemistry 29: 367-375.

Nauen W, Hartmann T (1980) Glutamate dehydrogenase from *Pisum sativum* L. Localization of the multiple forms and glutamate formation in isolated mitochondria. Planta 148: 7-16.

Nussabum S, Schmutz D, Brunold C (1988) Regulation of assimilatory sulfate reduction by cadmium in *Zea mays* L. Plant Physiol 88: 1407-1410.

Obata H, Umebayashi M (1993) Production of SH compounds in higher plants of different tolerance to Cd. Plant Soil 155/156: 533-536.

Osuji GO, Madu W (1997) Regulation of peanut glutamate dehydrogenase by methionine sulphoximine. Phytochemistry 46: 817-825.

Ouariti O, Boussama N, Zarrouk M, Cherif A, Ghorbel MH (1997) Cadmium and copper-induced changes in tomato membranes lipids. Phytochemistry 45: 1343-1350.

Ouzounidou G, Moustakas M, Eleftherou EP (1997) Physiological and ultrastructural effects of cadmium on wheat (*Triticum aestivum* L.) leaves. Arch Environ Contam Toxicol 32: 154-160.

Postius C, Jacobi G (1976) Dark starvation and plant metabolism. VI. Biosynthesis of glutamic acid dehydrogenase in detached leaves from *Cucurbita maxima*. Z Pflanzenphysiol 78: 133-140.

Pahlich E, Joy KW (1971) Glutamate dehydrogenase from pea roots: Purification and properties of the enzyme. Can J Biochem 49: 127-138.

Pahlich E, Ott W, Schad B (1980) Immunochemical investigations with highly purified glutamate dehydrogenase from pea seeds by means of the Ouchterlong test. J Exp Bot 31: 419-423.

Ratajczak L, Ratajczak W, Mazurowa H (1981) Effect of different carbon and nitrogen sources on the activity of glutamine synthètase and glutamate dehydrogenase in lupin embryonic axes. Physiol Plant 51: 277-484.

Sakakibara H, Fujii K, Sugiyama T (1995) Isolation and characterization of cDNA that encodes maize glutamate dehydrogenase. Plant Cell Physiol 36: 789-797.

Salt DE, Rauser WE (1995) MgATP-dependent transport of phytochelatins across the tonoplast of oat roots. Plant Physiol 107: 1293-1301.

Scheid HW, Ehmke A, Hartman T (1980) Plant NAD-dependent glutamate dehydrogenase. Purification molecular properties and metal ion inactivation of the enzymes from *Lemna minor* and *Pisum sativum*. Z Naturforsch 35: 213-221.

Schützendübel A, Polle A (2002) Plant responses to abiotic stresses: Heavy metal-induced oxidative stress and protection by mycorrhization. J Exp Bot 53: 1351-1365.

Skopelitis DS, Roubelakis-Angelakis KA (2002) Glutamate dehydrogenase genes are differentially regulated by the exogenous nitrogen environment. Congress of the Federation of European Societies of Plant Physiology. Crete Grece 2002. Poster 197. pp. 415.

Smart CM (1990) Gene expression during leaf senescence. New Phytol 126: 419-448.

Srivastava HS, Singh RP (1987) Role and regulation of L-glutamate dehydrogenase activity in higher plants. Phytochemistry 26: 597-610.

Suzuki A, Gadal P (1982) Glutamate synthase from rice leaves. Plant Physiol 69:848-852.

Syntichaki KM, Loulakakis KA, Roubelakis-Angelakis KA (1996) The amino acid sequence similarity of plant glutamate dehydrogenase to the extermophilic archaeal enzyme conforms to ist stress-related function. Gene 168: 87-92.

Tercé-Laforgue T, Mäck G, Hirel B (2004) New insights towards the function of glutamate dehydrogenase revealed during source-sink transition of tobacco (*Nicotiana tabacum* L.) plants grown under different nitrogen regimes. Physiol Plant 120: 220-228.

Thomas H (1978) Enzymes of nitrogen mobilization during senescence. Planta 142: 161-169.

Turano F, Dashner R, Upadhyaya A, Caldwell CR (1996) Purification of mitochondrial glutamate dehydrogenase from dark-grown soybean seedlings. Plant Physiol 112: 1357-1364.

Verklej JAC, Lolkema PC, De Neeling AL, Harmens H (1991) Heavy metal resistance in higher plants: biochemical and genetic aspects. In: Rozema J, Verkleij JC (ed.). Ecological Responses to Environmental Stresses, Kluwer Academic Publishers. Dordecht, The Netherlands, pp. 8-19.

Wallsgrove RM, Keys AJ, Lea PJ, Miflin BJ (1983) Photosynthesis, photorespiration and nitrogen metabolism. Plant Cell Environ 6: 301-309.

Wallsgrove RM, Turner JC, Hall NP, Kendall AC, Bright W (1987) Barley mutants lacking chloroplast glutamine synthetase. Biochemical and genetic analysis. Plant Physiol 83: 155-158.

Weatherburn MW (1967) Phenol-hypochlorite reaction for determination of ammonia. Anal Chem 39: 971-974.

Williams LE, Pittman JK, Hall JL (2000) Emerging mechanisms for heavy metal transport in plants. Biochem Biophys Acta 1465: 104-126.

Yamaya T, Oaks A, Matsumoto H (1984) Characteristics of glutamate dehydrogenase in mitochondria prepared from corn shoots. Plant Physiol 76: 1009-1013.

Yue SB (1969) Isoenzymes of glutamate dehydrogenase in plants. Plant Physiol 44: 453-457.

Zhu Y, Pilon-Smith EAH, Tarum A, Weber SU, Jouanin L, Terry N (1999b) Cadmium tolerance and accumulation in Indian mustard is enhanced by overexpressing γ-glutamylcysteine synthetase. Plant Physiol 121: 1169-1177.

Cadmium Toxicity and Tolerance in Plants
Editors: Nafees A. Khan and Samiullah
Copyright © 2006, Narosa Publishing House, New Delhi, India

Cadmium-A Metal-An Enigma: An Overview

C. Chatterjee[*] and B.K. Dube

Botany Department, Lucknow University, Lucknow – 226 007, India

1. INTRODUCTION

In recent times, several studies have been made on heavy metals in relation to their uptake and accumulation in crop plants. Several heavy metals such as cadmium (Cd), chromium (Cr), lead (Pb), mercury (Hg) are classified as pollutant elements as they are mainly concerned with the environmental pollution and detrimental effect in the food chain. In several cases these metals including Cd have direct effects on human health by high dietary intake. Acute Cd toxicity caused by food consumption is not very common, however, regular persistent exposure to high Cd levels in food could significantly increase the accumulation of Cd in certain body organs. The harmful effects associated with a specific Cd intake level; which can be modified or lessened by concentration of other interacting elements present in the diet such as Zinc (Zn), copper (Cu), iron (Fe) and selenium (Se) that influence the safe dietary level of Cd (Underwood, 1977). The World health Organization (WHO) (1972) has proposed a maximum tolerable intake not to exceed 400-500 µg Cd per week (Page et al., 1981)

Cadmium is naturally occurring metallic element, one of the components of the earth's crust and present everywhere in our environment. Its existence was revealed in 1817. It owes its name to *Cadmium fornacum* the 'zinc flowers' which formed on the walls of zinc distillation furnaces.

Its industrial applications were developed particularly during the first half of the 20th century based on its unique chemical and physical properties.

The naturally occurring presence of Cd in the environment results mainly from gradual phenomena such as rock erosion and abrasion or volcanic eruptions. Cd is therefore naturally present in air, water, soil and food stuffs.

2. PHYSICAL PROPERTIES OF CADMIUM

Chemical symbol : Cd, ; Form : Silvery white soft metal with a faint blue tinge ; Characteristics : malleable, ductile and flexible ; Melting point : 321° C ; Atomic number : 48 ; Atomic weight : 112.41; Density : 8.64 g cm^{-3}.

3. CHEMISTRY OF CADMIUM

Cadmium is the second member of group II b and triad (Zn, Cd, Hg) in the periodic classification of elements. The stable state of Cd in the natural environment is Cd (+2). It has a medium class b

---

[*] Corresponding Author (ccbot@sify.com)

character compared to Zn and Hg. This imparts moderate covalency in bonds and high affinity for sulphydryl groups, leading to increased lipid solubility, bioaccumulation and toxicity.

In animals, Cd accumulates in liver and kidney through its strong bindng with cysteine residues of metallothionein. Since the metabolism of Cd is closely related to Zn metabolism, metallothionein binds and transports both Cd and Zn. Cadmium seems to displace Zn in many vital enzymatic reactions, causing disruption or cessation of activity. Cd $(OH)_2$ is more basic than Zn $(OH)_2$, whereas Hg $(OH)_2$ is an extremely weak base. The halides of Zn and Cd are essentially ionic whereas $HgCl_2$ is covalent and almost undissociated in aqueous solution. The organometalic compounds of Zn and Cd are unstable in air and water.

Although they are metals, Zn and Cd are softer, have lower melting points and more electropositive than their neighbouring transition group metals. The chemistry of Cd is homologous to that of Zn and different from Hg in both the properties of the elements and its compounds.

Cadmium is a relatively rare metal in nature that is present in most ores of Zn, an element that it resembles. Cadmium is abundantly present in calamine soils.

4. OCCURRENCE AND SOURCES OF CADMIUM

Cadmium is widely distributed over the earth's surface (Krishnamurthi and Vishwanathan, 1991b). It is present in various types of rocks and soils, in water as well as in petroleum. Cd is not found in pure state in nature but occurs as sulphide along with Zn ores and in minor amounts in lead and copper ores.

The global production of Cd has increased steadily since 1910 when Cd electroplating was developed commercially. The average annual global production of Cd was 12000 tonnes in 1960-69 which increased by 15000-20000 tonnes in 1980-85 (Sandra Postel, 1986). Atmospheric emissions from man-made sources of Cd exceed those from natural sources. Natural sources are volcanic eruptions, ocean sprays and forest fires. The present level of Cd pollution in air, water and food samples in the fourteen states of India are surveyed in the late 1980s has been reported by Krishnamurthi and Vishwanathan (1991b). The principal uses of Cd are for the production of electroplating, pigments, chemicals and alloys. Pollution in air, soil and water has resulted from certain industrial processes involving Cd. Notably emissions by metallurgical industries have caused air and soil pollution and effluents from mines and plants concerned with the processing of ores have polluted water. Although there is no direct effect of Cd toxicity but the element is accumulated in the plants irrigated with this water, which makes potential hazard to human health consuming these food products containing elevated levels of Cd (Lepp, 1981; Woolhouse, 1983; Sigel, 1986).

Cadmium has a long biological half life (> 10 years) and its concentration in the body increases with age. The element reacts with several macro-molecules and organic compounds of biological importance e.g. purine, pyrimidines, nucleosides, nucleotides, RNA, DNA, enzymes etc. It competes with zinc to inhibit the SHO group of the thiol containing enzymes.

In some industrial units atmospheric cadmium concentration of several hundred micrograms (µg) per cubic meter of air have been reported. In surveys, cadmium burdens of 300-400 mg per person were found in exposed areas compared with the normal level of about 10 mg per person. The permissible limit of Cd for drinking water is about 5 mg L^{-1} and the maximum permissible discharge level for the effluents is about 2 mg L^{-1}. There have been a number of cases where ill health has been associated with chronic occupational exposure to dust, aerosole and fumes which contain the metal. Symptoms attributed to Cd poisoning were diverse and included disorders of the respiratory system, kidney and lungs. Some industrial workers exposed to excessive amounts of the metal showed a serious and continuous decline in health. A condition of chronic Cd poisoning which has occurred in the "Jintsu River Bank" in Japan is known as "Itai-Itai" disease. Excess Cd in the diet has been found to impair kidney function and hence disturb the metabolism of Ca and P and cause the bone disease.

Cadmium is a major environmental pollutant present in areas with heavy road traffic and near smelters and sewage sludge areas. Although not essential for plant growth, this metal is readily taken up by roots and translocated to aerial organs in many species (Tyler and McBride, 1982 ; Hardiman and Jacoby, 1984 ; Hendry et al., 1992).The Cd accumulated in plants is the major source of animal and human contamination (Bruwaene et al., 1984) and recent surveys have noticed the occurrence of alarming qualities of Cd in vegetables and food stuff from some polluted urban areas (Bulinski et al., 1990; Srikanth and Reddy, 1991).

5. NATURAL OCCURRENCE

Cadmium is present in soils, waters, plants and other environmental matrices and seldom found in pure state in the natural environment. The average concentration of Cd in the earth's crust is 0.15 - 0.20 ppm (Fleischer et al., 1974) but later the revised concentration was 0.098 ppm Cd. As Cd is closely related to Zn, it is mainly found in Zn, Pb-Zn and Pb-Cu-Zn ores.

The concentration of Cd in cultivated and non-cultivated soils are governed by the qualities of Cd found in the present material. The total Cd is lowest in soils derived from igneous rocks, would be intermediate in soils derived from metamorphic rocks and are the highest in soils derived from sedimentary rocks. The concentration of Cd in the surface soil could be elevated when in certain areas solid wastes containing Cd are disposed off regularly or recycled on the same soil and also Cd being largely immobile in soils.

The soil levels of Cd also depend on the Cd concentration present in native vegetation such as wild oats and mustard. Some reports suggest that residual soils developed shale parents materials had the greatest Cd concentration. An intermediate Cd concentration of 1.5 ppm has been found in association with alluvial soils with parent materials from mixed sources. Several reports on land disposal wastes have revealed that the concentration of Cd in plants depended not only upon the concentration of Cd in substrate but also upon a wide variety of plant and soil factors.

6. SOIL FACTORS INFLUENCING CADMIUM ABSORPTION BY PLANTS

Cadmium being a chemical element is dissolved in the soil solution, adsorbed in organic and inorganic colloidal surfaces, occluded into soil materials, precipitated with other compounds and incorporated into biological materials. A shift from solid-phase forms to that of the soil solution is essential to increase plant available chemical constituents in the soil. The factors governing the equilibrium between the solid and liquid phases of Cd in soils are complicated and not fully in the soil system. It is influenced by soil pH, temperature, organic matter content, oxidative-reduction potential, mineralogical composition and the type and concentration of other dissolved constituents.

It has been reported that amounts of Cd absorbed by plants tend to increase as the concentration of Cd in the soil increases. A large number of studies suggest that Cd accumulation by plant in relation to Cd concentration in soil deal with soil that has been amended with municipal sewage sludge.

Studies conducted separately by Page and Chang (1978) and by De Varies and Tiller (1978) demonstrate that plants grown on Cd enriched soils in containers in the green house absorb more Cd than the same plant grown on the same soil amended with identical amounts of Cd in the field. A reason has been suggested for this differential behaviour that most probably it is dependent on root development. Several observations show progressive increase in concentration of Cd in foliage and the edible part of carrot and radish and in the foliage of Swiss card (B*eta vulgaris* var. Cicla) and lettuce as the amounts of Cd added to the soil in the form of municipal sewage sludge is increased from 0.8 to 6.4 kg ha^{-1}.

The chemical form of Cd also influences the amount of Cd absorbed by plants. However, the uptake of Cd by corn (*Zea mays*) and rye (*Sceale cereale*) grown in soils amended with different types of sewage sludges and observed differences in Cd accumulation by plants where the concentration of Cd in soil was approximately the same but the sludge source of Cd differed.

Several chemical extractants have been tested to provide an index of Cd phytoavailability of Cd recovery from soils. They include weak acids, neutral salts and chelating agents. The Cd extracted by the NH_4OAC was regarded as an exchangeable form, whereas the HCl and HNO_3 extractions estimated closely the total amounts but that the exchangeable, complexed and HNO_3 – soluble fraction of Cd and the total amounts of NH_4OAC solution using oats as the test plant. Several workers (Bingham et al., 1976; Keeney and Walsh, 1975; Street et al., 1978) favour the use of DTPA extracting solution in predicting Cd uptake and yield by crops. The mobility and phytoavailability of Cd largely depends on its chemical form and speciation in soils. Cadmium solubility in soils is decreased as pH increased.

7. UPTAKE AND TRANSLOCATION OF CADMIUM IN PLANTS

Little is known about the mechanisms of Cd uptake by plants or animals at the tissue or cell level. Plants respond to heavy metal toxicity by involving different enzymes ion influx/efflux for ionic balance and synthesize small peptide such as poly (γ-glutanyl-cysteinyl) glycines called phytochelatins (PCS) mainly consisting of glutamate cystein and glycine. These peptides bind metal ions and reduce toxicity (Reddy and Prasad, 1990). The vacuole is a major compartment of Cd^{2+} accumulation in plants as has been suggested by Kortz et al. (1989) and Lange and Wagne (1990). The negative ecological effect of cellular uptake by aquatic plants acts as a biosorbant and influences considerably the levels of both free and bound Cd. Biosorption of Cd^{2+} generally comprises of:

1. Binding of cations to negatively charged groups on the cell surface.
2. Since Cd^{2+} is not required by plants, its entry into cell causes deleterious effect to the organisms at higher level in the food chain. Once Cd ions have entered the root they can either be stored or exported to other plant parts (Skowronski, 1984). However, the degree to which higher plants are able to take up Cd depends on its (a) concentration in the soil and its (b) bioavailability (c) organic matter (d) pH (e) redox potential (f) temperature (g) concentration of other elements.

Cadmium appeared to be absorbed passively (Cutler and Rains, 1974) and translocated freely for the same transmembrane carrier with nutrients such as K, Ca, Mg, Fe, Mn, Cu, Zn and Ni (Rivetta et al., 1997).

Cadmium is believed to penetrate the roots through the cortical tissue. As soon as Cd enters the roots, it can reach the xylem through an apoplastic and/or a symplastic pathway, complexed by several ligands such as organic acids and/or perhaps phytochelatins (Catalado et al., 1988). Normally Cd ions are retained in the roots and only small amounts are transported to shoots. In studies of Cd^{2+}, uptake and distribution in plants exposed to relatively low levels of metal Cd^{2+} is found to accumulate both in roots and shoots, the distribution depending on the species (Tunkendorf and Rauser, 1990).

In cowpea, the accumulation of Cd increased in different parts with an increase in Cd levels from 0.05 to 0.5 mM, and the increase was high in all parts from 0.2 to 0.5 mM Cd supply. This might be due to high concentration of Cd reducing the selectivity of the cell membrane thus allowing more rapid entry of the metal.

The distribution of Cd in each part at each level was almost of the same magnitude and was very high in inflorescence, much higher than reported for edible parts of soil grown field bean (3 ppm) and soybean (10-20 ppm) (Bingham and Page, 1975). The specific symptoms of excess Cd were associated with 16.3-21.1 ppm or more Cd present in different parts of cowpea. Cadmium concentration above these reflected very severe phytotoxicity of Cd. These values are somewhat higher than the reported for other plant species (Page et al., 1981; Adriano, 1986). In cowpea leaves, the values for threshold of toxicity and toxicity were 10.5 and 25 ppm respectively (Dube et al., 2003). These are almost similar to that reported for rice (Iimura and Ito, 1971), for spring barley Chino (1981a) and higher than that reported for bean leaves (Page et al., 1972).

The effect of Cd toxicity on growth was studied in different plant species and it was observed that under Cd stress, growth of plants was reduced and Cd accumulated more in root than shoot (Gussarsson, 1994; Cielinski et al., 1996; Hedva and Caspi, 1999; Romani et al., 2002).

The uptake and distribution of Cd in intact barley plants has also been determined. A large fraction of the Cd taken up by excised barley roots was apparently the result of exchange absorption and was displaced by subsequent desorption with unlabelled Cd. Another fraction of Cd which could not be displaced by desorption in unlabelled Cd was thought to result from strong irreversible binding of Cd, perhaps on sites of cell wall. The fraction of Cd taken up beyond that by exchange absorption by fresh roots was a linear function of temperature and inhibited by conditions of low oxygen and by the procedure of 2, 4 dinitrophenol. It was concluded that this fraction of Cd entered excised barley roots by diffusion, when followed by sequestering probably accounted for the accumulators of Cd observed in intact barley plants (Cutler and Rains, 1974).

The effect of various carboxylic and amino acids on the uptake and translocation of root-absorbed Cd by maize suggests the existence of Cd-organic acid interaction in soil rhizosphere environment of the plant (Nigam et al., 2002). It is important to study the chemical forms, mobility and distribution of toxic metals (Cd and Pb) in contaminated soils since these metals may eventually be translocated to plant tissues (Dudka et al., 1996; Zaman and Zareen, 1998).

Some field and green house experiments have demonstrated that the Cd concentration and pH of the soil are two important factors influencing the uptake of Cd by food crops.

8. SYMPTOMS OF EXCESS CADMIUM

The toxicity symptoms of Cd^{2+} are similar to Fe chlorosis and also necrosis, motling, red-orange leaf coloration and growth reductions have been described (Haghiri, 1973; Bingham and Page, 1975; Phalsson, 1989).

The characteristic symptoms of excess Cd in cowpea were development of marginal chlorosis on young and middle leaves, later affected leaves turned yellow and necrotic, dry and collapsed, no pods were produced at highest level of Cd due to toxicity of the element (Dube et al., 2003). At 0.4 mM Cd, the pods were immature, with very few empty, without any seeds (Table 1). In addition to the visible effects of excess Cd, the growth of plants was depressed markedly, the branches were thin, less in number with short internodes giving a bushy appearance to the plant, leaves were reduced in number as well as in size. The development of inflorescence was marked upto 0.2 mM Cd. Excess Cd not only retarded the growth of plants (Ernst, 1980) but also restricted formation of inflorescence. The reduction in root weight, lateral root formation, dwarfism of cowpea plants, early leaf drop and necrosis of foliage are common features in excess Cd.

9. CADMIUM EFFECTS ON CELL STRUCTURE

Cd induces disorders of cell wall microfibrils in *Chara vulgaris* (Heumann, 1987). Chronic exposure to low levels of Cd decreases a relative cell wall volume in *Clamydomonas bullosa* (Visviki and Rachlin, 1994) and degradation of the wall layer has been observed in *Anabaena flos-aquae* exposed to high Cd levels (Rachlin et al., 1984; Rai et al., 1990). Cd has also been found to cause a severe loss of cohesiveness of the outer polysaccharide layer of the heterocyst envelop of *Nostoc* and this effect was substantially ameliorated by Ca (Mateo et al., 1994).

Cadmium and chromium have also been found to be much more toxic to chloroplasts than to mitochondria in the brown alga *Cystoseira barbata* (Pellegrini et al., 1991) and in *Cyclotella meneghiniana*, respectively.

The more intense damage in chloroplasts may be related to metal-induced enhancement of the formation of free radicals.

Table 1: Influence of excess Cd on growth parameters, dry weight and pod and seed yield of cowpea.

DAMS	Days growth	Control	mM Cd supply					LSD P=0.05
			0.05	0.1	0.2	0.4	0.5	
29	52	\- Dry weight : g per plant \-						
		6.20	3.99	2.93	2.65	2.10	2.06	0.50
		\- Pod weight : g per plant \-						
		5.20	3.39	2.43	2.07	0.88	---	0.31
		\- Seed weight : g per plant \-						
		4.15	2.80	2.05	1.75	---	---	0.08
		\- Weight of 100 seeds : g \-						
		7.40	6.46	4.27	3.88	---	---	0.63

DAMS = Days after metal supply
Dube et al., 2003

10. METABOLISM

Cadmium is one of the most toxic metals in plants active at concentrations much lower than those of other heavy metals (Hardiman and Jacoby, 1984). It has inhibitory effect on plant growth (Greger et al., 1991; Godbold, 1991; Snehlata, 1991), chlorophyll synthesis (Stobart et al., 1985; Padmaja et al., 1990; Somashakhariah et al., 1992) and photosynthetic activity (Baszynski, 1986).

10.1. Photosynthesis

Cd is a potent inhibitor of photosynthesis (Bazzaz et al., 1974), particularly the O_2 evolving reaction of photosystem II (Bazzaz and Govindjee, 1974) and respiration (Clijsters and Van Assche, 1985). Bazzaz and Govindjee (1974) observed that Cd^{2+} is known to affect the PS II activity. The reduction in dry matter yield could be due to the disorganization of photosynthetic apparatus.

Baszynski et al. (1980) showed that the effect of Cd on photosynthesis was secondary as compared to chlorophyll synthesis in tomato plants. Cd reduced chlorophyll level (Imai and Seigel, 1973) by interacting with protochlorophyllide reductase and δ-amino-levulinic acid (ALA) formation (Stobart et al., 1985), Cd inhibits chlorophyll synthesis, decreases the chlorophyll a/b ratio, and causes disorganization of grana (Baszynski et al., 1980).

Cadmium inhibits the organization of chlorophyll by interfering with the organization of pigment protein complexes that are essential for optimal function of PS II.

Cadmium at low concentration e.g. 2-9 ppm reduced dry matter production upto 50% in some field crops (Baszynski, 1986).

Keshan and Mukherji (1992) studied the effect of Cd toxicity on chlorophyll content, Hill Activity and chlorophyllase activity in *Vigna radiata* leaves. They found that leaf chlorophyll concentration and Hill Activity decreased with increased chlorophyllase activity. Cd interacts with the water balance (Costa and Morel, 1994) and damages the photosystem apparatus, in particular the light harvesting complex II and the photosynthesis I and II (Siedlecka and Krupa, 1996).

Mungbean (*Phaseolus vulgaris* L.) seedlings were treated with different concentrations of cadmium acetate (Cd) under both light and dark growth conditions. Cd inhibited δ-amino-levulinic acid (ALA) synthesis and ALA dehydratase activity significantly whereas it had no effect on protoporphyrin IX, Mg-protoporphyrin ester and protochlorophyllide contents. Chlorophyll (Chl) (a and b) levels were decreased by Cd in both light and dark grown seedlings. The *in vitro* inhibition of ALA dehydrogenase by Cd and the comparison of *in vivo* and *in vitro* activity of ALA-dehydratase suggested that inactivation of the enzyme probably by reacting with sulphydryl groups present at the active site of the enzyme. Thus chlorophyll synthesis is regulated by Cd at the level of rate limiting enzymes of porphyrin biosynthesis viz., ALA synthesis and ALA dehydratase in germinating seedlings (Padmaja et al., 1990). Similar studies have also been made in other plant species (Stobart et al., 1985).

Kalita et al. (1993) during their investigation noted that Cd affects seed germination, early seedling growth and chlorophyll content of *Triticum aestivum*. Germination decreased with increasing Cd concentration from 9.1% for the control to 18.89% for 1000 ppm Cd. Chlorophylls a and b content decreased from 0.19 to 0.10 mg and from 0.11 to 0.04 mg g^{-1} fresh weight, respectively.

Cadmium causes a reduction in net photosynthetic rate (Krupa et al., 1993) as it inhibits the activity of photosystem II (PS II) by affecting the water splitting system at the level of magnoprotein (Van Duijvendik-Malteoli and Desmeta, 1975). The inhibition of root Fe (III) reductase induced by Cd led to Fe deficiency and it seriously affected photosynthesis (Alcantara et al., 1994). Photosynthetic rate and chlorophyll content was more pronounced in third leaf from the top in maize. While investigating Cd phytotoxicity, Prasad (1995) observed that it inhibits the chlorophyll and carotenoid contents, gas exchange, decrease in fresh weight and biomass. Cd inhibited the oxidative mitochondrial phosphorylation, probably increasing the passive permeability of the mitochondrial inner membrane (Kessler and Brand, 1995). Cd also actively inhibits the stomatal opening but the mechanism has not been established.

In *Medicago sativa*, Becerril et al. (1989) found that Cd inhibited transpiration and CO_2 assimilation in a drastic manner, whereas in *Picea abies* seedlings decreased CO_2 assimilation was mainly due to stomatal closure (Schlegel et al., 1987).

In plants exposed to Cd-containing nutrient solution increased stomatal resistance without a reduction in leaf pressure potential has been observed (Poschenrieder et al., 1989).

In Cd treated *Phaseolus vulgaris* (bush bean) plants, the decrease in leaf area was not only due to the reduced cell size but also due to decreased intracellular spaces. Cd-reduced photosynthesis and transpiration were attributed to stomatal closure (Bazzaz et al., 1974; Greger and Johansson, 1992).

In *Brassica napus*, Cd lowered total chlorophyll and carotenoid contents and increased the non-photochemical quenching (Larsson et al., 1998). During the investigation of physiological responses of barley plants (*Hordeum vulgare*) to Cd contamination in soil during ontogenesis, Vassilev and Tsone (1999) observed that there was slight decrease in photosynthesis. The rate of transpiration as well as plastid concentration did not change significantly.

In soybean seedlings Cd^{+2} affected growth and inhibited photosynthesis. Both the length and fresh mass decreased more in roots than in shoots. Cd^{2+} stress caused an increase in ratio of chlorophyll (a + b) /b by 1.3 fold and ratio of total xanthophylls β-carotene by 3 fold as compared to the control. A reduced activity of photosystem II by about 85% measured in Cd^{2+} treated chloroplasts was associated with a dramatic quenching of fluorescence emission intensity, with a band shift of 4 nm. A major suppression of absorption was accompanied with shift in peaks in the visible region of the spectrum. In Cd^{2+} treated chloroplasts a selective decline in linolenic acid (18:3), the most unsaturated fatty acid of chloroplasts was found parallel with the 10 fold enhancement in ethylene production. A three fold increase in peroxidase activity was found in chloroplasts treated with Cd^{2+} compared to the control. Addition of 1 mM glutathione (GSH) counteracted all the retardation effects in soybean seedling growth induced by Cd^{2+}. Thus GSH may control the Cd^{2+} growth inhibition as it detoxifies Cd^{2+} by reducing its concentration in the cytoplasm and removing hydrogen peroxidase generated in chloroplasts (El-shinitinaway, 1999).

Cd, Ni and Cu have been reported to inhibit photosystem II, ATP synthatase and various Calvin cycle enzymes (Lucero et al., 1976; Mostowska, 1977; Krupa et al., 1993). This is likely to disturb the balance between electron pressure in photosynthetic electron transport chain and availability of electron acceptors.

Cadmium specially inhibits chlorophyll and carotenoid biosynthesis (Dubey, 1997). The resulting decrease in pigments causes deficiency in light harvesting capacity (Ouzoinidou, 1996; Moustakas et al., 1997) and consequently decreases photosynthetic activity of the cells.

Cd^{2+} in addition to other metals directly influence the photosynthetic electron transport processes and inhibit CO_2 fixation. The effects are due to three major processes:
(1) Oxidative breakdown of chlorophylls and carotenoids
(2) Oxidative damage of proteins and membrane structures and
(3) Substitution of metal co-factors.

Cadmium affects photosynthesis by inhibition of different reaction steps of the Calvin cycle, and not by interaction with photosynthetic reactions in the thylakoid membranes (Weigel, 1985 a:b). The membrane bound photosynthetic reactions in isolated mesophyll protoplasts were not impaired by Cd concentrations which drastically inhibited CO_2 fixation.

Cadmium inhibits electron flow on the reducing side of PS I. According to Siedlecka and Baszynski (1993) the site of Cd inhibition in PS I is between primary electron acceptor X and NADP Cd treatment causes Fe deficiency indicating that the light phase of photosynthesis was affected in the treated plants due to Cd-induced Fe deficiency.

Cadmium (200 ppm) applied through the rooting medium to 30 day old wheat plants decreased net CO_2 exchange and PS II activity (Malik et al., 1992). Cadmium causes permanent stomatal closure and increased ethylene production results in senescence (Fuhrer, 1988). Cadmium exposure also decreased thylakoid membrane total lipids like MGDG (monogalachosyl diaacylglycerol), DGDG (digalaehosyl diacyl glycerol) and SQDG (Sulphoquinouosyl diacylglycerol) resulting in thylakoid degradation and dissociation of oxygen evolving complex peptides.

Cadmium exerts its toxicity through membrane damage and inactivation of enzymes, possibly through reaction with sulphydryl groups of proteins (Mathys 1975; Fuhrer, 1988), sulphydryl inactivation was suggested to explain the inhibitory effects of Pb and Cd on the activity of the chloroplast enzyme Rubisco and phosphoribulokinase *in vitro*.

Sheoran et al. (1990) investigated the effect of Cd and Ni on photosynthesis and enzymes of the photosynthetic carbon reduction (PCR) cycle in pigeon pea (*Cajanus cajan*) and concluded that the reduction in photosynthesis was through a decrease in chlorophyll content and effects on stomatal conductance and the electron transport system.

Cadmium treatment induced symptoms of iron deficiency. Cadmium is an effective inhibitor of plant metabolism.

10.2. Cadmium and Enzyme Activity

In germinating seedlings of mungbean (*Phaseolus vulgaris* cv. K 16) treated with different concentrations of cadmium acetate (10, 50 and 100 µM), Cd^{2+} lowered the chlorophyll and haem levels. The level of lipid peroxidases was higher on day 3 than on day 6. However, the Cd^{2+} treatment significantly enhanced the level of lipid peroxidases. Similarly, a dose dependent induction of lipoxygenase (EC 1.13.11.12) activity was observed with Cd^{2+} treatment. Further the activities of antioxidant enzymes such as superoxide dismutase (EC 1.15.1.1) and catalase (EC 1.11.1.6) were decreased. The results suggest that lipoxygenase mediated accumulation of lipid peroxidases on one hand and inhibition of free radical scavenging enzymes like superoxide dismutase and catalase on the other hand caused a pronounced reduction in the chlorophyll and heme levels of the seedlings. The experiments conducted on the effect of Cd^{2+} on dark grown seedlings did not conform with the result of light grown seedlings. Though chlorophyll and heme levels decreased in a dose dependent manner, no accumulation of lipid peroxidases was observed suggesting that the inhibition of chlorophyll synthesis by Cd^{2+} is achieved both by reaction with constituent biosynthetic enzymes as well as peroxidase mediated degradation (Somashakharaiah et al., 1992).

The suitability of nitrate reductase activity and the level of some metabolites as an *in vivo* test system for Cd toxicity was evaluated in submerged macrophyte *Hydrilla vesticillata*. Cadmium concentration ranging from 0.01-80 µM affected nitrate reductase activity in a different way. It had stimulatory effect upto 1.0 µM Cd, while higher concentrations inhibited the enzyme activity significantly. The protein synthesis inhibitor cyclohexamide inhibited nitrate reductase activity during *in vivo* and *in vitro* assays. However the effect of Cd on nitrate reductase activity under *in vitro* assay was more pronounced. Low Cd exposures had no effect but high metal exposure augmented nitrate uptake.

The Cd induced NO_3 uptake did not result in recovery of inhibited enzyme activity *in vivo*. It appears that nitrate reductase activity is more sensitive to Cd toxicity than the evaluated products of nitrate assimilation such as total organic nitrogen and soluble proteins. Cd has a differential response to chlorophyll levels, lower concentrations enhanced the pigment level while higher ones reduced it. Cadmium exposure always enhanced the levels of carotenoids (Rai et al., 1998).

Cadmium toxicity also causes elongation inhibition in legume seedlings. The germination decreased with an increasing concentration of Cd. In *Phaseolus aureus* Cd influences the activity of hydrolytic enzymes such as alpha amylase and proteases which seem to inhibit the seedling growth. The effect of the element seems to be more on the roots (Kalita et al., 1993).

Toxic doses of Zn and Cd inhibit shoot growth but increase the capacity of several leaf enzymes in dwarf beans (*Phaseolus vulgaris* L.). Both effects were studied as a function of the metal concentrations applied to the plant. There was a linear relationship between the metal content of the primary leaf and the nutrient solution. When leaf metal content exceeded toxic threshold values, shoot growth became inhibited and an increase in capacity of the following enzymes was measured in the leaf, glucose-6-phosphate dehydrogenase, glutamate dehydrogenase, isocitrate dehydrogenase, malic enzyme glutamate oxaloacetate transaminase, peroxidase. The threshold values were similar for growth inhibition as well as for enzyme capacity induction. Both effects were strongly correlated to each other, specially under conditions of toxic zinc treatment. Measurement of enzyme capacity might therefore provide a useful criterion for the evaluation of the phytotoxicity of soils contaminated by Zn and/or Cd (Van Assche et al., 1988).

Cd^{2+} enhanced the activity of cytochrome oxidase (Vallee and Ulmer, 1972). Kesseler and Brand (1995) explained the decrease in phosphorylation efficiency under elevated Cd^{2+} levels by an increase in the permeability of the mitochondrial inner membranes to protons in the presence of Cd. Cadmium is known to reduce the activity of isocitrate dehydrogenase (ICDH) and malate dehydrogenase (MDH).

10.3. Oxidative Stress

The content of carboxyl groups in leaf extracts of pea was two fold higher in plants treated with Cd than in the control plants. By using different antibodies some of the oxidized proteins were identified as rubisco, glutathione reductase, Mn-superoxide dismutase and catalase. The incubation of leaf crude extracts with increasing H_2O_2 concentration showed a progressive enhancement in carboxyl content and the pattern of oxidized proteins was similar to that found in Cd treated plants. Oxidized proteins were more efficiently degraded and the proteolytic activity increased 20% due to the metal treatment. In peroxisomes purified from pea leaves a rise in the carboxyl content similar to that obtained in crude extracts from Cd treated plants was observed but the functional ability of the peroxisomal membrane was not apparently affected by Cd. Results show the participation of both oxidative stress, probably mediated by H_2O_2 and the proteolytic degradation in the mechanism of Cd toxicity in leaves of pea plants and they appear to be involved in the Cd induced senescence previously reported in these plants (Romero-Puertas et al., 2002).

To study the relationship between heavy metal ion toxicity and oxidative stress in plant cells, leaf segments from 14 day old sunflower seedlings were incubated in solution containing 0.5 mM Fe (II), Cu (II) or Cd (II) ions for 12 h in the light. Treatment with metal ions produced a decrease in chlorophyll and glutathione (GSH) contents as well as an increase in lipid peroxidase and lipoxygenase activity. Free radical scavengers such as sodium benzoate and mannitol prevented the decrease in chlorophyll and GSH content and the lipid peroxidation and lipoxygenase increased. While Fe (II) and Cd (II) ions caused a decrease in superoxide dismutase activity, Cu (II) ions raised its level. However all three metal ions caused a decrease in other antioxidant enzymes (catalase, ascorbate peroxidase, glutathione reductase and dehydroascorbate reductase). Free radical scavengers protected these enzymes against inactivation. No effect of these scavengers was observed on superoxide dismutase activity. These results indicate that excess Fe (II), Cu (II) or Cd (II) ion produce oxidative damage in plant leaves (Gallego *et al.*, 1996). Glutathione synthetase activity was increased during Cd treatment (Ruegsegger et al., 1990). Cd produces oxidative stress (Hendry et al., 1992) but in contrast with other heavy metals such as Cu, it does not seem to act directly on the production of reactive oxygen species (via Fenton and/or Heber Weiss reaction) (Salin, 1988).

Membranes are considered as the main target of Cd^{2+} toxicity (Hendry et al., 1992). Cd^{2+} exerts its toxicity by producing membrane damage due to lipid peroxidation mediated by activated oxygen radicals that is quenched by antioxidant enzymes (Reddy and Prasad, 1992). Glutathione (GSH) and associated antioxidant systems have been implicated in adaptation of plants to various oxidative stresses (Alscher, 1989; Verkleij et al., 1990).

In *Phaseolus aureus* Cd ions produced lipid peroxidation, decrease of catalse activity and increase of guaiacol peroxidase and ascorbate peroxidase (Shaw, 1995). In roots and leaves of

Phaseolus vulgaris, 5 μM Cd enhanced activities of guaiacol and ascorbate peroxidase and increased lipid peroxidation (Chaoui et al., 1997), in pea plants Cd treatment notably increased lipid peroxidation (Lozano Rodriquez et al., 1997); whereas no peroxidation was noted in Cd exposed plants and hairy roots of *Daucus carota* (di Toppi et al., 1998). Varying responses to Cd induced oxidative stress are probably related both to the levels of Cd supplied and to the concentration of thiolic groups already present or induced by Cd treatment. Thiol posses strong antioxidative properties and they are consequently able to counteract oxidative stress (Pichorner et al., 1993).

Although Cd is the most toxic among the heavy metals, it has been reported to stimulate the growth of zinc limited marine diatom, *Thalossiosira weissfligii*, by substituting Zn in certain macro molecules (Price and Morel, 1990). This shows that in extreme conditions toxic metals such as Cd can also act as micronutrients.

The relationship between Cd and Zn phytotoxicities and oxidative reactions in bean plants has been observed. The ten day old bean plants were subjected to 5 μM Cd and 100 μM Zn separately showed some reduction of growth. In response to each metal lipid peroxidation was enhanced in all plant organs and catalase activity was decreased in both roots and leaves but not in stems. However, Cd and Zn stimulated the activity of guaiacol-dependent peroxidase only in stems, where native electrophoresis revealed at least two new anionic isoenzymes. The induction of one of the isogluaiacol peroxidase was Zn specific. The exposure of metals did not modify the activity of ascorbate peroxidase either in roots or in stems.

10.4. Respiration

In terrestrial plants Cd treatment results in increased respiration, probably directly affecting the cytoplasm and damage to the mitochondrial structures.

Cd-membrane interactions seem to involve the binding of Cd^{2+} to sulphydryl groups of the mitochondrial membrane in *Zea mays* as suggested by the effects of sulphydryl protecting agent dithiothreitol (Miller et al., 1973). Cd^{2+} shows a higher affinity to mitochondrial structures and therefore, exhibits a stronger effect on mitochondrial metabolism viz., electron transport and phosphorylation (Bittell and Miller, 1974), but has no effect on mitochondrial volume as measured by the light transmission properties of a mitochondrial suspension (Kesseler and Brand, 1995; Lorimer and Miller, 1969).

10.5. Water Stress

Cadmium also inhibits abscisic acid accumulation during drying of excised leaves. In clover and lucerne, Cd and Pb induced several changes in gas exchange and water relations (Becenil et al., 1989). Cadmium induces a decrease in water stress resistance in bush bean plants (*Phaseolus vulgaris* cv. Contender) by affecting endogenous abscisic acid, water potential, relative water content and cell wall elasticity (Barcelo et al., 1986 a, b; Poschenrieder et al., 1989).

10.6. Growth

Growth and ultrastructural changes of Cd-treated wheat plants (*Triticum aestivum*) were described by Ouzounidou et al. (1997). Young plants were grown in nutrient solutions with Cd concentrations upto 1 mM. At these concentrations, elongation of roots and of above ground parts was reduced to 28% and 40% of the control, respectively. Increased Cd concentrations in leaves of plants under Cd stress were accompanied by declining growth and concentration of Fe, Mg, Ca and K. The author suggested Cd induced premature senescence in such plants.

10.7. Toxicity of Cadmium and Nutrient Concentrations

In Cd-stress leaf weight was reported reduced more than root weights (Cieslinski et al., 1996). They further conclude that leaf dry weights were the best indicators of Cd toxicity. Increased Cd concentrations in leaves of plants under Cd stress were accompanied by declining concentration of Fe, Mg, Ca and K. The growth inhibition, reduced chlorophyll content and an inhibition of photosynthesis in upper plant parts may have resulted from Cd effects on the plant content of essential nutrients. In Cd-treated plants the structure of chloroplasts is changed. This may be due to Cd-induced premature senescence (Ouzounidou et al., 1997).

The possible interaction of two stresses UV-B radiation and Cd when applied simultaneously to *Brassica napus* L. cv. Paroli, the changes in chlorophyll fluorescence, growth and uptake of selected elements were observed. Exposure to Cd significantly increased the amount of Cd in both roots and shoots, increase also occurred in the concentration of Fe, Zn, Cu and P in roots while K was reduced.

Significant increase in Mg, Ca, P, Cu and K occurred in plants exposed to Cd and UV-B radiation together, Mn decreased significantly under combined exposure treatment. The rise in S content may have been due to stimulated glutathione and phytochelation synthesis. Cd exposure significantly increased root dry weight, leaf area, total chlorophyll content, carotenoid content and phytochemical quantum yield of photosynthesis. The chlorophyll a:b ratio showed a reduction with UV-B at no or low Cd concentrations used (2 and 5 µM $CdCl_2$) (Larsson et al., 1998).

10.8. Interaction of Cadmium with Other Elements

The toxicity of Cd^{2+} *in vivo* during the early phases of radish (*Raphanus sativus* L.) seed germination and the *in vitro* effect on radish calmodulin (CaM) were studied. Cd was taken up in the embryo axes of radish seeds, and the increase in fresh weight of embryo axes after 24 h of incubation was inhibited significantly in the presence of 10 m mol m^{-3} Cd^{2+} in the external medium, when the Cd content in the embryo axes was 1.1 µ mol g^{-1} fresh weight. The re-absorption of K which characterizes germination was inhibited by Cd^{2+} suggesting that Cd affected metabolic reaction: The slight effect of Cd on the transmembrane electric potential of the cortical cells of the embryo axes excluded a generalized toxicity of Cd at the plasma membrane level. After 24 h of incubation Cd induced no increase in total acid soluble thiols whereas Cd binding peptides were able to reduce Cd toxicity. Calcium added to the incubation medium partially reversed the Cd induced inhibition of the increase in fresh weight of embryo axes and concomitantly reduced Cd uptake (Rivetta et al., 1997).

The effect of organic acids viz., citrate, malate, succinate and calcium was observed on Cd toxicity and it was found that organic acids normally increase the growth of the algae *Shigeoclonium tenue*, Kutz in terms of dry weight, total chlorophyll and protein content, succinate was most effective followed by malate in reducing the toxic effects of Cd. The addition of organic acids resulted in the increase of internal Cd level but decreasd the toxic effect of Cd on growth. In contrast to the effect of organic acids, calcium not only reduced the toxic effect of Cd but also decreased the uptake of Cd by the algae (Vanaja et al., 2000).

The uptake of Cd, Se (IV) and Zn by the fresh water alga *Scenedesmus oldiquitis* and the subsequent transfer and release budget in *Daphnia magna* were investigated under different nutrient additions and cell incubation conditions. An increase in ambient phosphate concentrations from 0.5 µmol L^{-1} significantly increased the intracellular accumulation of Cd. The percentage of Cd distributing in the intracellular pool of algae also increased substantially with increasing ambient P concentrations. The assimilation efficiency (AE) of Cd was generally independent of the nutritional conditions. Responses of trophic transfer *Daphnia* to nutrient enrichment was metal specific (Yu and Wang, 2004).

Low concentration of Cd^{2+} significantly inhibited pollen germination and germ tube elongation in red pine (*Pinus resinosa*) (Chaney and Strickland, 1984).

11. PHYTOREMEDIATION

Heavy metal pollution of soils and waters mainly caused by mining and burning of fossil is a major environmental problem. Heavy metals unlike organic pollutants cannot be chemically degraded or biodegraded by micro-organisms. Several methods for reducing soil trace element availability to plants have been reported. Sorption, ion exchange precipitation, attenuate and increasing soil pH have been proposed (Johnson et al., 1977; Sims, 1986; Shuman, 1986; Sims and Kline, 1991).

An alternative biological approach to deal with this problem is phytoremediation i.e. the use of plants to clear up polluted waters and soils (Black, 1995; Salt et al., 1995). Heavy metals metalloids can be removed from polluted sites by phytoextraction which is the accumulation of the pollutants in the plant biomass (Kumar et al., 1995). Compared with other remediation technologies, phytoremediation is less expensive (1000 fold less) than excavation suitable for treatment of large volumes of substrate with low concentration of heavy metals.

However the presence of heavy metals inhibits plant growth, limiting the application of phytoremediation. Therefore, one trait that is of great significance to phytoremediation is the ability of plants to tolerate the toxic metals that are being extracted from the soil.

The proposed remediation method involves the use of trace-elements tolerant plant species that are able to hyperaccumulate trace elements in plant shoots (Baker et al., 1991). This method known as phytoremediation involves successive croppings of hyperaccumulator plants to translocate potentially polluting trace elements from soils to plant shoots (Chaney, 1983; Ebbs et al., 1997). Recently many plants have been reported to accumulate high level of toxic metals in their sinks in the aquatic and terrestrial ecosystems. Such plants can remove the pollutant metals from soils and water (Cunninghan, et al., 1995; Ensley et al., 1997). EDTA and citric acid can also be used for remediation of Cd, Cr and Ni from soil using sunflower (Targut et al., 2004).

Therefore it appears that selecting a more appropriate plant species for the phytoextraction (phytoremediation) i.e. removal of metals from the contaminated soils, sludges and sediments in the vicinity of the industrial unit having specific metal contaminant using a potential plant species/cultivar is yet a task ahead.

Instead of going to the root cause of this metal toxicity problem, most widely acceptable solution of biological remediation is by using plant species having capacity to accumulate the toxic metals. Such bioremediation process has potential advantage over conventional treatment process in terms of low cost and is environmentally compatible.

The metal hyper accumulators are defined as having concentrations of more than 100 ppm for Cd^{2+} (on dry weight basis) (Brown et al., 1994). The high amounts of metals in the hyperaccumulator plants tissue suggests the existence of defence mechanisms to avoid the harmful effects caused by the metal. However, this mechanism of accumulation of metals and their effects on the metabolism of submerged plants are not fully understood (Guilizzoni, 1991) and there is an immediate need to unravel these problems for specific need of pollution abatement.

Several species of aquatic plants usually endemic to metalliferous soils are known for their ability to accumulate large quantities of metals (Baker and Brooks, 1989). They possess tremendous capacity to concentrate metals in their tissues at levels much higher than their surrounding habitat.

Indian mustard (*Brassica juncea* L.), Italian serpentine plant (*Alyssun beztolonii*) and *Thlaspi caerulescens* are some of the most potential terrestrial plant species which were recently been used to extract toxic metals from soil and sediments and translocate these metals to the roots, harvestable stalks and leaves etc. of the plants (Baker and Brooks, 1989; Reeves, 1992; Baker et al., 1994; Brown et al., 1994; Ensley et al., 1997; Schiekler and Caspi, 1999). All aquatic plants are not equally effective for removal of heavy metals. Plants such as *Phragmites communis, Seirpus lacustris, Eichhoraia crassipes, Elodea cariadensis, Egeria densa, Hydrilla spp., Bacopa monnieri, Limnanthemum cristamum* and the algal macrophyte, *Hydrodictyon reticulatum* are suitable for removal of different metals (Wolverton et al., 1975; Rai and Chandra, 1989; Sinha and Chandra, 1990; Rai and Chandra, 1992).

Using mRNA differential display, identification of sequenced and quantified the induction of a number of transcripts that are up-regulated by a brief (2 h) exposure to 25 µM $CdCl_2$ including one transcript which is also highly responsive to iron (Fe) deficiency. These transcripts represent both nuclear and chloroplast – encoded genes and include both suspected functions. The magnitude of induction and functional analysis suggest possible utility for these genes in the study of metal stress sensing in green plants and development of novel Fe acquisition and phytoremediation strategies (Peter et al., 2002).

A consideration of phytoremediation of trace elements is the disposal or utilization of the harvested plant material. Another potential alternative for disposal of trace element laden plant material is utilizing the harvested plants as an organic source for the eventual production of electricity in biomass fired plants. In this scenario, the plant is fuelled by agricultural prunings such as harvested plant material. Combustion is efficient because of a circulating fluidized bed boiler which results in a reduced amount of air pollutants.

Air emissions are scrubbed clean by a large cyclone, multiclones and an electrostatic precipitator. Trace element i.e. Cd which would eventually accumulate over time in residual ash may be extracted depending upon the cost.

References

Adriano DC (1986) Cadmium. In: Adriano DC (ed.). Trace Elements in the Terrestrial Environment. Springer Verlag, New York, pp. 106-155.

Alcantara E, Romera FJ, Cañnete M, De La Guardia MD (1994) Effects of heavy metals on both induction and function of root Fe (III) reductase in Fe-deficient cucumber (*Cucumis sativus* L.) plants. J Exp Bot 45: 1893-1898.

Alscher RG (1989) Biosynthesis and anti-oxidant function of glutathione in plants. Physiol Plant 77: 457-464.

Baker AJM, Brooks RR (1989) Terrestrial higher plants which accumulate metallic elements: a review of their distribution, ecology and phytochemistry. Biorecovery 1: 81-126.

Baker AJM, Reeves RD, Mc Grath SP (1991) In Situ *Bioreclamation*. Butterworth –Heinemann, Boston, MA. pp. 600-605.

Baker AJM, Reeves RD, Hajar ASM (1994) Heavy metal accumulation and tolerance in British populations of the metallophyte *Thlaspi caerulescens* J & C Presl (Brassicaceae) New Phytol 127: 61-68.

Barcelo J, Cabot C, Poschenrieder C (1986a) Cadmium-induced decrease of water stress resistance in bush bean plants (*Phaseoluv vulgaris* L. cv. Contender) II. Effects of Cd on endogenous abscisic acid levels. J Plant Physiol 125: 27-34.

Barcelo J, Poschenrieder C, Andreu I, Gunse B (1986b) Cadmium-induced decrease of water stress resistance in bush bean plants (*Phaseoluv vulgaris* L. cv. Contender) I. Effects of Cd on endogenous abscisic acid levels. J Plant Physiol 125: 17-25.

Baszynski T (1986) Interference of Cd^{2+} in functioning of the photosynthetic apparatus of higher plants. Acta Soc Bot Pol 55: 291-304.

Baszynski T, Wajda L, Krol M, Wolinska D, Krupa Z, Tukendorf A (1980) Photosynthetic activities of Cd-treated plants. Physiol Plant 48: 365-370.

Bazzaz FA, Carlson RW, Rolfe GL (1974) The effect of heavy metals on plants. Environ Poll 7: 241-246.

Bazzaz MB, Govindjee (1974) Cd^{2+} is known to effect the PS II activity. Environ Lett 6: 1-12.

Becernil JM, Gonzalez-Murua C, Munoz-Rueda R, DeFelipe MR (1989) The changes induced by cadmium and lead in gas exchange and water relations in clover and Lucerne. Plant Physiol Biochem 27: 913-918.

Bingham FT, Page AL (1975) Cadmium. In: Adriano (ed.). Trace Elements in the Terrestrial Environment. Springer Verlag, New York, pp. 106-155.

Bingham HT, Page AL, Mahler RJ, Ganje TJ (1976) Cadmium availability to rice in sludge-amended soil under 'flood' and 'non-flood' culture. Soil Sci Soc Amer J 40: 715-719.

Black H (1995) Absorbing possibilities: phytoremediation. Environ Health Perspect 103: 1106-1108.

Brown SL, Chaney RL, Angle JS, Baker AJM (1994) Phytoremediation potential of *Thlaspi caerulescens* and bladder campion for Zn and Cd – contaminated soils. J Environ Qual 23: 1151-1157.

Bruwaene RV, Kirchmann R, Impens R (1984) Cadmium contamination in agriculture and zootechnology. Experientia 40: 42-52.

Bulinski R, Kot A, Bloniarz J, Wiszogrodzka L (1990) Studies on some trace elements content in food stuffs of home growth. Part XI: Evaluation of contamination with harmful metals of crop products. Bromatol Chem Toksycol 23: 105-108.

Catalado DA, Mc Fadden, KM, Garland TR, Wildung RE (1988) Organic constituents and complexation of nickel (II), iron (III), cadmium (II), and plutonium (IV) in soybean xylem exudates. Plant Physiol 86: 734-739.

Chaney RL (1983) Land treatment of hazardous wastes. In: Parr JF, Marsh PB, Kla JM (eds.). Plant Uptake of Inorganic Waste Constituents Noyes Data Corp, Park Ridge, NJ, pp. 50-76.

Chaney WR, Strickland RC (1984) Relative toxicity of heavy metals to red pine pollen germination and germ tube elongation. J Environ Qual 13: 391-394.

Chaoui A, Mazhoudi S, Ghorbal MH, El Ferjani E (1997) Cadmium and zinc induction of lipid peroxidation and effects on antioxidant enzyme activities in bean (*Phaseolus vulgaris* L.). Plant Sci 127: 139-147.

Chino M (1981a) Cadmium. In: Adriano DC (ed.). Trace Elements in the Terrestrial Environment. Springer Verlag, New York, pp. 106-155.

Cielinski G, Neilsen GH, Hogue EJ (1996) Effect of soil cadmium application and pH on growth and cadmium accumulation in roots, leaves and fruit of strawberry plants (*Fragaria ananossa* Duch). Plant Soil 180: 267-276.

Clijsters H, Van Assche F (1985) Inhibition of photosynthesis by heavy metals. Photosynthesis Res 7: 31-40.

Costa G, Morel LJ (1994) Water relations, gas exchange and amino acid content in Cd-treated lettuce. Plant Physiol Biochem 32: 561-570.

Cunninghan SD, Bnerti WR, Huang JW (1995) Phytoremediation of contaminated soils. Trends Biotech 13: 393-397.

Cutler MJ, Rains WD (1974) Characterization of Cd uptake by plant tissue. Plant Physiol 54: 67-71.

De Varies MPC, Tiller KG (1978) Sewage sludge as a soil amendment with special reference to Cd, Cu, Mn, Ni, Pb and Zn – comparisons of results from experiments conducted inside and outside a glass house. Environ Poll 16: 231-240.

di Toppi SL, Lambardi M, Pazzagli L, Cappugi G, Durante M, Gabbrielli R (1998) Response to cadmium in carrot *in vitro* plants and cell suspension cultures. Plant Sci 137: 119-129.

Dube BK, Pandey VN, Sinha P, Chatterjee C (2003) Cadmium phytotoxicity and disturbances in cowpea physiology. Poll Res 22: 105-111.

Dubey RS (1997) Photosynthesis in plants under stressful conditions. In: Pessarakli M (ed.). Handbook of Photosynthesis. Marcel Dekker, New York, pp. 859-876.

Dudka S, Piotrowska M, Terelak H (1996) Transfer of cadmium, lead, and zinc from industrially contaminated soil to crop plants: a field study. Environ Poll 94: 181-188.

Ebbs SD, Lasat MM, Brady DJ, Cornish J, Gordon R, Kochiar LV (1997) Heavy metals in the environment-phytoextraction of cadmium and zinc from a contaminated soil. J Environ Qual 26: 1424-1430.

El-Shinitinaway F (1999) Glutathione counteracts the inhibitory effect induced by cadmium on photosynthetic process in soybean. Photosynthetica 36: 171-179.

Ensley BD, Raskin I, Salt DE (1997) Phytoremediation applications for removing heavy metal contamination from soil and water. In: Sayler (ed.). Biotechnology in the Sustainable Environment. Plenum Press, New York, pp. 59-64.
Ernst WHO (1980) Biochemical aspects of cadmium in plants. In: Nriagu JO (ed.). Cadmium in the Environment. Wiley Inter Science, New York, pp. 639-654.
Fleischer M, Sarofim AF, Fassett DW, Hammond P, Shacklette HT, Nisbet IC, Epstein S (1974) Environ Health Perspec 7: 253-323.
Fuhrer J (1988) Ethylene biosynthesis and cadmium toxicity in leaf tissue of beans plants *Phaseoluv vulgaris* L. Plant Physiol 70: 162-167.
Gallego SM, Benavides MP, Tomaro ML (1996) Effect of heavy metal ion excess on sunflower leaves: evidence for involvement of oxidative stress. Plant Sci 121: 151-159.
Godbold DL (1991) Cadmium uptake in Norway spruce (*Picea abies* L. Karst.) seedlings. Tree Physiol 9: 349-357.
Greger M, Johansson M (1992) Cadmium effects of leaf transpiration of sugar beet (*Beta vulgaris*). Physiol Plant 86: 465-473.
Greger M, Brammer E, Lindberg S, Larsson G, Idestam AJ (1991) Uptake and physiological effects of cadmium in sugar beet (*Beta vulgaris)* related to mineral provision. J Exp Bot 42: 729-737.
Guilizzoni P (1991) The role of heavy metal and toxic materials in physiological ecology of submerged macrophytes. Aq Bot 87: 87-109.
Gussarsson M (1994) Cadmium-induced alterations in nutrient composition and growth of *Betula pendula* seedlings: the significance of fine roots as a primary target for cadmium toxicity. J Plant Nutr 17: 2151-2163.
Haghiri F (1973) Cadmium uptake by plants. J Environ Qual 2: 93-96.
Hardiman RT, Jacoby B (1984) Absorption and translocation of cadmium in bush beans (*Phaseolus vulgaris*) Physiol Plant 61: 670-674.
Hedva S, Caspi H (1999) Response of anti oxidative enzymes to Ni and Cd stress in hyperaccumulator plants of the genus *Alyssum*. Physiol Plant 105: 39-44.
Hendry GAF, Baker AJM, Edwart CF (1992) Cadmium tolerance and toxicity, oxygen radical processes and molecular damage in Cd tolerant and Cd sensitive clones of *Holcus lanatus* L. Acta Bot Nether 40: 271-281.
Heuman HG (1987) Effect of heavy metal on growth and ultra structure of *Chara vulgaris*. Protoplasma 136: 37-48.
Iimura K, Ito H (1971) Cadmium. In: Adriano DC (ed.). Trace Elements in the Terrestrial Environment. Springer Verlag, New York, pp. 106-155.
Imai I, Seigel S (1973) A specific response to toxic cadmium levels in red kidney bean embryos. Physiol Plant 29: 118-120.
Johnson MS, Mc Neillyu T, Putwain PD (1977) Revegetation of metalliferrous mine soil contaminated by lead and zinc. Environ Poll 12: 261-277.
Kalita MC, Devi P, Bhattacharya I (1993) Effect of cadmium on seed germination, early seedling growth and chlorophyll content of *Triticum aestivum*. Indian J Plant Physiol 36: 189-190.
Keeney DR, Walsh LM (1975) In: Proceedings of International Conference on Heavy Metals in the Environment. Toronto, Ontario, Canada, pp. 379-401.
Keshan U, Mukherji S (1992) Effect of cadmium toxicity on chlorophyll content, Hill activity and chlorophyllase activity in *Vigna radiata* L. leaves. Indian J Plant Physiol 35: 225-230.
Kesseler A, Brand MD (1995) The mechanism of the stimulation of state 4 respiration by Cd in potato tuber (*Solanum tuberosum*) mitochondria. Plant Physiol Biochem 33: 519-528.
Kortz RM, Evangelou BP, Wagner GJ (1989) Relationships between cadmium, zinc, Cd-peptides and organic acid in tobacco suspension cells. Plant Physiol 91: 780-787.
Krishnamurthi CR, Vishwanathan P (1991) Cadmium in the Indian environment and its human health implication. In: Krishnamurthi CR, Vishwanathan P (eds.). Toxic Metals in the Indian Environment. Tata McGraw Hill Publishing Company Ltd., New Delhi, pp. 75-95.
Krupa Z, Oquist G, Hunter NPA (1993) The effects of cadmium on photosynthesis of *Phaseolus vulgaris*: A fluorescence analysis. Physiol Plant 88: 626-630.

Kumar PBAN, Dushenkov V, Motto H (1995) Phytoextraction: the use of plants to remove heavy metals from soils. Environ Sci Technol 29: 1232-1238.

Lange RV, Wagne GJ (1990) Subcellular localization of cadmium and cadmium binding peptides in tobacco leaves. Plant Physiol 92: 1086-1093.

Larsson EH, Bornman JF, Asp H (1998) Influence of UV-B radiation and Cd^{2+} on chlorophyll fluorescence, growth and nutrient content in *Brassica napus*. J Exp Bot 49: 1031-1039.

Lepp NW (1981) Effect of Heavy Metal Pollution on Plants. Vol. I. Applied Science Publishers, London and New Jersey.

Lorimer GH, Miller RJ (1969) The osmotic behaviour of corn mitochondria. Plant Physiol 44: 839-844.

Lozano-Rodriguez E, Hernandez LE, Bonay P, Carpena-Ruiz RO (1997) Distribution of Cd in shoot and root tissues of maize and pea plants: physiological disturbances. J Exp Bot 48: 123-128.

Lucero HA, Andreo CS, Vallejos RH (1976) Sulphydryl groups in photosynthetic energy conservation. II. Inhibition of phosphorylation in spinach chloroplasts by $CdCl_2$. Plant Sci Lett 6: 309.

Malik D, Sheoran IS, Singh R (1992) Lipid composition of thylakoid membranes of cadmium treated wheat seedlings. Indian J Biochem Biophys 29: 350-354.

Mateo P, Fernandez-Pinas F, Bonilla I (1994) O_2-induced inactivation of nitrogenase as a mechanism for the toxic action of Cd^{2+} on *Nostoc* UAM 208. New Phytol 126: 267-272.

Mathys W (1975) Enzymes of heavy-metal-resistant and non-resistant populations of *Silene cucubalus* and their interactions with some heavy metals in vitro and in vivo. Physiol Plant 33: 161-165.

Miller RJ, Bittell JE, Koeppe DE (1973) The effect of cadmium on electron and energy transfer reactions in corn mitochondria. Physiol Plant 28: 166-171.

Mostowska M (1977) Environmental Factors affecting chloroplasts. In: Pessarakli M (ed.). Handbook of Photosynthesis. Marcel Dekker, New York, pp. 407-426.

Moustakas M, Lanaras T, Symeonidis I, Karataglis S (1997) Growth and some photosynthetic characteristics of field grown *Avena sativa* under copper and lead stress. Photosynthetica 30: 389-396.

Nigam R, Srivastava S, Prakash S, Srivastava MM (2002) Plant availability of cadmium in presence of organic acids: an interactive aspect. J Environ Biol 23: 175-180.

Ouzoinidou G (1996) The use of photoacoustic spectroscopy in assessing leaf photosynthesis under copper stress: correlation of energy storage to photosystem II fluorescence parameters and redox change of P700. Plant Sci 111: 229-237.

Ouzounidou G, Moustakas M, Eleftheriov EP (1997) Physiological and ultrastructural effects of cadmium on wheat (*Triticum aestivum* L.) leaves. Arch Environ Contam Toxicol 32: 154-160.

Padmaja K, Prasad DDK, Prasad ARK (1990) Inhibition of chlorophyll synthesis in *Phaseolus vulgaris* L. seedlings by Cd acetate. Photosynthetica 24: 399-405.

Page AL, Chang AC (1978) Trace elements impact on plants during cropland disposal of sewage sludges. In: Proceedings-Fifth National Conference on Acceptable Sludge Disposal Techniques. Information Transfer Inc., Rockville, Maryland, pp. 91-106.

Page AL, Bingham FT, Chang AC (1981) Cadmium. In: Lepp NW (ed.). Effect of Heavy Metal Pollution on Plant. Applied Science Publishers, London, pp. 77-109.

Page AL, Bingham FT, Nelson C (1972) Cadmium absorption and growth of various plant species as influenced by solution cadmium concentration. J Environ Qual 1: 288-291.

Pellegrini L, Pellegrini M, Delivopoulos S, Berail G (1991) The effects of cadmium on the fine structure of brown alga *Cystoseirs barbata* forma repens Zinova et Kaliguna. Br Phycol J 26: 1-8.

Peter R, Siripornadulsil S, Rubinelli GF, Sayre RT (2002) Cadmium and iron stress inducible gene expression in the green alga *Chlamydomonas reinhardtii*: evidence for H_{43} protein function in iron assimilation. Planta 215: 1-13.

Phalsson AB (1989) Toxicity of heavy metals (Zn, Cu, Cd, Pb) to vascular plants: a literature review. Water Air Soil Poll 47: 287-319.

Pichorner H, Koroi SAA, Thur A, Ebermann R (1993) The two and the four electron transfer to molecular oxygen mediated by plant peroxidase in the presence of thiols. In: Welinder KG, Rasmussen SK, Penel C, Greppin (eds.). Plant Peroxidases: Biochemistry and Physiology, University of Geneva, pp. 131-136.

Poschenrieder C, Gunse B, Barcelo J (1989) Influence of cadmium on water relations, stomatal resistance and abscisic acid content in expanding bean leaves. Plant Physiol 90: 1365-1371.

Prasad MNV (1995) Inhibition of maize leaf chlorophylls, carotenoids and gas exchange functions by cadmium. Photosynthetica 31: 635-640.

Price NM, Morel FMM (1990) Cadmium and cobalt substitution for zinc in a marine diatom. Nature 344:658-660.

Rachlin JW, Jensen TE, Warkentine B (1984) The toxicological response of the alga *Anabaena flos-aquae* (Cyanophyceae) to cadmium. Arch Environ Contam Toxicol 13:143-151.

Rai LC, Jensen TE, Rachlin JW (1990) A morphometric and X-ray energy dispersive approach to monitoring pH altered cadmium toxicity in *Anabaena flos-aquae*. Arch Environ Contam Toxicol 19:479-487.

Rai UN, Chandra P (1989) Removal of heavy metals from polluted waters by *Hydrodictyon reticulatum* (Linn.). Lagerheim Sci Total Environ 87-88: 509-515.

Rai UN, Chandra P (1992) Accumulation of copper, lead, manganese and iron by field population of *Hydrodictyon reticulatum* (Linn.). Lagerheim Sci Total Environ 116: 203-211.

Rai UN, Gupta M, Tripathi RD, Chandra P (1998) Cadmium regulated nitrate reductase activity in *Hydrilla verticillata* (l.f.) Royale. Water Air Soil Poll 106: 171-177.

Romani S, Shaikh MS, Suseelan KN, Kumar SC, Joshua DC (2002) Tolerance of *Sesbania* species to heavy metals. Indian J Plant Physiol 7: 174-178.

Reddy GN, Prasad MNV (1990) Heavy metal binding proteins/peptides: occurrence, structure, synthesis and functions: a review. Environ Exp Bot 30: 251-264.

Reddy GN, Prasad MV (1992) Cadmium induced peroxidase activity and isozymes in *Oryza sativa*. Biochem Arch 8: 101-106.

Reeves RD (1992) The hyperaccumulation of Ni by serpentine plants. In: Baker AJM, Proctor J, Reeves RD (eds.). The vegetation of Ultramafic (Serpentine) Soils: Proceedings of the First International Conference on Serpentine Ecology. Intercept. Andover, pp. 253-277.

Rivetta R, Negrini N, Cocucci M (1997) Involvement of Ca^{2+} calmodulin in Cd^{2+} toxicity during the early phases of radish (*Raphanus sativus* L.) seed germination. Plant Cell Environ 20: 600-608.

Romero-Puertas MC, Palma JM, Gomez M, Del Rio LA, Sandalio LM (2002) Cadmium causes the oxidative modification of proteins in pea plants. Plant Cell Environ 25: 677-686.

Ruegsegger A, Schmutz D, Brunold C (1990) Regulation of glutathione synthesis by cadmium in *Pisum sativum* L. Plant Physiol 93: 1579-1584.

Salin ML (1988) Toxic oxygen species and protective systems of the chloroplasts. Physiol Plant 72: 681-689.

Salt DE, Blaylock M, Kumar NPBA, Dushenkov V, Ensley BD, Chet I, Raskin I (1995) Phytoremediation: a novel strategy for the removal of toxic metals from the environment using plants. Biotechnology 13: 468-474.

Sandra P (1986) Altering the earth's chemistry assessing and risk. World Watch paper 71, July 1986, World Watch Institute.

Schiekler H, Caspi H (1999) Response of anti oxidative enzymes to nickel and cadmium stress in hyper accumulation plants of the genus *Alyssum*. Plant Physiol 105: 39-44.

Schlegel H, Godbold DL, Huttermann A (1987) Whole plant aspects of heavy metal induced changes in CO_2 uptake and water relations of spruce (*Picea abies*) seedlings. Physiol Plant 69: 265-270.

Shaw BP (1995) Effects of mercury and cadmium on the activities of anti oxidative enzymes in the seedlings of *Phaseolus aureus*. Biol Plant 37: 587-596.

Sheoran IS, Singal HR, Singh R (1990) Effect of cadmium and nickel on photosynthesis and the enzymes of the photosynthetic carbon reduction cycle in pigeonpea (*Cajanus cajan*). Photosynth Res 23: 345-351.

Shuman LM (1986) Effect of liming on the distribution of manganese, copper, iron and zinc among soil fractions. Soil Sci Soc Am J 50: 1236-1240.

Siedlecka A, Baszynski T (1993) Inhibition of electron flow around photosystem I in chloroplasts of Cd-treated maize plants is due to Cd-induced iron deficiency. Physiol Plant 87: 199-202.

Siedlecka A, Krupa Z (1996) Interaction between cadmium and iron and its effects of photosynthetic capacity of primary leaves of *Phaseolus vulgaris*. Plant Physiol Biochem 35: 833-841.

Sigel H (1986) Metal Ions in Biological Systems: concept on Metal Ion Toxicity. Vol 20, Marcel Dekker Inc., New York.

Sims JT (1986) Soil pH effects on the distribution and plant availability of manganese, copper and zinc. Soil Sci Soc Am J 50: 367-373.

Sims JT, Kline JS (1991) Chemical fractionation and plant uptake of heavy metals in soil amended with-co-

composted sewage sludge. J Environ Qual 20: 387-395.

Sinha S, Chandra P (1990) Removal of Cu and Cd from water by *Bacopa monneiri* L. Water Air Soil Poll. 51: 271-276.

Skowronski T (1984) Energy dependent transport of cadmium by *Stichococcus bacillaris*. Chemosphere 13: 1379-1384.

Snehlata (1991) Phasic pre-treatment effects of cadmium on seedling growth and activity of hydrolytic enzymes in *Phaseolus aureus* cultivars. J Environ Biol 12: 299-306.

Somashakhariah BV, Padmaja K, Prasad ARK (1992) Phytotoxicity of Cd ions on germinating seedlings of mungbean (*Phaseolus mungo*): Involvement of lipid peroxidases in chlorophyll degradation. Physiol Plant 85: 85-89.

Srikanth R, Reddy SRP (1991) Lead, cadmium and chromium levels in vegetable grown in urban sewage sludge: Hyderabad, Indian Fed Chem 40: 229-234.

Stobart AK, Griffiths WT, Ameen Bukhari I, Sherwood RP (1985) The effect of Cd^{2+} on the biosynthesis of chlorophyll in leaves of barley. Physiol Plant 63: 293-298.

Street JJ Sabey BR, Lindsay WL (1978) Influence of pH, phosphorus, cadmium, sewage sludge and incubation time on the solubility and plant uptake of cadmium. J Environ Qual 7: 286-290.

Targut C, Pepe MK, Cutright TJ (2004) The effect of EDTA and citric acid on phytoremediation of Cd, Cr, and Ni from soil using *Helianthus annuus*. Environ Poll 131: 147-154.

Tunkendorf A, Rauser WE (1990) Changes in glutathione and phytochelatins in roots of maize seedlings exposed to cadmium. Plant Sci 70: 155-166.

Tyler LD, Mc Bride MB (1982) Influence of Ca, pH and humic acid on cadmium uptake. Plant Soil 64: 259-262.

Underwood EJ (1977) Trace Elements in Human and Animal Nutrition. 4th ed. Academic Press, New York.

Vallee BL, Ulmer DD (1972) Biochemical effects of mercury, cadmium and lead. Rev Biochem 41: 91-128.

Van Assche F, Cardinaels C, Clijsters H (1988) Induction of enzyme capacity in plants as a result of heavy metal toxicity: dose response relations in *Phaseolus vulgaris* L. treated with zinc and cadmium. Environ Poll 52: 103-115.

Van Duijvendik-Malteoli MA, Desmeta GM (1975) On the inhibitory site of PS II in isolated chloroplasts. Biochim Biophys Acta 408: 164-169.

Vanaja M, Charyulu NVN, Rao KVN (2000) Effect of some organic acids and calcium on growth, toxic effect and accumulation of cadmium in *Stigeoclonium tenue* Kutz. Indian J Plant Physiol 7: 163-167.

Vassilev A, Tsone V (1999) Physiological responses of barley plants (*Hordeum vulgare*) to cadmium contamination in soil during ontogenesis. Environ Poll 103: 289-293.

Verkleij JAC, Koevoets P, Van'triet J, Bank R, Nijdam Y, Ernst WHO (1990) Poly (γ-glutamylcysteinyl) glycines or phytochelatins and their role in cadmium tolerance of *Silene vulgaris*. Plant Cell Environ 13: 913-921.

Visviki I, Rachlin JW (1994) Acute and chronic exposure of *Dunaliella salina* and *Chlamydomonas bullosa* to copper and cadmiu: effects on ultrastructure. Arch. Environ. Contam Toxicol 26: 154-162.

Weigel HJ (1985a) Inhibition of photosynthetic reactions of isolated intact chloroplast by cadmium. J Plant Physiol 119: 179-189.

Weigel HJ (1985b) The effect of cadmium on photosynthetic reaction of mesophyll protoplast. Physiol Plant 63: 192-200.

Wolverton BC, Mc Donald RC, Gordon J (1975) Water hyacinth and alligator weeds for final filtration of sewage NASA. Tech Memo TM – X – 72724.

Woolhouse HW (1983) Toxicity and tolerance in the responses of plants to metals. In: Lange OL, Noble PS, Osmond CB, Ziegier H (eds.). Encyclopedia of Plant Physiology. Vol 12, Springer Verlag, Berlin, pp. 245-289.

World Health Organization (1972) Evaluation of certain food additives and contaminants. Mercury, lead and cadmium. WHO Technical Report Series N.O. 505, Sixteenth Report of the Joint FAO/WHO Expert Committee on Food Additives, WHO, Geneva.

Yu RQ, Wang WX (2004) Biokinetics of cadmium, selenium and zinc in fresh water alga *Scenedesmus obliquus* under different phosphorus and nitrogen conditions and metal transfer to *Daphnia magna*. Environ Poll 129: 443-456.

Zaman MS, Zareen F (1998) Growth responses of radish plants to soil Cd and Pb contamination. Bull Environ Contam Toxicol 61: 44-50.

Index

α-tocopherol	18
β-alanine	81
β-carotene	169
β-oxidation	11
β-mercaptoethanol	141
γEC peptides	128
γ-glutamylcysteine synthetase	101
δ-amino-levulinic acid	167
γ-glutamylcysteine synthetase	21, 112
γ-glutamyl cysteine dipeptidyl transpeptidase	24
A. thaliana	36, 44, 45, 47, 50
ABA	67, 77
Abiotic stresses	32, 124
Abiotic	11, 15, 46, 71, 74
Abrasion	159
Abscisic acid	174
Abscission	109
Absorption	92, 98
Accumulation	8, 9, 49, 65, 67, 71, 92, 98, 121, 143, 159, 174, 177
Accumulator	92
Acer	77
Acid invertase	72
Acid phosphatase	72
Acid rains	64
Acid soils	64
Acidic environment	116
Acrylamide	142
Active absorption	90
Active metabolism	91
Adaptable agriculture	104
Adaptations	16
Additive	111
Adsorption	35, 36
Aerobic	10
Aerosole	161
Aerotolerant anaerobes	32
Agricultural	3, 5
Agriculture	2, 64
Agricultural activities	4
Agricultural soils	6, 103, 105
Agrobacterium	49
Agroecosystem	5
Agronomic practices	92
Agronomist	4
Agropyron repens	38
Agrostis	65, 92
Air	159
Air emissions	178
Air pollutants	178
ALA dehydratase	71, 72
ALA synthase	71, 72
Aldose	48
Aldehyde reductase	48
Algae	32, 115, 176
Alkaline phosphatas	72
Alkaloids	25
Alkyl peroxides	45
Alluvial soils	162
Alpha amylase	171
Alternative oxidases	27
Alyssum	36, 44
Alyssun	178
Amaranthus	26, 36
Amaranthus lividus	38
Aminating	151
Amination	151, 156
Amine oxidases	12
Amino acids	70, 80, 128, 165
Amino-terminal targeting sequences	34
Ammonia	138, 147
Ammonia assimilation	140, 151, 156, 157
Ammonium	75, 138, 154
Ammonium accumulation	144
Ammonium detoxification	145
Ammonium sulphate	109
Amylase	95
Amylose	99
Anabaena	166
Anabolic	157
anaerobic	10
Animal	4
Anion	91
Ankistrodesmus	67
Antagonistic	36, 63, 65, 111
Anthesis	91
Anthocyanin	47
anthropogenic	35, 64, 115
Antigenic properties	139
Antioxidant	8, 17, 25

Antioxidant defences	50	Biomass accumulation	96
Antioxidant enzymes	2, 171	Biomin	114
Antioxidative	73	Biomolecules	71, 75
Antioxidative enzyme	108	Bioremediation	177
Apo-phytochelatin	82	Biosphere	4
Apoplast	19, 27	Biosynthesis	45, 49
Apoplastic	7, 10, 65, 164	Biotic	32, 46, 124
Apoplastic isoenzymes	39	Biotoxicity	99
Aquatic	5, 87, 177	*Brassica*	24, 85, 87, 95, 110, 169, 178
Aquatic plants	65	*Brassica chinensis*	97
Arabidopsis halleri	85	*Brassica juncea*	49, 95, 108, 112
Arabidopsis thaliana	37, 81, 101	*Brassica napus*	108, 175
Arabidopsis	9, 24, 67, 80, 101, 105, 139, 154	*Brassica oleraceae*	112
Aromatic electron donors	46	Bright-yellow 2 cells	40
Arsenate reductase gene	50	*Bz-2*	47
Arsenic	4	*C. reinhardtii*	138
ascomycete	24	Ca 56	
Ascorbate	17, 25, 32, 38, 42, 46, 111	Cabbage	35, 36
Ascorbate peroxidase	13, 26, 38, 72, 73, 79, 173	Cadmium	2, 4, 35, 36, 63, 87, 88, 91, 115, 137, 159, 168, 171
Ascorbate-glutathione cycle	21, 27, 43, 44, 111	*Cajanus*	170
Ascorbic acid	79	Calamine soils	160
Asparagine	71, 154	Calcareous	107
Asparatic acid	71	Calcium	4, 78, 176
Aspergillus	8	Callose	80
Assimilation	75, 138	Calmodulin	175
Assimilation efficiency	176	Calvin	7, 12, 32, 76
Atmosphere	64, 90	Calvin cycle,	170
Atmospheric	35, 88, 103	Cambium	67
Atomic	4	*Camellia sinensis*	46
ATP	10, 76	Cancer	5
ATPase	65	*Candida glabrata*	129
Atrazine	46	Carbohydrate	46, 74
Autoradiograms	143	Carbon	13
Auxin	70	Carbon oxides	3
Azolla	38	Carbonate rich soils	109
Bacopa monnieri	178	Carbonic anhydrase	71
Barley	37, 92	Carcinogenic	5, 103
Beta vulgaris	163	*Cardaminopsis*	39
Betaine	7	Carotene	25
Bioaccumulation	2, 160	Carotenoid	76, 168, 169
Bioavailability	6, 41, 61, 65, 63, 164	Carrot	92
Biochemical	127	Casparian strips	67
Bioconcentration	38	*CAT1*	37
Bioconcentrations	5	*CAT2*	37
Biogeochemical	63	*CAT3*	37
Biomagnification	2	Catabolic	157
Biomass	5, 45, 85, 93, 104, 107, 110, 143, 168	Catabolic enzyme	138

Index

Catalase	13, 26, 36, 72, 73, 79, 94, 171, 172, 173
Catalytic cycle	34
Cations	77, 91, 96
Cation exchange	90
Cation exchange capacity	42, 90
Cd chelation	134
Cd complexation	130
Cd concentration	132
Cd detoxification	126
Cd mobility	104
Cd phytoavailability	109
Cd phytoextraction	104, 105
Cd sequestration	116, 130
Cd tolerance	112
Cd toxicity	88, 94
Cd	9, 26
Cd-contaminated soils	113
Cd-ligand complexes	43
Cd-S crystallite	129
CDTA	40
CEC	90
Celery	36
Cell	71
Cell death	14
Cell division	120
Cell membrane	97
Cell wall	12, 65, 67, 68, 77, 82, 93, 97, 165
Cell wall elasticity	174
Cellular	13
Cement factories	63
Ceratophyllum demersum	40
Cercospora	13
Cereal	5, 36
Chara	166
Chelating agents	41
Chelation	23, 82, 90
Chelators	133
Chemical	4
Chemical pollutants	3
Chicory	39
Chlorella	67
Chloride	65
Chlorophyll	6, 7, 11, 13, 25, 76, 94, 95, 112, 167, 168, 171, 175
Chlorophyll degradation	71
Chlorophyll fluorescence	113
Chlorophyll synthesis	167
Chlorophyllase	167
Chloroplast	14, 21, 43, 45, 73, 76
Chlorosis	7, 68, 86, 165
Chromatographic profiles	127
Chromium	4, 159
Chronic phytotoxicity	111
Chronic	15
Cichorium intybus	39
Citrate	82, 176
Citric acid	91
Clamydomonas	166
Clay	35
Clones	9, 10
Clover	76
Cobalt	4
Codex committee	103
Compartmentalization	82, 97, 111
Compartmentation	67
Complexation	133
Concentrations	4, 6, 172
Contaminants	103
Contaminated	165
Contaminated soil	109
Contamination	2, 4, 87, 88, 99
Copper	4, 33, 65, 159
Corn	35, 36, 163
Cortex	67, 80
Cowpea	164
Critical	89
Crop	63
Crotalaria	44
Cu	7
Cu/Zn protein	32
Cu/Zn-SOD	33, 34
Cu/Zn-SODs	33, 34
Cucumber	19, 33, 92
Cultures	8
Cuscuta	24
Cyclohexamide	171
Cyclone	178
Cyclotella meneghiniana	166
Cys	118
Cysteine	22, 23, 25, 43, 47, 71, 80, 101
Cystoseira	166
Cytochrome c	19
Cytochrome c peroxidase	39
Cytochrome oxidase	172
Cytoplasm	92
Cytosol	21, 27, 32, 45, 66, 111, 112, 129
Cytosolic	39
Cytotoxic	12, 118
Daphnia	176

de novo	6, 75, 140, 156, 157	Electrokinetic	38
DEAE-cellulose system	147	Electron	10, 17, 27, 78
Deamination	151, 156	Electron transport	170
Decontamination	87, 114	Electron transport chain	12, 77
Deficiencies	97	Electrophoresis	141
Deficit	71	Electrophylic	46
Degradation	171	Electroplating	64, 160
Dehydratase	168	Electrostatic precipitator	178
Dehydroascorbate	19, 29, 42	Elemental sulphur	109
Dehydroascorbate reductase	26, 28, 42, 173	*Elodea cariadensis*	178
Dehydrogenases	17	Emission	35
Deionized	59	Endodermis	10, 67, 80
Deleterious	93	Endoperoxides	13
Demineralization	5	Endoplasmic reticulum	45
Desorption	165	Energy metabolism	90
Detoxification	9, 17, 20, 27, 41, 48, 50, 65, 71, 79, 81, 101, 116, 117, 120, 126, 130, 133, 138	Environment	3, 4, 38, 63, 64, 88, 162
		Environmental	15
		Environmental pollutants	5
Dianthus chinensis	41, 42, 50	Environmental pollution	3
Diethyltriamine pentaacetic acid	65	Environmental stress	37
Diffusion	65, 165	Enzymes	7
Dihydric anions	91	Epidermis	10
Dioxygen	10	Epiphyte	99
Disorganisation	7	*Eremosphera*	67
Disposal	104	Erosion	38
Distribution	94, 165	Erythrocuprein	32
Disulfide	21	Erythrocytes	32
Dithiothreitol	77, 174	*Escherichia coli gshII*	49
DNA	8, 47, 161	*Escherichia coli*	112
DNAase	95	*Escherichia*	32
DNA synthesis	74	Ethanol	10
Dry biomass	45	Ethylene	82, 169, 170
Dry matter yield	96	Ethylenebis (oxyethylenetrinitrilo) tetraacetic acid	41
Dry weight	70		
DTPA	40	Ethylene-diamine-tetraacetic acid	108
Duckweed	38	Eukaryotes	15, 32
Dust	161	Europe Union	88
Dwarfism	166	Evapotranspiration	107
Ecophysiological	22	*ex situ*	104
Ecotype	105	Exchangeable	90
EDTA	40, 41, 52, 56, 108	Excluder	65, 92
Efflux	97	Exodermis	10
Egeria densa	178	Exogenous nitrogen environment	138
Eggplant	92	Exosmosis	94
EGTA	40	Exposure	3
Eichhoraia crassipes	178	Exudation	16
Eichhornia crassipes	67	Fatty acids	36
Eichhornia	38	Fe	7, 56

Feedback	22	Glutathione reductase	13, 26, 43, 72, 73, 172, 173
Fenton	8	Glutathione S-transferase	20
Fermentation	10	Glutathione synthesis	111
Ferredoxin	42, 74	Glutathione-ascorbate cycle	26
Fertilizers	6, 35, 63, 64, 87, 115	Glutathione-S-transferase	43, 46
Fe-SOD	33, 34	Glycine	21, 70, 80, 81, 163
Flavonoid-binding proteins	47	Glycolate	36
Flavoprotein	43	Glycolipids	73
Fluorescence	76	Glycolysis	77
Food chain	2, 5, 103, 133	Glyoxylate	12, 36
Food stuffs	159	Glyoxysomes	36
Forest fires	3, 161	*Gossipium*	85, 108
Fossil	6, 176	*gr1*	43
Fossil fuels	3	*gr2*	43
Fragrance	104	Grain	103
Free amino acids	152	Grana	7
Free ions	4	Grasses	36
Free metal ions	111	Groundwater	59, 108
Free radical	171	Growth inhibition	118
French	36	Growth reductions	165
Fruit	5	Growth	5, 63, 68, 165, 171
Fruiting	94	GS/GOGAT cycle	75, 138
Fungicides	99	GS/GOGAT	137
Fv/Fm	95	GSH synthase	120
Galactose	18	GSH	8, 118
Gas exchange	113, 168	GST25	47
Gaseous pollutants	27	GST26	47
Gasoline	64	Guaiacol	46
GDH isoenzymes	155	Guaiacol peroxidase	45, 72
gdh mRNA	157	*H. annuus*	36, 44
Gene	11, 93	H^+-ATPase	72
Genes expression	120	H_2O_2	11, 12, 25
Genetic engineering	112	H_2S	98
Genotoxic	137	Haber-Weiss reaction	33
Genotoxicity	74	Haber-Weiss	13
Genotypes	50, 94, 100, 103, 107, 108	Halliwell-Foyer-Asada cycle	28
Germination	176	Heat shock proteins	81
Gibberellins	114	Heavy metals	3, 41, 47, 87, 93, 104, 105, 107, 115, 118, 151, 177
Glucose	10, 18	Heavy metal contamination	89
Glucose-6-phosphate dehydrogenase	172	Heavy metal detoxification	111
Glutamate dehydrogenase	72, 75, 137, 138, 156, 172	HEDTA	40
Glutamate	81, 138	*Helianthus annus*	79
Glutamate oxaloacetate	172	*Helianthus annuus*	38
Glutamic	80	*Helianthus*	39, 85, 87, 108
Glutamine synthatase	72, 75	Heme peroxidase	39
Glutamine	81, 154	Herbicide	46
Glutathione	13, 17, 32, 49, 66, 78, 79, 82, 100, 111, 169, 173	Heterogeneous	110
Glutathione peroxidase	26, 27	Hexadecanoic acid	76

Hexameric protein	139	Indophenol oxidase	32
hGSH	118, 133	Industrial	2
hGSH/GSH ratios	120	Industrial emission	87
hGSH synthesis	124	Industrial processes	6, 115
Higher plants	26	Industrial waste	88
Higher weighted complex	116	Inflorescence	166
Histidine	66	Inorganic pyrophosphatas	72
Histograms	127	Inorganic	38
HMW complex	130, 132	International food standards organisation	103
HMW complexes	131	Intracellular Cd	131
HMW	127	Invertase	74
Homeostasis	13, 18	Ion diffusion	90
Homeostatic processes	9	Ion exchange	177
Homodimer	33, 40	Ion radii	77
Homogenous	110	Iron	4, 6, 33, 159
Homoglutathione	20	Isocitrate dehydrogenase	172
Hordeum	36, 140	Isoenzymes	34, 43, 146
Hordeum vulgare	169	Isoleucine	71
Hormone	7	Itai-Itai	5, 87, 103, 161
Horticultural	5	Jintsu River Bank	161
hPC	123	K^+ ions	14
Human health	3, 64, 85	K^+-ATPase	72
Humic acid	90	Kinases	14
Hybridisation	143	Kinetin	140
Hydrilla	171, 178	Kreb's cycle	77
Hydrodictyon reticulatum	178	Lactate dehydrogenase	95
Hydrogen peroxidase	169	*Lactuca*	26
Hydrogen peroxide	10, 82, 140	Leaching	38, 59
Hydroperoxides	13, 44	Lead	4, 159
Hydroperoxy fatty acid	45	Leaf area	70
Hydrophobic	46	Leaf gas-exchange	108
Hydroponic	96	Leaf	64
Hydroquinones	25	Leak	92
Hydroxyl radical	34	Leakage	14
Hydroxyl radicals	10	Legumes	36
Hydroxylase	25	*Lemna minor*	38
Hyperaccumulating	8	Lesions	13
Hyperaccumulation	9, 83, 105, 111	Lettuce	35, 36, 92
Hyperaccumulator	65, 67, 83, 105, 116, 177	Leucine	71
Hyperoxia	13	L-galactose dehydrogenase	18
Igneous rocks	162	Ligand	17, 65
Immobilization	80	Lignification	8, 82, 86
in situ	75	Lignin	45
in vitro	73, 75, 168, 175	*Limnanthemum cristamum*	178
in vivo	168, 171	*Linum*	85, 108, 156
Indian mustard	49, 108	Lipid	6, 8, 20, 45, 94, 160, 171
Indicator	65, 92	Lipid hydroperoxides	44
Indoleacetic acid	45	Lipid peroxidation	14, 94, 114, 173

Lipid peroxides	13
Lipoxygenase	171, 173
Liquid nitrogen	141
LMW	131
LMW complex	132
Loam	35
Localization	64
Low molecular weight complex	116
Lucern	76
Lumen	42
Lycopersicon esculentum	146, 152, 155, 156
Lycopersicon	81, 123, 140
Macro	7
Macromolecular complexes	92
Macromolecules	13, 92
Macronutrients	113
Macrophyte	178
Macrophytes	5
Magnaporthe	24
Magnesium	4
Maize	20, 37, 46, 47, 92, 95, 121, 127, 139, 152
Malate	82, 176
Malate dehydrogenase	172
Malic acids	109
Malic enzyme	172
Malondialdehyde	14, 73, 78, 94
Mammalian	11
Manganese	4, 33
Mannitol	25
Mannose	18
Maturation	133
Medicago	168
Membrane lipids	73
Membrane	6
Membrane permeability	73, 138
Mentha	85, 108
Mercaptide bonds	81
Mercury	4, 159
Mesophyll	170
Metabolic	13, 71
Metabolic absorption	90
Metabolism	3, 12, 68, 70, 74
Metabolomics	51
Metal	8, 16, 63
Metal accumulation	128
Metal chelation ability	131, *See*
Metal complexation	121, 126, 131
Metal detoxification	126
Metal stress	123
Metal tolerance	112
Metal transporters	112
Metalliferous	9, 105, 178
Metalloids	177
Metalloprotein	146
Metallothioneins	24, 80, 82, 86, 111
Metallothiopeptide	67
Metamorphic rocks	162
Methionine	71
Mg	56
Michaelis constant	74
Michaelis-Menten kinetics	65
Micronutrients	7, 77
Microorganisms	82, 98, 99, 103
Middle lamellae	70
Mineral nutrition	111
Mineralogical composition	162
Mining	2, 88
Mitochondria	10, 11, 12, 14, 17, 21, 27, 32, 43, 45, 67, 73, 138
Mitochondrial electron transport	11
Mn	57
Mn-SOD	33, 34, 48
Mn-superoxide dismutase	172
Mobility	64
Molecular	82
Molybdenum	4
Monodehydroascorbate	19, 29
Monodehydroascorbate reductase	26, 28
Monothiol	119
Morphogenesis	93, 120
Motling	165
mRNA	178
Mucilage	80
Multiclones	178
Multigene	37
Mungbean	168
Municipal waste	87
Mustard	26, 67
Mutagenic	5
Mutation	13
Mycorrhizal fungus	99
NAD (P)H	12
NAD(H)-GDH	155
NADH-GDH pathway	75
NADH-GDH	144
Native PAGE	43
Natural sources	161
Natural	3
Necrosis	8, 165, 166

Nematodes	115	Organelles	6, 34
Ni	26	Organic	38, 63
Nickel	4	Organic acid	98, 101
Nicotiana	35, 85, 87	Organic matter	6, 35, 65, 85, 90, 92, 162, 164
Nicotiana glauca	49	Organic nitrogen	171
Nicotiana plumbagnifolia	48	*Oryza*	25, 66, 70
Nicotiana tabaccum	108	Osmoregulation	74
Nitrate assimilation	74	Osteomalacia	87
Nitrate reductase	72, 75, 95, 137171	Overexpression	112
Nitrate	74, 78, 152, 171	Oxalate	19, 82
Nitric oxide synthase	12	Oxidases	12
Nitrilotriacetic acid	109	Oxidation	36
Nitrite reductase	137	Oxidative damage	35, 170
Nitro-cellulose membranes	142	Oxidative stress	13, 14, 35, 43, 47, 51, 94, 111, 112, 123, 172
Nitrogen	3, 7, 10, 22, 78, 94, 114, 137, 138, 152	Oxidative	2, 6, 10, 12, 13, 63, 73, 78, 79
Nitrogen assimilation	152	Oxidative-reduction potential	162
Nitrogen metabolism	138, 155	Oxides	35
Nitrogen remobilization	152	Oxidized glutathione	79
Nitrogenase	95	Oxoglutarate	138
Nodules	10, 19, 46	Oxohistidine	34
Non essential element	3	Oxygen	76
Non-cultivated soils	162	Oxygen reactive species	2
Non-food crops	104	Oxyradical	17, 35
Non-metabolic absorption	90	Ozone	13, 20
Nonphotosynthesizing plant cells	11	*P. sativum*	120, 123, 127, 128, 132, 133, 134
Nonphotosynthetic	18	Paraquat	13
Non-protein thiols	128	Parsley	22, 92
Non-redox metal	78	Passive absorption	90
Non-woody biomass	108	Pathogens	10, 27, 45
NT103	47	Pb 26	
NT107	47	$Pb(NO_3)_2$	52
Nuclear	34	PC complex	127
Nucleoli	73	PC synthase	120
Nucleosides	161	PC synthesis	131, 133, 134
Nucleotides	74	PC-Cd complexes	128
Nucleus	10, 73	PC-coated Cd	130
Nutrient deficiency	113	Pea	35, 36, 152
Nutrient	63	Pectins	70
Obligate anaerobes	32	Pennycress	105
Ocean sprays	161	Pentose phosphate pathway	77
Oilseed	112	PEP Carboxylase	72
Oligomeric	128	PEPC	76
Olygomers	132	Pepper	36, 38, 44, 46, 92
Omega	47	Peptidases	70
Omics	51	Peptides	2, 24, 80
Onion	92	Peptidoglycan	99
Ontogenesis	93, 169	Pericycle	92

Permeability	63	Phylogenetic	34
Peroxidase activity	169	Physiological	96
Peroxidase	39, 73, 82, 94	Phytoaccumulation	114
Peroxidation	8, 13, 14, 73, 78, 111	Phytoavailability	85, 109, 163
Peroxides	73	Phytochelatins	2, 20, 67, 66, 81, 111, 115, 125, 163
Peroxisomal	39		
Peroxisome biogenesis genes	37	Phytochelation	175
Peroxisomes	10, 11, 12, 27, 33, 36, 172	Phytodegradation	38
Pesticides	6	Phytoextraction	38, 40, 41, 82, 86, 87, 104, 107, 108, 109, 110, 113, 177
Petunia	47		
pH	6, 35, 63, 65, 90, 92, 98, 162	Phytoextraction potential	111
Phaeodactylum tricornutum	44	Phytofiltration	38
Phaseolus aureus	38, 40, 46, 171, 173	Phytometallothionein	67
Phaseolus vulgaris	38, 40, 46, 48, 123, 169, 170, 172	Phytoremediation	3, 9, 38, 48, 50, 38, 85, 107, 177, 178
Phaseolus vulgaris,	173	Phytoremediation capacity	23, 49
Phenylalanine ammonia-lyase	82	Phytostabilization	38
Phloem	5, 67, 91	Phytostabilization	38, 65
Phosphatases	75	Phytotoxic effects	133
Phosphate	6, 35, 63, 87, 103, 115	Phytotoxic	2, 25, 119
Phosphate fertilisers	2, 104	Phytotoxicity	4, 5, 79, 113, 114
Phosphatic	82	Phytotron	42, 50
Phosphatidyl choline	73	Phytovolatilization	38
Phosphatidyl glycerol	73	*Picea*	168
Phosphoenolpyruvate carboxylase	76	Ping-pong mechanism	39, 46
Phospholipid hydroperoxide GPX	45	*Pinus resinosa*	176
Phospholipids	73	*Pinus sylvestris*	43
Phosphoribulokinase	170	*Pisum sativum*	38, 125, 126, 133, 156
Phosphorite	89	*Pisum sativum's*	133
Phosphorites	64	Plant growth	2, 124, 167
Phosphorus fertilizer	89	Plant height	94
Phosphorylation	76, 77, 174	Plant metabolism	170
Phosphorylation efficiency	172	Plant morphology	113
Photochemical	4	Plasma membrane	7, 42, 65, 67, 98
Photoinhibition	14	Plasmalemma	14, 65
Photooxidation	14	Plasmamembrane	86
Photoreduced ferredoxin	41	Plasticity	68
Photo-reduced	11	Plastids	49
Photorespiration	36	Plastoquinone	76
Photosynthesis	5, 15, 26, 63, 73, 74, 76, 95, 111, 137, 167, 169, 170, 175	Pollen germination	176
		Pollutant	63, 137, 144, 159
Photosynthetic pigments	108	Pollution	2, 4, 64, 138, 139, 161
Photosynthetic rate	114	Polyacrylamide gel electrophoresis	146
Photosynthetic	11, 18	Polyacrylamide	139, 141
Photosystem II	169	Polyamines	7
Photosystem	95, 167	Polyclonal antibodies	25
Phragmites australis	48	Polymorphic	9
Phragmites communis	178	Polymorphisms	9
Phragmitis australis	38	Polypeptides	12
Phseolus	154	Polyphenoloxidase	95

Polyphosphate	99	Redox potential	7, 21, 35, 36, 63, 65, 164
Polysaccharides	68	Redox reactions	98
Polythiol accumulation.	120	Redox state	18
Polythiol production	126	Redox-regulated gene expression	14
Polythiol synthesis	118, 120, 125	Reduced sulphur	43
Polyunsaturated lipids	14	Relative growth rate	70
Poplar	23, 36, 42, 107	Relative water content	174
Populus	36, 85, 107	Remediation	64
Porphyrin	See	Renal disfunctions	5
Potassium	4	Reproduction	6
Potato	36	Resistance mechanism	119
Power stations	63	Respiration	6, 10, 15, 77, 86, 174
Precipitation	177	Rhizodermis	80
Productivity	6, 63	Rhizosphere	65, 86, 97, 98, 99, 165
Programmed cell death	12, 40	Ribonuclease	73
Prokaryotes	32	Rice	36, 38, 64, 66, 73, 78, 92
Proline	71	RNA	47, 161
Protease	72, 81	RNAase	95
Proteasome	81	Rock erosion	159
Proteins	17, 47, 63, 70, 71, 80, 124, 141	Rock mineralization	64
Proteolysis	70, 144, 153	Roots	6, 66
Proteolytic degradation	172	Root growth	124
Proteomics	51	ROS	7
Protochlorophyllide	167	ROS-detoxification	43
Protonation	94	Rubber tyres	64
PSI	11	Rubisco	114, 152, 170
Pumpkin	92	RUBISCO	76
Purine	161	RUBP Carboxylase	72
Putrecine	7	Rye	163
Pyrimidines	161	S deficiency	113
Pyrogallol	46	*S. alfredii*	10
Pyrometallurgical	38	*S. tuberosum*	36, 44
Pyrophosphatase	75	S/SH ratios	130
Quanta	13, 26	*Saccharomyces*	101
Quantum yield	175	Salicylic acid	39
Quenching	169	Salinity	92, 139
Radiation	13	*Salix*	9, 107, 109
Radish	92	Salt	27
Rainbow pink	41	Sandy	35
Rpe	92	Scavenging	3
Raphanus sativus	38, 175	S-Cd crystallites	134
Raphanus	44	*Scenedesmus*	176
Ratio of variable fluorescence to maximal fluorescence	95	*Schizosaccharomyces*	101, 127
		SDS-PAGE	43, 141
Rauwolfia	100	Secondary antioxidant	18
Reactive oxygen species	71	Secondary metabolism	8
Recuperation	104	Sedimentary rocks	162
Redox	7	Sedimentation	88

Index 189

Sedum	10, 105
Seirpus lacustris	178
Selenium	4, 159
Semi-metals	4
Semiquinone form	41
Senescence	7, 70, 93, 139, 140
Sensitive type	92
Sequestering	8
Sequestration	9, 15, 20, 67, 65, 67, 81, 86, 113, 132
Serine	81
Sesamum	70
Sesamum	70
Sewage sludge	2, 6, 35, 82, 87, 103, 163
Sewage	88
S-glutamylcysteinyl	47
SH groups	71
Shigeoclonium	176
Shoot	7
Silene	65, 123, 128
Silvicultural	5
Singlet oxygen	10
Smelters dust	103
Smelting	88
S-nitrosoglutathione	12
Sod1	34
Sod2	34
Sod3	34
Sod3.3	34
Sod3.4	34
Sod4	34
Sod4A	34
Sod5	34
Sodium dodecyl sulphate gel	146
Sodium	4
Soil	159
Soil pollutants	138
Solidago	67
Soluble protein	46
Soluble stromal enzymes	40
Sorghum	92
Soybean	19, 38, 44, 47, 35, 36, 76, 91, 164
Specific water content	77
Spectrophotometry	141
Spermine	7
Spinach	36, 76
Stimulation	94
Stomatal	13
Stress adaptive reaction	101
Stress	12
Stress proteins	81
Stroma	42
Stromal	39
Subcellular	92
Succinate dehydrogenase	77
Succinate	77, 176
Sucrose phosphate synthase	72
Sucrose synthase	72, 74
Sugar	7
Sugar beet	92
Sugarcane	35, 44
Sulfate	20
Sulfhydryl groups	43
Sulphate fertilizers	98
Sulphate	7
Sulphide accumulation	116, 126, 130
Sulphide content	117
Sulphide incorporation	132, 133
Sulphide ions	130
Sulphur rich peptides	23
Sulphur	3, 20, 22, 47, 109, 113, 125, 130
Sulphur-oxidising bacteria	109
Sulphydryl	160, 174
Sulphydryl groups	168
Sunflower	20, 36, 38, 46, 77, 95, 105, 108, 114
Superoxide dismutase	13, 26, 73, 79, 94, 171
Superoxide dismutase	32, 72
Superoxide	10
Superoxide radicals	94
Surface/ground-water	89
Swiss chard	35, 92
Symplasm	17, 80
Symplast	96
Symplastic	7, 67, 164
Symptoms	7
Syndrome	16
Synergistic	36, 96, 97, 111
T. caerulescens	105, 110
Taraxacum officinale	39
Temperature	7, 63, 139, 162, 164
Teratogenic	5
Terrestrial	87, 177
Tetrameric Fe porphyrin protein	36
Tetrazolium	32
Thalossiosira	173
Thermostable	33
Thiol	32, 115
Thiol formation	124
Thiol groups	125
Thiol production	133

Thiol/disulphide exchange	21	Vessels	7
Thiolate peptides	100	*Vicia*	123
Thlaspi	9, 38, 39, 83, 92, 105, 107, 116, 178	*Vicia faba*	44
Threshold	6	Vitamin	25
Thylakoid	25, 39, 73, 76, 170	*Vitis vinifera*	156
Thylakoid membrane	42, 73	Volcanic eruptions	159, 161
Thylakoid membrane-bound	39	Volcanic	3
Tillering	91	Vulcanization	98
Tissues	34	Waste incinerators	63
Tobacco	19, 22, 92	Wastewater	52
Tocopherol	17	Water potential	174
Tolerable concentrations	119	Water relations	6, 111
Tolerance	3, 8, 9, 20, 79, 80, 108, 111, 116, 118, 120, 121, 133	Water table	4
		Water-holding capacity	42
Tolerance mechanisms	88	Watermelon	33
Tomato	22, 33, 91, 92, 95	Water-water cycle	27
Tonoplast	24, 81	Weeds	5
Topsoil	4	Wheat	46, 92
Toxic	2, 3, 38, 41, 89, 171	White clover	96
Toxicity	4, 11, 63, 65, 71, 90, 103, 165, 167, 173	Wild type plants	49
		Willow	107
Toxicology	4	World health Organization	159
Trans-1,2-cyclohexylenedinitrilotetraacetic acid	40	Xanthine oxidase	12
		Xanthophyll	25
Transcription	11	Xenobiotics	20
Transduction	19	Xylem	7, 91
Transgenic CAT deficient tobacco line	38	Xylem transport	91
Transgenic plants	48	Yeast	20, 47, 115
Translocation	9, 49, 66, 165	Yellow-brown soil	89
Transpiration	68, 76, 77, 91, 152, 169	Yields	4
Transport	9, 64, 92	*Zea mays*	123, 163, 174
Trichomes	68	Zinc	4, 6, 33, 159
Tripeptide	120	Zincification	89
Triticum	39, 65, 70, 101, 168	Zn transporter system	86
Turnip	92	Zn-ligand complexes	43
Tyre production	89	ZnO	52
Ubiquitin	81	Zymogram	146
Unsaturated fatty acid	169		
Uptake	9, 36, 40, 64, 65, 77, 90, 92, 143, 159, 163, 165		
Urban traffic	63		
Vacuolar	17		
Vacuolation	9		
Vacuole	82, 129		
Valerianella locusta	107		
Valine	71		
Vascular	67		
Vegetable	5		